Advances in Herpetological Medicine and Surgery

Advances in Herpetological Medicine and Surgery

Guest Editor

Tom Hellebuyck

MDPI

Basel • Beijing • Wuhan • Barcelona • Belgrade • Novi Sad • Cluj • Manchester

Guest Editor
Tom Hellebuyck
Department of Pathobiology,
Pharmacology and Zoological
Medicine
Ghent University
Merelbeke
Belgium

Editorial Office
MDPI AG
Grosspeteranlage 5
4052 Basel, Switzerland

This is a reprint of the Special Issue, published open access by the journal *Animals* (ISSN 2076-2615), freely accessible at: https://www.mdpi.com/journal/animals/special_issues/Advances_in_Herpetological_Medicine_and_Surgery.

For citation purposes, cite each article independently as indicated on the article page online and as indicated below:

Lastname, A.A.; Lastname, B.B. Article Title. *Journal Name* **Year**, *Volume Number*, Page Range.

ISBN 978-3-7258-2655-1 (Hbk)
ISBN 978-3-7258-2656-8 (PDF)
https://doi.org/10.3390/books978-3-7258-2656-8

Cover image courtesy of Tom Hellebuyck

Contents

About the Editor

Tom Hellebuyck

Tom Hellebuyck is the head of clinic at the Division for Exotic Companion Animals and Wildlife of the Department of Pathobiology, Pharmacology and Zoological Medicine (Faculty of Veterinary Medicine of the Ghent University, Belgium). In 2006, he obtained his master's degree as a veterinarian, and in 2012, he finished his PhD in veterinary sciences with a focus on dermatological diseases in lizards. He became a diplomate of the European College of Zoological Medicine (Herpetological Medicine & Surgery) in 2013. At present, his main research projects focus on infectious diseases, reproductive medicine, and oncology in reptiles. He is an author and co-author of several scientific publications and book chapters. He is a frequent speaker at national and international conferences and symposia and is a member of several organizations concerned with veterinary medicine and herpetological societies. In addition, he acts as an advisor to authorities with regard to reptile legislation, medicine, and welfare.

Preface

Herpetological medicine constitutes an area of specialty within the broader field of zoological medicine and encompasses the veterinary care of extremely diversified vertebrate taxa with highly variable biological properties. Despite the fact that the field of herpetological medicine is evolving in a myriad of ways, and that clinicians who work with reptiles are able to rely on a constantly increasing array of possibilities, there is a continuing need to advance herpetological medicine by elaborating the scientific knowledge that allows the development and optimization of diagnostic and therapeutic possibilities in reptile species. This Special Issue focuses on disorders of the spectacle and subspectacular space in squamates, the evaluation of anesthetic protocols in different reptile species, the comparison and evaluation of surgical procedures for the treatment of reptile reproductive disorders, novel insights in reptile oncology, and the various aspects of viral as well as bacterial diseases in reptiles.

Tom Hellebuyck
Guest Editor

animals

Review

Anatomy, Physiology, and Disorders of the Spectacle, Subspectacular Space, and Its Lacrimal Drainage System in Squamates

Tom Hellebuyck * and Ferran Solanes Vilanova

Department of Pathobiology, Pharmacology and Zoological Medicine, Faculty of Veterinary Medicine, Ghent University, Salisburylaan 133, B-9820 Merelbeke, Belgium
* Correspondence: Tom.Hellebuyck@UGent.be

Simple Summary: Various snake and lizard species have transparent and fused eyelids that make up a spectacle. From an evolutionary point of view, it is assumed that ancestral snakes developed this structure to protect the eye as an adaptation to their burrowing, underground existence. Between the spectacle and the eye lies a narrow subspectacular space that is filled with a tear-like fluid that drains to the roof of the mouth via the lacrimal duct. Although disorders of the eye are relatively uncommon in squamates with a spectacle, disorders of the spectacle, the subspectacular space, and its drainage system are frequently encountered in veterinary practice. As the spectacle is an integral part of the skin it is renewed with each shedding cycle. Retention of the old layers of the spectacle may occur because of infectious as well as non-infectious causes. Besides inflammation and trauma of the spectacle, overfilling due to blockage of the drainage system (pseudobuphthalmos) as well as bacterial and parasitic infection of the subspectacular space (subspectacular infection) are undoubtedly the most frequently observed disorders in snakes and spectacled lizards. An adequate diagnostic approach often allows the establishment of a successful treatment for these conditions.

Abstract: Various squamate species have completely fused eyelids that make up a transparent spectacle. The spectacle is a continuation of the integument that is renewed with each shedding cycle and creates a narrow subspectacular or corneospectacular space that is filled with lacrimal fluid. The latter is considered as the analogue of the conjunctival sac in other vertebrates. Almost all reptiles that have a spectacle lack a nictitating membrane, bursalis muscle, and lacrimal glands. The lacrimal fluid in the subspectacular space is secreted by the Harderian gland. The features of the spectacle and its lacrimal drainage system are an illustration of the enormous variation of the morphological adaptations that are seen in reptiles and one of the most distinguishable traits of snakes and most gecko species. Whereas ocular disease in squamates with a spectacle is infrequently seen in practice, disorders of the spectacle and the subspectacular space are commonly encountered. In order to apply an adequate diagnostic and therapeutic approach for these conditions, a sound knowledge and understanding of the anatomical and physiological peculiarities of the spectacle, subspectacular space, and lacrimal drainage system are fundamental.

Keywords: lizards; pseudobupthalmos; snakes; spectacle; subspectacular abscess

Citation: Hellebuyck, T.; Solanes Vilanova, F. Anatomy, Physiology, and Disorders of the Spectacle, Subspectacular Space, and Its Lacrimal Drainage System in Squamates. *Animals* **2023**, *13*, 1108. https://doi.org/10.3390/ani13061108

Academic Editor: Volker Schmidt

Received: 26 February 2023
Revised: 16 March 2023
Accepted: 20 March 2023
Published: 21 March 2023

1. Introduction

It is generally accepted that moveable eyelids are the primitive condition for squamata [1–4]. In several squamate lineages, however, the eyelids may show a varying degree of modification ranging from thinning of the dermal layers to the presence of a single transparent scale or multiple transparent scales creating a window in the lower eyelid (Figure 1) or covering the eye. In various taxa, moveable eyelids have been completely replaced by a truly transparent and rigid spectacle (Table 1) [2–6]. The reptilian spectacle is considered the most sophisticated type of spectacle (tertiary spectacle) that is posteriorly lined with conjunctival epithelium bordering a narrow subspectacular or corneospectacular space filled with a fluid analogous to the tear film of lidded vertebrates allowing the eye to rotate freely [2,3,6].

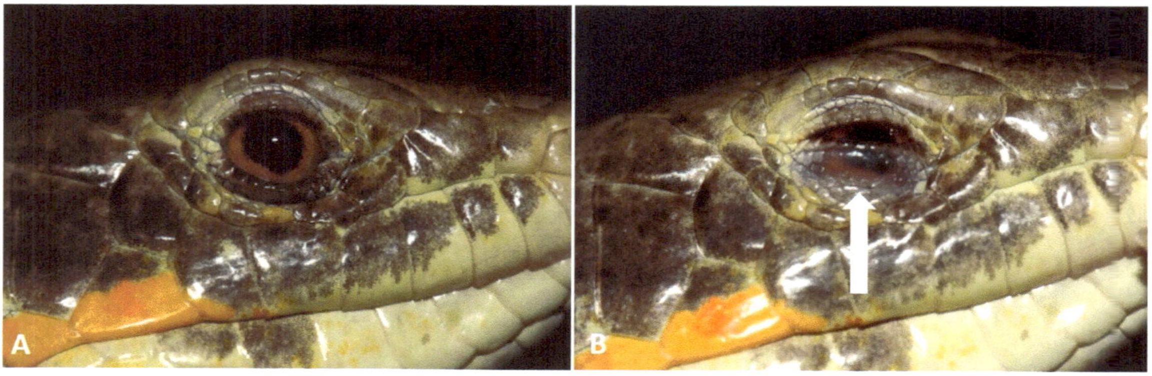

Figure 1. (**A**) Schneider's skink (*Eumeces schneiderii*); this species has a window in the lower eyelid (arrow) made up by multiple transparent scales (**B**).

Table 1. Overview of squamate lineages that show a varying degree of modification of the eyelids or possess a truly transparent and rigid spectacle.

Clade/Superfamily	Family	Comments
Amphisbaenia (Worm lizards)	Amphisbaenidae Bipedidae Blanidae Cadeidae Rhineuridae Trogonophidae	Reduced eyes. Moveable eyelids have been replaced by a transparent ocular scale that covers the eye (Figure 2A,B). Sometimes the ocular scale is not thinner than the skin and pigmentation might be present [2,6].
Dibamidae (Blind lizards)	Dibamidae	Reduced eyes. Moveable eyelids have been replaced by transparent scales that cover the eye [4,6].
Gekkota	Pygopodidae Diplodactylidae Gekkonidae Phyllodactylidae Sphaerodactylidae	Only the clade of the eublepharid geckos (*Eublepharidae*) has fully moveable eyelids. A true transparent and rigid spectacle is present in all other Gekkota [1,5–9].
Scincomorpha	Scincidae (skinks)	Many species belonging to various genera have a transparent window in the lower eyelid made up by multiple transparent scales (Figure 1). Genera *Leiolopisma* and *Trachylepsis* (formerly *Mabuya*): single large transparent scale in the lower eyelid large enough to cover the cornea [6,10–12]. Genera *Ablepharus* (Figure 2C), *Morethia*, *Proablepharus*, and *Cryptoblepharus:* true spectacle [5,12]. Genus *Ablepharus*: only genus in which lacrimal glands are present [4–6].
	Xantusiidae (Night lizards)	True spectacle (Figure 2D) [2,4,13,14].

Table 1. *Cont.*

Clade/Superfamily	Family	Comments
Lacteroidea	Gymnophtalmidae (Spectacled lizards, microteiids)	Most species: moveable upper eyelids and immobile lower eyelids with a transparent window [3,4,14–17]. Monophyletic tribe Gymnophtalmini: true spectacle [15,18,19].
	Lacertidea	Genus *Eremias*: simplest eyelid modification involving a thinning of the layers of multiple scales creating an 'eyelid window'. Genus *Ophisops:* true spectacle.
Typhlopoidea (Scolecophidia, Blind snakes)		Reduced eyes. Transparent ocular scale rather than a spectacle. The scale extends well beyond the margins of the globe, in some cases to the mouth (oculolabial scale) (Figure 2E,F) [2,3,20,21].
All other snakes than Blind snakes (Alethinophidia)		True spectacle [2,3,20,21].

Figure 2. A varying degree of modification of the eyelids is present in several squamata lineages. In worm lizards (Amphisbaenia), such as the Turkish worm lizard (*Blanus strauchi*) (**A**) and the checkerboard worm lizard (*Trogonophis wiemanni*) (**B**), reduced eyes are covered by an ocular scale. (**C**) The Budak's snake-eyed skink (*Ablepharus budaki*) and other members of its genus have a true spectacle and are the only known reptile species that possess a lacrimal gland. (**D**) In all members of the family Xantusiidae, such as the Costa Rican tropical night lizard (*Lepidophyma reticulatum*), moveable eyelids have been replaced by a true spectacle. Ocular scales that extend well beyond the margins of the globe are present in blind snakes such as the long-nosed worm snake (*Myriopholis macrorhyncha*) (**E**) and the Eurasian blind snake (*Xerotyphlops vermicularis*) (**F**). ©Frank Pasmans.

Outside of the squamata, modification of the eyelids is only seen in a few turtle species (e.g., the Indian flapshell turtle (*Lissemys punctata*) and the eastern long-necked turtle (*Chelodina longicollis*)) that have a single large transparent scale in the lower eyelid large enough to cover the cornea [13]. Except for some burrowing snakes (Scolecophidia: Typhlopoidea, blind snakes) that possess an ocular or oculolabial scale and species of the eublepharid geckos that possess fully moveable eyelids (Figure 3), snakes and geckos (Figure 4) are the only taxa where the spectacle occurs in all species regardless of habitat and ecology (Table 1) [2,3,20–23]. The phylogenetic position of eublepharid geckos suggests that having functional eyelids is a re-aquisited feature of more primitive forms, although spectacles may have also evolved independently in non-eublepharid geckos and their sister taxa [7–9,22].

Figure 3. Members of the eublepharid clade, such as this leopard gecko (*Eublepharis macularius*), are the only Gekkota that possess fully moveable eyelids.

Figure 4. The spectacle in a New Caledonian giant gecko (*Rhacodactylus laechianus*) is bordered by a extra-brillar ring that resembles eyelids but is a separate, fixed structure.

Although we might never fully understand what prompted the ancestors of snakes to develop a spectacle, it is generally accepted that snakes evolved from burrowing lizards that lived underground and the reduced eyes and spectacle are adaptations to this burrowing existence to primarily protect the eye. It would not have been until snakes re-emerged from their underground existence that they re-developed functional eyes that were initially adapted to nocturnal conditions [2,3,6,23,24].

Although it was first hypothesized that the spectacle evolved from a modification of the nictitating membrane [13], studies in natricid and viperid snakes demonstrated that the spectacle is derived from the concentric fusion of the mesenchymal tissues that form the eyelids at the level of the pupil during the embryonic stage. The fused eyelids become transparent and are voided of all glands and muscles [6,10,24–26]. The spectacle is well developed at the time of birth in ovoviviparous and oviparous snakes, but in some species, fusion of the eyelids and formation of the spectacle occurs post-oviposition [3,5]. Most species belonging to the Gymnophthalmidae family, also referred to as the spectacled lizards or microteiids, have moveable upper eyelids and immobile lower eyelids with a transparent window [4]. Within this family, the members of the monophyletic tribe Gymnophtalmini have a true spectacle and embryonic fusion of the eyelids takes place above the pupil region [3,4,15–19]. It remains uncertain, however, if functional eyelids can be either considered as a primitive condition or as a re-acquisition of the character in *Tretrioscincus* species belonging to the single genus of the tribe where functional eyelids are present [4]. The spectacle of other saurian species and the ocular scale of amphisbaenians and blind snakes are also considered to be formed by fusion of the eyelids, but it is unknown if the embryonic development pattern equals that of snakes or Gymnophthalmini [1,4]. In amphisbaenians, the thickness of the ocular scale and its degree of pigmentation is variable, suggesting that the emphasis placed on vision varies significantly among these fossorial squamates [2].

2. The Structure and Function of the Spectacle in Squamates

The periocular scales form a discrete rim that continues in a transition zone of dermis that gradually becomes thinner and connects with the spectacle (Figure 5) [27–29].

Figure 5. The transition zone between the periocular scales and the spectacle can be clearly distinguished in this West African Gaboon viper (*Bitis gabonica* rhinoceros).

Histologically, the spectacle consists of three layers and has been best studied in snakes [30,31]. The outer epithelial layer (3.5–10.5 μm) has one (germinal layer) or more layers of basal cells and is covered by keratin made up of 2 to 4 alternating layers of alpha and beta keratin depending on the stage of the shedding cycle. Beta keratin composition seems to differ between species and bears a relationship with taxonomy, suggesting that optical transparency is not restricted to a few isoforms [32]. The inner epithelial layer (1.6–3.6 μm) is a single germinal layer that is considered as the continuation of the palpebral conjunctiva and is separated from the underlying stroma layer by a basement membrane. The free border of the inner epithelium that delineates the lacrimal-fluid-filled subspectacular space is considered as the homologue of the corneal endothelium [31]. The stroma (9–132 μm) contains parallel layers of evenly spaced lamellar of collagen fibrils with an alternating course, similar to what is seen in the vertebrate cornea. In contrast to the cornea, however, the stroma layer contains blood vessels, fibroblasts, and nerve fibers [2,31].

In snakes, spectacular thickness seems to reflect evolutionary adaptation and development to their natural habitat and ranges from 74 to 244 μm. The thinnest spectacles are found in arboreal and terrestrial species, whereas the thickest occur in aquatic and fossorial or burrowing species. In general, colubrids have thinner spectacles than boas and pythons. The thinnest spectacles are found in viperids and the thickest in pipe snakes [33].

In addition to the protective role of the spectacle, it also constitutes the main refractive ocular surface (refractive index n $\geq$ 1.5) and plays a crucial role in the quality of vision [2,31,32]. The refractory potential of the spectacle together with the lacrimal fluid in the subspectacular space equals that of the lens in species that do not possess a spectacle and renders the role of the cornea in refraction irrelevant [34]. The spectral transmittance has been studied in snakes as well as lizards and a wide range of wavelengths from red light to ultraviolet (UV) (280–315 nm) are transmitted; however, the spectacle also acts as an optical filter that reduces the passage of UV irradiation with short wavelengths in diurnal species [35,36].

Although the spectacle is optically transparent, similar to other parts of the integument it is well vascularized [37]. As spectacled reptiles are nearly the only vertebrates that possess non-retinal blood vessels in the visual fields, adaptations to minimize loss of clarity in the optical transmissive regions of the eye are necessary. In addition to transparent blood vessel walls and a specific spatial layout and density in the spectral regions that serve the foveal and binocular visual fields, it has been shown that these vessels undergo cycles of dilation and constriction to impact visual clarity as minimally as possible in snakes [2,37]. During activity or whenever a snake is visually stimulated, spectacle vessels remain constricted for longer periods than during a resting phase. During the renewal phase of the shedding cycle, spectacle vessels remain dilated and blood flow remains strong and continuous [2,34,37]. The mechanism and patterns of blood flow in the spectacle seem to guarantee visual acuity when snakes are active or presented with a visual stimulus and promote physiological shedding by avoiding desiccation of the spectacle layers during ecdysis [35,38]. Similar vascularization patterns of the spectacle or ocular scale have also been described in geckos, xantusiid lizards, and amphisbaenids [37,39].

Clinically, it is important to differentiate between a normal vascularization pattern, also during different stages of the shedding cycle, and neovascularization of the cornea, as well as physiological dilatation of iris vessels (Figure 6) or reactive dilation in case of uveitis. Engorged spectacular vessels might be seen in the chronic stages of pseudopubhthalmos or subspectacular abscesses (Figure 7) [31].

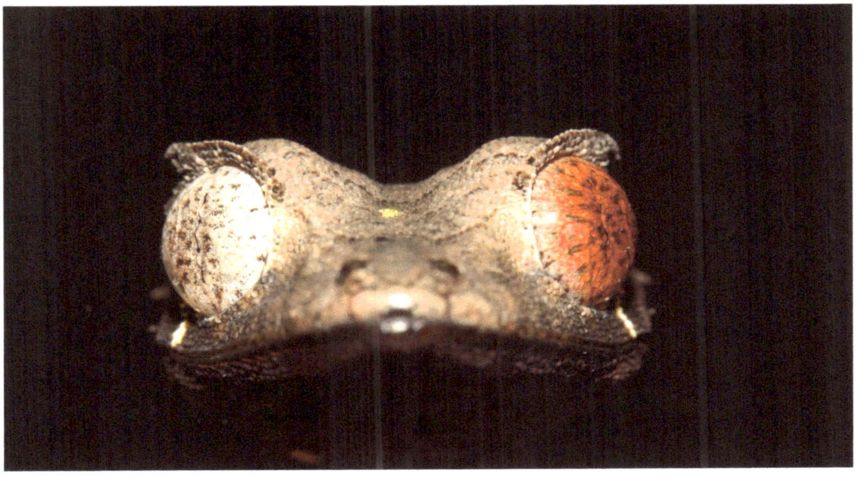

Figure 6. Unilateral dilation of iris vessels in the left eye of a satanic leaf-tailed gecko (*Uroplatus phantasticus*) can occur as a transient physiological phenomenon.

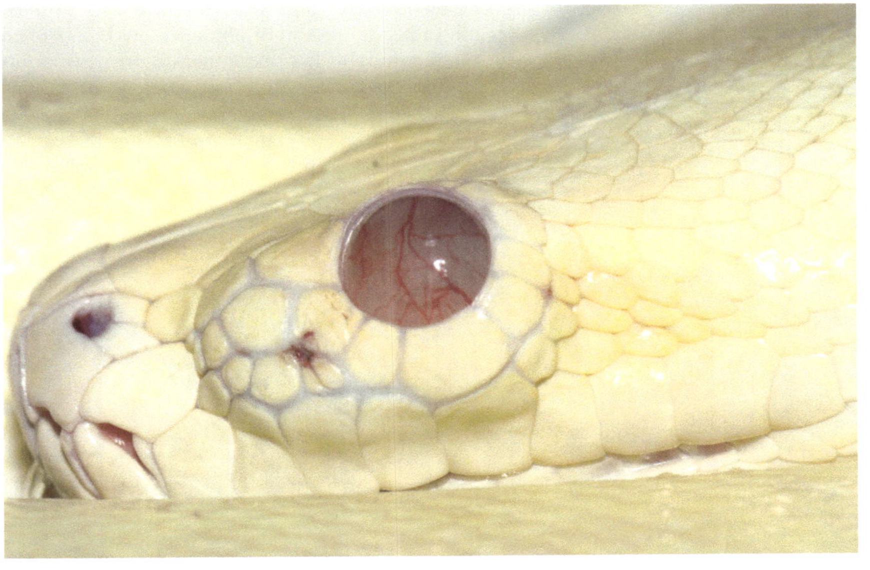

Figure 7. Engorged spectacular vessels in a *Burmese python* (*Python bivittatus*) with chronic pseudopubhthalmos. Besides overfilling of the subspectacular space, prominent swelling of the periocular and facial region can be noted.

3. The Lacrimal System in Squamates

In snakes, lacrimal glands are absent; however, non-Harderian glandular structures that empty into the subspectacular space have been identified in boid species and parts of the epithelium of the subspectacular space might be secretory [6,40]. The serous Harderian gland of snakes is located in the dorsonasal orbit, medial and posterior to the eye and contains several ductules that drain into a lacrimal sac or ampulla that communicates with the ventral portion of the subspectacular space [5,41,42]. In general, saurian species with spectacles also lack lacrimal glands (except for members of the *Ablepharus* genus; Table 1) and their Harderian gland is located in the nasal aspect of the orbit. In geckos, it has been demonstrated that it is a seromucous gland that secretes proteins and lipids [3,26]. The Harderian gland is primarily responsible for the production of the tear film but might also play a role in vomeronasal olfaction [26,43].

In snakes, a single punctum is located in the ventromedial medial fornix of the subspectacular space and continues in the lacrimal duct [5,6,40,41]. After the duct passes through the lacrimal foramen, lateral to the prefrontal bone it shows a circuitous course rostrally and medially between the caudal aspect of the maxillary and palatine bones. Next, it runs between the vomer and the hypochoanal cartilage and exits in the oral cavity, medial to the opening of the duct of the vomeronasal organ [6,41]. Based on a computed tomographic anatomical study of the ophidian lacrimal system, the complexity of the tortuous course of the lacrimal duct varies between snake species [41]. This variation may be pertinent to predispositions of certain snake taxa towards the development of disorders of the subspectacular space, in particular pseudobuphthalmos and subspectacular infections, and also has implications towards the treatment and prevention of these disorders [41].

4. Ecdysis and Healing of the Spectacle

Ecdysis is the periodical process of renewal of the outer layer of the epidermis of the entire skin, including the spectacle [44–46]. In most squamates, this is a discontinuous, cyclic process. During the cleavage phase, the old skin separates from the newly formed layer by enzymatic activity of the lymphatic fluid. In snakes, the skin often appears dull or has a blueish hue during the renewal phase. This is most obvious in the spectacle as the result of accumulation of old and new layers of the epidermis causing opacification [44–49]. Curiously, this phenomenon does not occur or is almost unremarkable in lizards that have spectacles [2]. Shortly before the actual shed, the spectacle and skin of snakes regain their normal appearance as the lacunae of the stratum intermedium are broken down [47]. If a normal shedding cycle occurs, snakes shed their entire skin in a single piece (Figures 8 and 9). Most lizards that have spectacles show a more gradual renewal process of the skin in comparison to snakes but will eventually also slough the old skin in a short period of time [48,49].

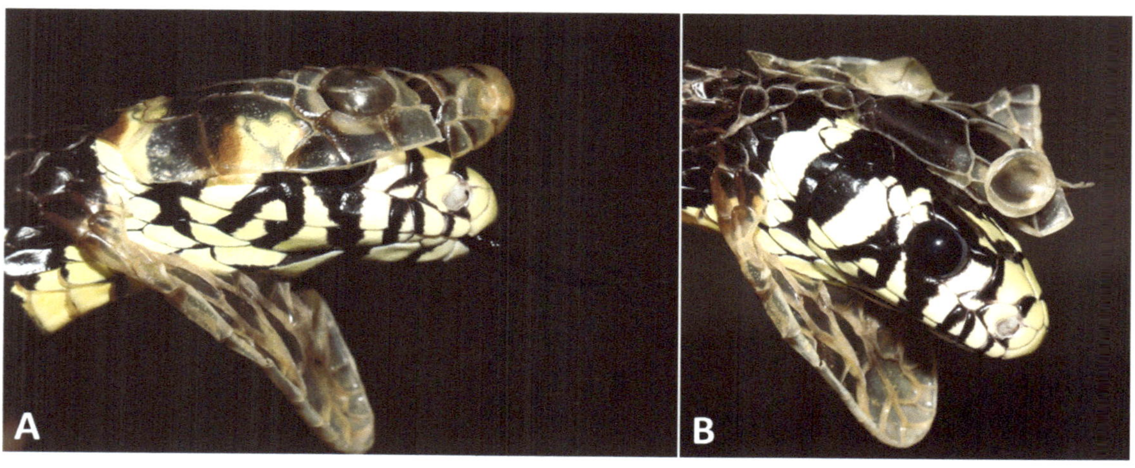

Figure 8. Ecdysis in a chicken snake (*Spilotes pullatus*). (**A**) Sloughing of the skin in a healthy snake should always occur in a single piece and include the spectacles. (**B**) The shed spectacle and its transition to the sloughed skin of the head are clearly visible and resemble a cast of the newly formed spectacle.

Figure 9. In most spectacled lizards, the renewal of the skin is a more gradual process in comparison with snakes, but the actual sloughing of the skin occurs in a short period of time. The spectacle should always be released together with the shed of the head as seen in (**A**) a satanic leaf-tailed gecko (*Uroplatus phantasticus*) and (**B**) a Madagascar day gecko (*Phelsuma madagascariencis*).

The clinical and histological healing of the spectacle was studied in ball pythons (*Python regius*) [50]. Twenty-four hours after experimental removal of a quadrant of the spectacle, infiltration of inflammatory cells and the development of a proteinaceous plug are noted at the cut edges of the spectacle. After 7 to 10 days, a proteinaceous crust bridges the spectacle defect. As the cellular inflammatory component gradually decreases, the deeper germinal epithelium becomes hyperplastic and covers the defect, whereas a prominent vascular response is present. After 21 days, the regenerated spectacle shows a normal vascular pattern beneath the crust. Initially, the thickness and transparency of healed spectacles may show considerable variation, and a normal appearance with a restored subspectacular space is mostly regained after one or more shedding cycles (Figure 10) [50]. The development of the proteinaceous crust is considered as a highly important step in the healing process as it acts as a scaffold for the regenerating epithelium of the spectacle [50].

Figure 10. Appearance of the spectacle in a Burmese python (*Python bivittatus*) following a first shed after performing a partial spectaculectomy.

Regeneration of the spectacle has been described in two snake species for the treatment of a near-complete spectaculectomy or spectacle avulsion by use of porcine small intestinal submucosal or amniotic membrane grafting combined with application of topical ointments [51,52]. In both cases, full regeneration of the spectacle following two or more shedding cycles was observed; based on these results, grafts are considered to serve as a scaffold for the healing of larger or complete spectacular defects [51,52].

5. Clinical Examination of the Spectacle and Subspectacular Space

Assessment of the spectacle and subspectacular space should always be included as a part of the general clinical examination of spectacled squamates [46,48]. Evidently, the presence of the spectacle excludes the application of fluorescein dye stain of the cornea, measuring lacrimal secretion or intraocular pressure [31].

Although many abnormalities can be detected with the naked eye, the detection and in-depth assessment of moderate to discrete changes often require the use of a magnifying glass. Slit-lamp biomicroscopy is highly useful to obtain a magnified three-dimensional view of the spectacle and the subspectacular space, as well as the anterior segment of the eye and the lens if a normal opacity of the spectacle and lacrimal fluid in the subspectacular space is present [31]. Although the normal microvasculature of the spectacle can vary greatly, it can be assessed using a slit lamp [31,35,38,53].

Comparison of the left and right spectacle often allows the detection of mild distention of the subspectacular space or discrete changes in the opacification and the transparency of the lacrimal fluid (Figure 11). Any abnormality detected at the level of the spectacle or spectacular space should warrant appropriate testing and sampling. Systemic disease should be considered in case of bilateral lesions.

Figure 11. Bilateral opacification of the spectacle in a Burmese python (*Python bivittatus*). The blueish hue of the left spectacle is related to the renewal phase of the skin, whereas opacification of the right spectacle is caused by the presence of flocculent exudate in the ipsilateral subspectacular space due to bacterial infection.

Optical coherence tomography (OCT) and scanning laser ophthalmoscopy have been used to visualize the spectacle, assess the thickness of the spectacle, and evaluate the subspectacular space [33,53]. In particular, OCT proves to be a valuable tool to assess the anterior segment structures of the spectacle [53,54].

During the clinical examination, and especially during the sampling and treatment of lesions of the spectacle and subspectacular space, appropriate local and/or general analgesic and anesthetic protocols should be used that are tailored on an individual basis and take into account the involved reptile species, anatomical structure, and lesion.

6. Retention of the Spectacle

Especially in snakes, retention of the spectacle is a commonly observed disorder. As the transition zone of the spectacle to the periocular zone is one of the body sites with the highest mechanical resistance, retention may also occur in healthy reptiles. If retained spectacles are part of generalized dysecdysis, this might be attributed to non-infectious causes, such as inappropriate environmental conditions (mainly inappropriate humidity levels or a lack of objects in the captive environment to rub against), as well as infectious causes, including ectoparasitosis (Figure 12) and bacterial, mycotic, or viral dermatitis. Malnutrition or generalized disorders, especially those resulting in dehydration and/or hypoproteinemia, may also result in impaired shedding, including spectacular retention [44–46,48]. The diagnosis and treatment of the primary causes of retained spectacles are indispensable parts of the treatment of this disorder.

Figure 12. Extreme infestation with snake mites (*Ophionyssus natricis*) in a deceased ball python (*Python regius*).

Retention of the spectacle is easily diagnosed if several layers of shed spectacle have accumulated in contrast to retention of a single layer, especially shortly after sloughing, as this results in minimal changes in opacification. Whenever possible, the shed skin should be examined for the presence of the spectacles (Figure 13).

In some cases, the retained spectacle may be released spontaneously with the next shed (Figure 14).

Figure 13. Examination of the shed skin from a northern spiny-tailed gecko (*Strophurus ciliaris*) confirms complete shedding of the spectacle.

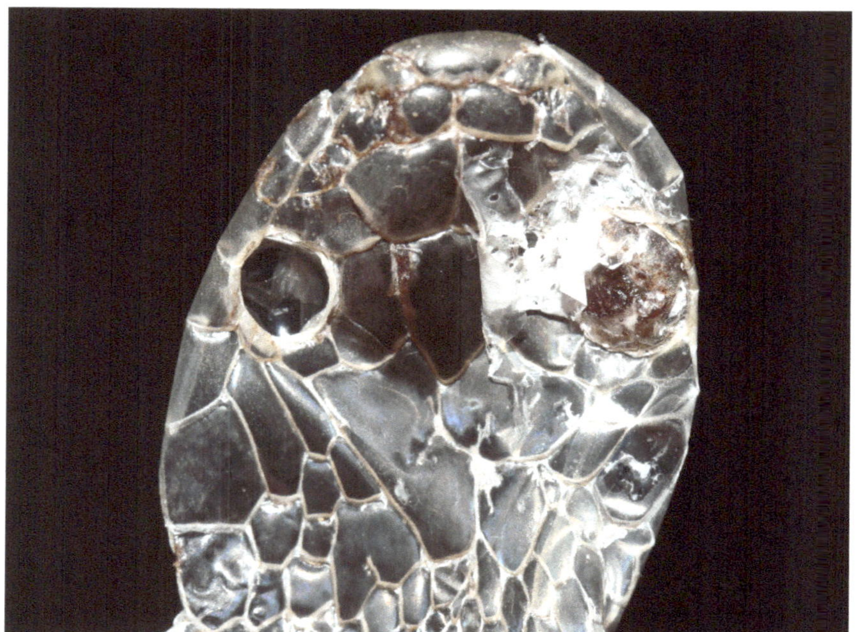

Figure 14. Fresh shed skin from the head of a corn snake (*Pantherophis guttatus*) detached a retained spectacle at the right side.

Multiple layers of retained shed often become very rigid and highly difficult to remove. Premature removal or aggressively peeling back the retained spectacle may damage the normal epidermal layer and in severe cases the subspectacular space and/or cornea might be exposed with the development of corneal ulceration, exposure keratitis, or even panophthalmitis [44–46,48]. Repeatedly putting the affected snake or lizard in lukewarm water baths for 10 to 15 min in combination with optimization of the environment and/or applying topical eye droplets, tear preparations, or ointments multiple times a day mostly allows the safe removal of retained spectacles (Figure 15) [44–46].

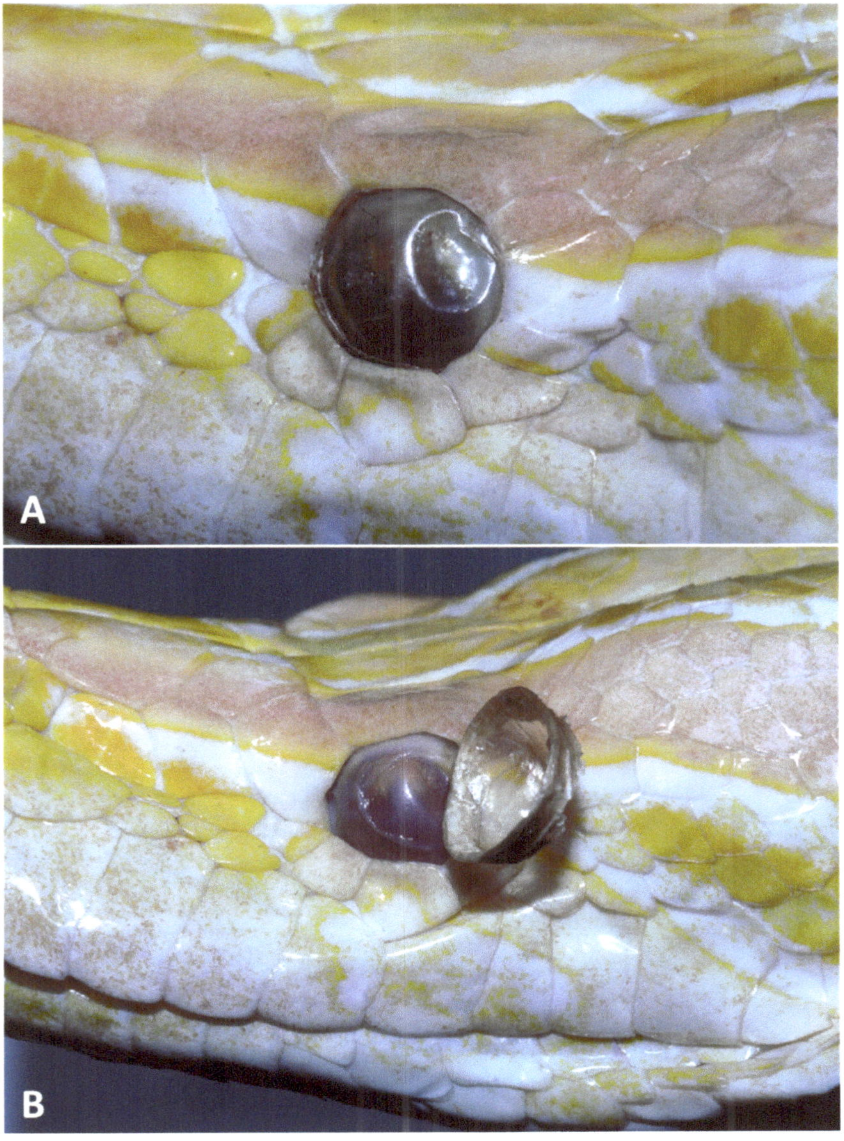

Figure 15. (**A**) Multiple retained layers of the spectacle in a Burmese python (*Python bivittatus*). (**B**) Release of the retained layers after a few days of topical treatment with ophthalmological ointments.

Hydrodissection is a technique that consists of irrigating saline between the spectacle layers using a 23-to-27-gauge catheter that seems to be quite effective to rapidly delineate and remove retained layers without the risk of removing the basal layer of the spectacle [31].

7. Indentation and Trauma of the Spectacle

Bilateral or unilateral indentations of the spectacle can be regularly seen in various snake species, notably ball pythons. In most cases these indentations are physiological, although they may also be associated with mild trauma, generalized skin disorders, or systemic diseases. Indentations of the spectacle will resolve quickly with the next shed unless underlying disease is present [44–48].

Injuries of the spectacle are most frequently seen as a consequence of iatrogenic trauma (e.g., inappropriate removal of retained spectacle layers), trauma from the environment, especially in snakes or lizards that show repetitive rubbing or explosive flight behaviour, or biting lesions caused by prey items (Figure 16) [48]. Often, this coincides with more extensive rostral trauma, dermal granulomas, and/or stomatitis. Although mild trauma and abrasions of the spectacle will often heal quickly with the next shedding phase, secondary bacterial or fungal infection might occur if the deeper layers of the spectacle are exposed [44–48].

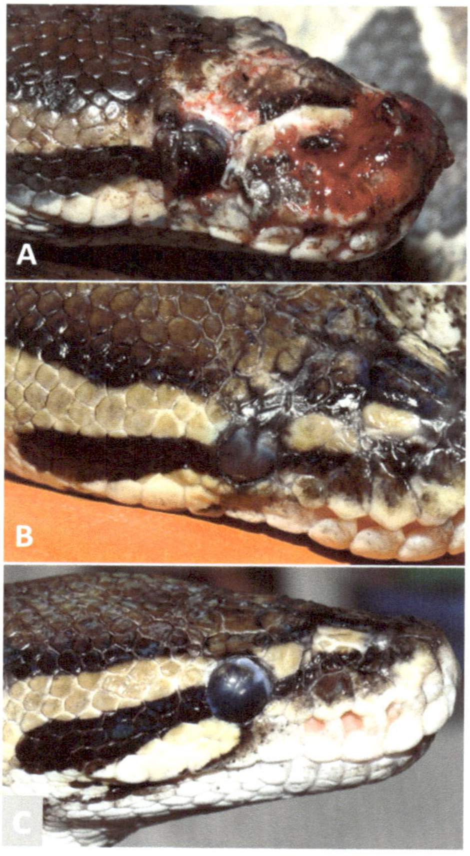

Figure 16. (**A**) Rat bite induced rostral trauma involving the spectacle in a ball python (*Python regius*). (**B**) Healing of the skin and spectacle following topical wound treatment and shedding (week 4). (**C**) Mild residual scarification of the skin and spectacle after a second shedding phase (week 15).

It has been postulated that in cases of traumatic near-complete spectaculectomy or avulsion of the spectacle, spontaneous regeneration will not occur and intensive treatment, including the use of grafting, is necessary [31,51,52]. Based on our experiences, however, topical application of antimicrobial ointments until a first shed has occurred has a fair chance of resulting in complete regeneration of the spectacle and subspectacular space if the basal layers at the transition zone are still intact. Although the regenerated spectacle may initially appear wrinkled, it mostly regains its normal appearance after one or more additional shedding cycles.

8. Spectaculitis

Spectaculitis may be seen as a localized lesion (Figure 17) or be part of a generalized dermatoses, including those that are associated with systemic disease [29,31,55]. If spectaculitis coincides with dermatitis, a mutual etiology can often be revealed (Figure 18) [44–48].

Figure 17. Spectaculitis associated with bacterial infection in a crested gecko (*Correlophus ciliatus*).

Figure 18. Dermatitis of the preocular and nasal skin extending to the medial part of the spectacle associated with opportunistic mycosis in a great lakes bush viper (*Atheris nitschei*).

Gram-positive bacteria are more commonly encountered in snakes with spectaculitis than in other diseases [28,56]. *Ophidiomyces ophiodiicola* has been reported as a cause of spectaculitis in both wild and captive snakes [57,58]. Rather anecdotally, the presence of spectacular urate depositions as well as cases of disseminated lymphoma and myeloproliferative disease with involvement of the spectacle have been reported in snakes [31]. The latter disorders may be associated with localized or generalized inflammation and opacification of the spectacle. Local or diffuse opacification of the spectacle may develop following exposure to excessively short UV-B wavelengths [31,59].

Histological features of spectaculitis include serocellular crust formation, epithelial and stromal edema, stromal neovascularization, and infiltration of inflammatory cells. In most cases, the stroma and outer epithelial layers are involved, whereas the inner epithelium only seems to be involved in severe cases of deep bacterial or mycotic infections [28,29]. If the inner epithelial layer is involved, cellular hypertrophy and hyperplasia may occur and necrotic epithelial cells may enter the subspectacular space [28,29].

Correlophus and *Rhacodactylus* species (Diplodactilydae) frequently present with localized, ulcerative lesions in the center of the spectacle (Figure 19). Although the exact etiology is unknown, the development of these lesions might be associated with opportunistic infection caused by bacteria that are part of the oral microbiota. Although most geckos clean and moisturize the spectacle by use of the tongue, *Correlophus* and *Rhacodactylus* species seem to be predisposed to such infections if the integrity of the epidermal barrier of the spectacle is impaired due to microtrauma, lacerations, inappropriate environmental conditions, and/or secondary changes due to nutritional deficiencies. Especially during the vitellogenic phase of the reproductive cycle, female crested geckos seem to be particularly prone to the development of these typical lesions of the spectacle. In most cases, lesions resolve following application of topical vitamin A and/or antimicrobial ointments followed by renewal of the spectacle, although fibrosis might develop and recurrent episodes are commonly observed.

Figure 19. Ulcerative lesion located at the center of the spectacle in a crested gecko (*Correlophus ciliatus*).

9. Pseudobuphthalmos

Pseudobuphthalmos, also referred to as bullous spectaculopathy, results from obstruction of the lacrimal drainage system and is seen in snakes (Figures 20 and 21) as well as lizards (Figure 22) [11,31,47].

Figure 20. Pseupdobuphthalmos in a neonate Dumeril's boa (*Acrantophis dumerili*) that is presumably related to developmental abnormalities in the lacrimal drainage system.

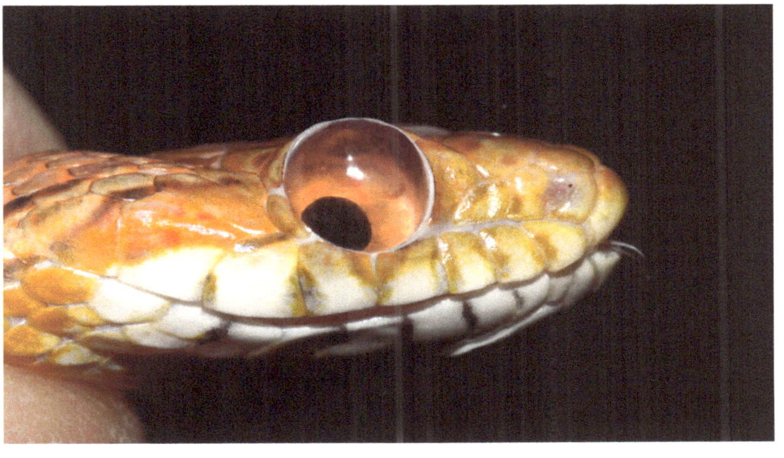

Figure 21. Pseudobuphthalmos presenting as a severe distention of the subspectacular space related to obstruction of the lacrimal duct in a corn snake (*Pantherophis guttatus*).

In comparison to subspectacular infection, pseudobupthalmos is more frequently seen in young animals and it may occur uni- or bilaterally. Typically, the excessive lacrimal fluid in the subspectacular space is clear and the spectacle as well as the ocular globe show a normal appearance. In severe cases, swelling of the facial and medial periocular region (Figure 23) may occur [11,47,60].

Congenital pseudobupthalmos in neonates is possibly associated with developmental abnormalities of the lacrimal duct [60]. In ovoviviparous snakes, the authors have witnessed cases where up to 60% of the neonates from a single clutch were affected (Figure 24). In some cases, development into subspectacular abscesses was observed.

Figure 22. Pseupdobuphthalmos in a juvenile crested gecko (*Correlophus ciliatus*) resulting from developmental abnormalities of the lacrimal drainage system.

Figure 23. Pseudobupthalmos in a juvenile boa constrictor (*Boa constrictor*) combined with swelling of the facial region associated with ascending *Pseudomonas aeruginosa* infection via the lacrimal duct. If left untreated, subspectacular infection and abscessation might develop.

Mild pseudobupthalmos can be seen shortly before the actual sloughing of the shed in snakes. This might be attributed to obstruction of the drainage at the lacrimal punctum caused by the accumulation of spectacle layers in the ventromedial subspectacular space [31]. Although those cases mostly resolve spontaneously, recurrence might be observed with each shedding cycle [61]. Other causes of pseudobuphthalmos also include direct or indirect blockage along the course of the lacrimal drainage such as caseous stomatitis with debris obstructing the drainage opening (Figure 25) or ascending infection of the lacrimal duct (Figure 26), abscesses, granulomas, foreign objects (e.g., substrate), trauma, or neoplasia located in the roof of the mouth or the facial tissues [11,31,47]. Especially in lizards, hypovitaminosis A with the development of metaplasia of the lacrimal duct and subsequent (sub)obstruction of the duct should be considered [48]. Similar predispositions and etiologies are commonly associated with dacryocystitis in, e.g., rabbits and cats in which the lacrimal duct also shows pronounced tortuosity [31,41].

Figure 24. Idiopathic congenital pseudobupthalmos in a neonate Malayan pit viper (*Calloselasma rhodostoma*) shortly after the first shed after hatching. A morbidity of approximately 50% was witnessed within the clutch.

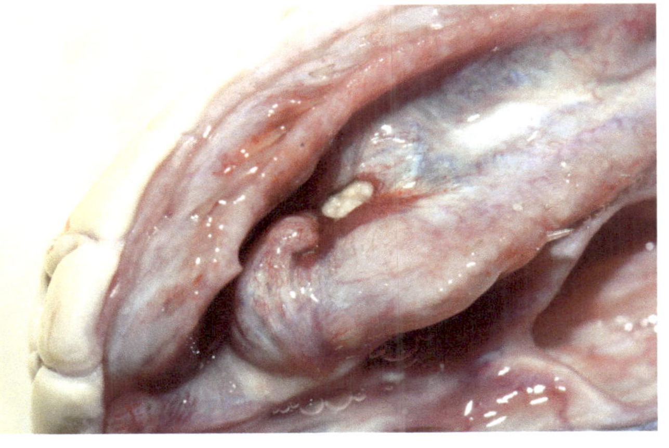

Figure 25. Debris obstructing the combined orifice of the lacrimal duct and vomeronasal organ in the roof of the mouth in a Burmese python (*Python bivittatus*) with pseudo-buphthalmos.

In case of pseudobuphthalmos, sampling of the clear fluid from the subspectacular space mostly yields negative results following microbiological examination [31]. Pseudobuphthalmos, however, can be an early presentation of ascending infection of the lacrimal duct from the oral cavity and might eventually lead to the development of subspectacular abscessation once infection has reached the subspectacular space. Secondary infection may occur if the distended spectacle is traumatized or perforated, mostly in cases where affected snakes or lizards show reactive and repetitive rubbing behavior. Detection and elimination of the primary etiology is fundamental towards the successful treatment of pseudobophtalmos. Congenital pseudobuphthalmos that is related to developmental disorders has a poor prognosis and, theoretically, conjunctivolateralostomy is often the single treatment option [11,31,47,60].

As the fluid in the subspectacular space is often pure lacrimal fluid, percutaneous aspiration of the fluid by inserting a needle between the periocular scales or aspiration following puncture through the spectacle (Figure 27) is highly feasible. Repeated aspiration of the fluid and flushing of the subspectacular space may be necessary. Unusually, a partial spectaculectomy needs to be performed as described for treating subspectacular abscesses.

If the spectacle is persistently and severely distended, it may be considerably stretched and become wrinkled and collapse once the excessive volume of subspectacular fluid is alleviated [31].

Figure 26. Medial peri-ocular and facial swelling associated with ascending bacterial infection of the lacrimal duct in a northern spiny-tailed gecko (*Strophurus ciliaris*).

Figure 27. (**A**) Pseudobuphthalmos in a gopher snake (*Pituophis catenifer*). (**B**) Drainage of the subspectacular space via punction through the spectacle.

Successful treatment of pseudobuphthalmos can be confirmed by assuring restored patency of the lacrimal duct. The latter may be performed by injection of a 1:5 balanced salt dilution of 10% fluorescein into the subspectacular space, either percutaneously (Figure 28) or through the spectacle [41]. If these routes seem ineffective, it might be useful to attempt retrograde flushing via the drainage orifice of the lacrimal duct in the roof of the mouth, although this might induce iatrogenic infection from the oral cavity.

Figure 28. Boa constrictor (*Boa constrictor*) following percutaneous injection of a 1:5 balanced salt dilution of 10% fluorescein into the subspectacular space to ascertain patency of the lacrimal drainage system following treatment for pseudobuphthalmos.

10. Subspectacular Infection

Subspectacular infection is presumably the most common disorder of the spectacle and its drainage system in reptiles. Boa constrictors (*Boa constrictor*) (Figure 29) and Burmese pythons (*Python bivittatus*) (Figure 30) seem to be especially predisposed to the development of subspectacular infection, although the disorder is also diagnosed in colubrids and to a lesser extent in viperids and lizards (Figure 31). Although subspectacular infection might resemble pseudobuphthalmos in the earliest stages, the fluid in the subspectacular space mostly becomes nontransparent with the appearance of flocculent exudate as seen in hypopyon of the anterior eye chamber [11,31,44,47,61,62].

If bilateral infection is present (Figure 32), systemic infection should be considered as a primary etiology. The amount of exudate, thickening, and opacification of the spectacle as well as engorgement of the blood vessels (Figure 33) are correlated with the severity and/or chronicity of the disorder. In chronic stages or cases of severe distension of the spectacle, the spectacle may show a whitish appearance and become extremely fragile because of impaired vascularization (Figure 34). In those cases, the spectacle might collapse and wrinkle or might even detach at the level of the transition zone when manipulated. When left untreated, subspectacular infection might extend to the periocular and facial region [11,31,44,47,61,62].

Except for rare cases where infection enters the subspectacular space through perforation of the spectacle or results from systemic infection, ascending infection from the oral cavity via the lacrimal duct is the predominant cause of subspectacular infection [11,44,45]. It is of fundamental importance to diagnose and eliminate facilitating or primary causes of subspectacular infection in reptiles, such as upper respiratory tract and oropharyngeal infection, in particular caseous stomatitis (Figure 35) and esophagitis. Especially in boa constrictors, these infections are commonly seen as a comorbidity to inclusion body disease caused by divergent reptarenaviruses [31,63].

Figure 29. Subspectacular infection associated with bacterial infection in a boa constrictor (*Boa constrictor*). The subspectacular space is filled with flocculent exudate.

Figure 30. Flocculent exudate causing partial opacification of the subspectacular space and mild engorgement of the blood vessels of the spectacle in a Burmese python (*Python bivittatus*).

Figure 31. Thick-tailed gecko (*Underwoodisaurus milii*) with subspectacular abscess associated with bacterial infection.

Figure 32. Bilateral subspectacular abscesses as seen in this Burmese python (*Python bivittatus*) is a relatively unusual presentation and should warrant the inclusion of possible systemic bacterial infection as a primary etiology.

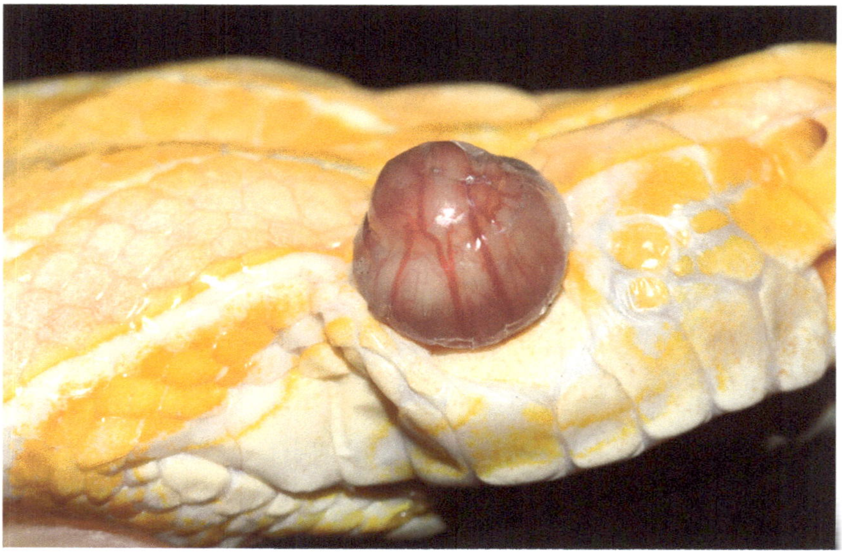

Figure 33. Prominent engorgement of blood vessels as well as severe distention and deformation of the spectacle as observed in a Burmese python (*Python bivittatus*), indicating chronic subspectacular abscessation.

Figure 34. Extremely chronic subspectacular abscess in a Schultze's pit viper (*Trimeresurus schultzei*). Avulsion of the spectacle may occur if manipulated because of the chronic inflammatory changes.

In almost all cases, bacterial isolates can be cultured from subspectacular abscesses and generally comprise Gram-negative bacteria, such as *Pseudomonas aeruginosa*, *Aeromonas* species, and *Salmonella* species, but infection caused by *Clostridium perfrigens* or methicillin-resistant *Staphylococcus* sp. infection has also been reported [11,61,62,64,65]. Interestingly, combined infection of the subspectacular space with flagellated protozoan parasites, presumably *Tritrichomonas* and *Monocercomonas* species, are frequently seen in snakes with bacterial subspectacular abscesses and flagellated protozoan infection has also been demonstrated in geckos with subspectacular abscesses [11,66]. Although the role in the etiopathogenesis of subspectacular abscesses remains unclear, it is certain that these protozoal parasites also reach the subspectacular space via ascending infection from the oral cavity, as the parasites can also be detected in the oropharyngeal cavity in most of these cases. Less frequently, *Serpentirhabdias*-species eggs and nematode worms have been detected in mostly bilateral subspectacular lesions in snakes [67,68].

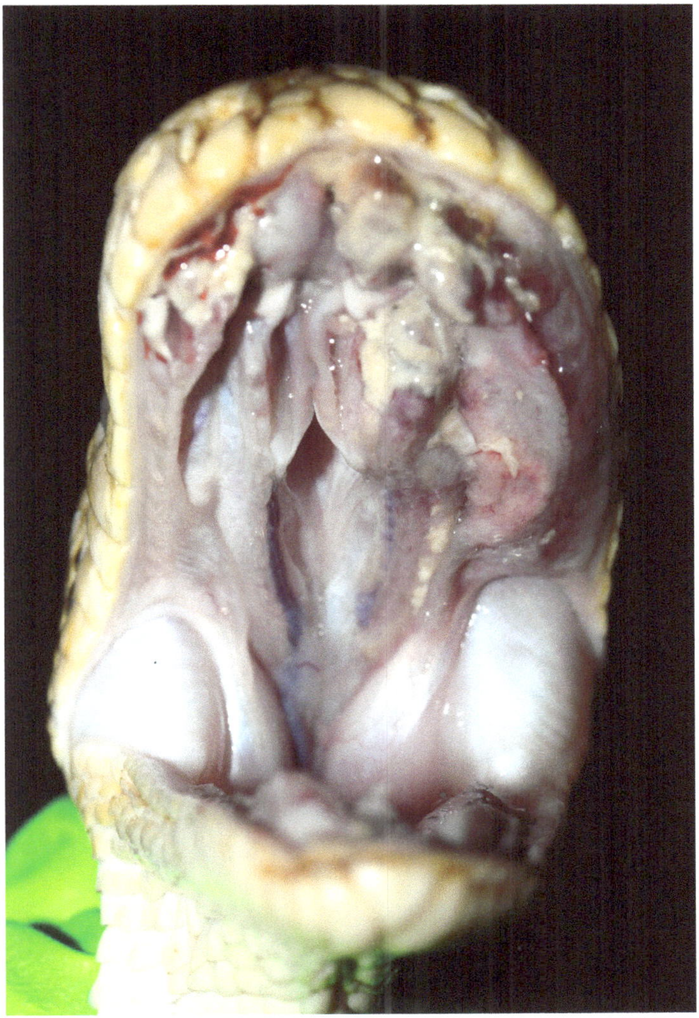

Figure 35. Caseous stomatitis as seen in this reticulated python (*Malayopython reticulatus*) is commonly seen in snakes with subspectacular infection as associated bacteria as well as flagellated protozoa may reach the subspectacular space following ascending infection via the lacrimal duct.

Although subspectacular infection is easily diagnosed based on the clinical appearance, the management of this disorder should always include the search for possible primary systemic infection or causes of generalized or localized immunosuppression. The collection of exudate from the subspectacular space (Figures 36 and 37) for cytology (wet preparations, Gram staining) and microbiological culture (including mycology), including susceptibility testing, is of fundamental importance. Molecular testing might be necessary to demonstrate fastidious organisms (e.g., *Chlamydia* spp. and *Chlamydia*-like organisms, *Mycobacterium* spp.). As the flocculent exudate mostly prevents the collection of qualitative samples by FNA (Figure 34), it is advisable to collect samples when creating a partial speculatectomy opening, as almost in all cases this is an integral part of the treatment [11,31,47,61,62,64].

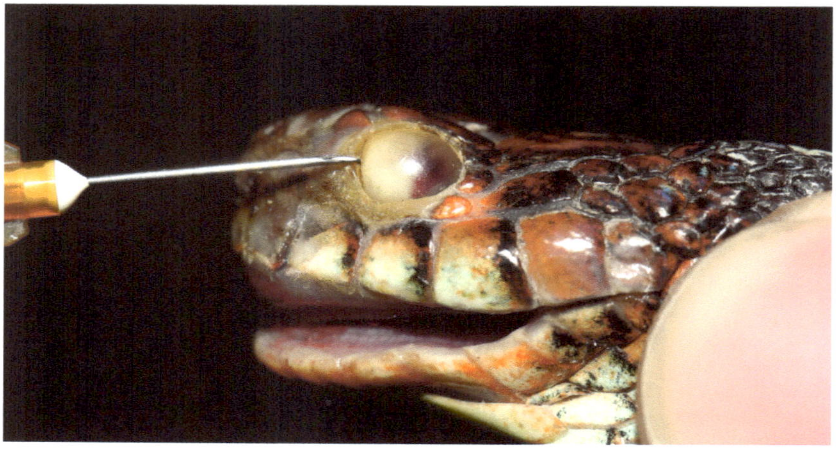

Figure 36. Sampling of a subspectacular abscess in a red-spotted garter snake (*Thamnophis sirtalis concinnus*) via fine needle aspiration through the spectacle.

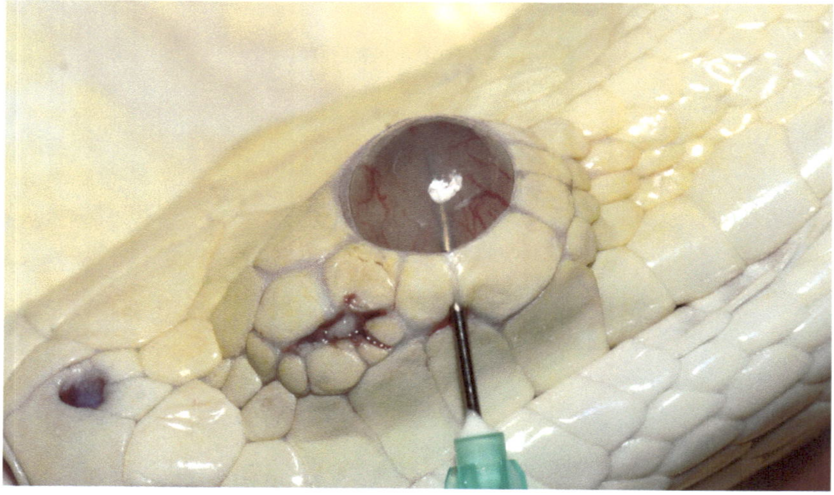

Figure 37. Sampling of a subspectacular abscess in a Burmese python (*Python bivittatus*) via fine needle aspiration by insertion of the needle in between the periocular scales.

Conventionally, performing partial spectaculectomy entails the creation of a small 25-to-30-degree wedge in the ventrolateral quadrant of the spectacle, as this minimizes the risk of damaging the drainage punctum [41]. In small-sized reptiles, it might be advisable to use an operation microscope to avoid iatrogenic damage to the cornea. Introduction of a needle into the subspectacular space is recommended, followed by the use of Wescott tenotomy scissors to excise the spectacle window (Figures 38 and 39).

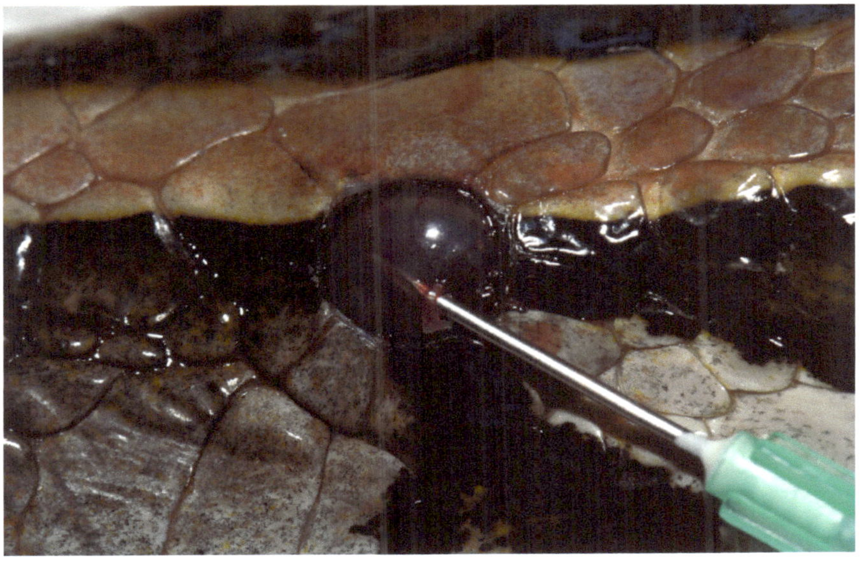

Figure 38. Insertion of a 21-gauge needle into a Burmese python (*Python bivittatus*) through the spectacle may allow aspiration of exudate from the subspectacular space and is also used as a first step in performing a partial spectaculectomy.

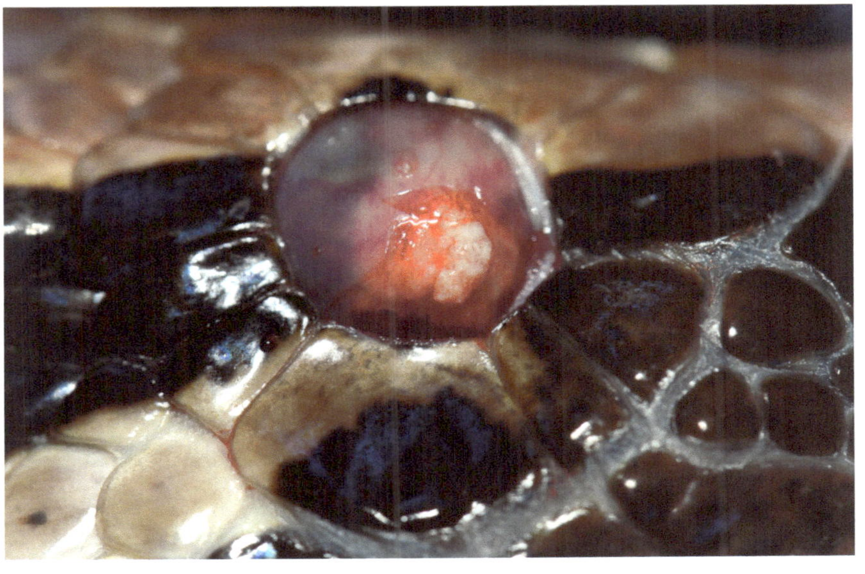

Figure 39. Partial spectaculectomy in a Burmese python (*Python bivittatus*) with a subspectacular abscess allows removal of the exudate from the subspectacular space.

Once the wedge has been created, the debris can be removed if complete aspiration was not feasible and topical antimicrobial administration can be performed. Treatment should be based on antimicrobial susceptibility testing and irrigation of the subspectacular space for an average 5 to 7 days is recommended [11,31,63,64]. Although crust formation is considered as an integral part of the healing process of the spectacle [50], it will prevent topical applications from reaching the subspectacular space. Consequently, it is recommended to avoid total dissolution of the crust during repeated irrigation of the subspectacular space or to perform this percutaneously [31,50].

As geckos moisten and clean their spectacles using their tongue, ingestion of excessive amounts of topical applications should be avoided. If keratomalacia is present, topical sodium EDTA may be useful to reduce collagenase activity [31]. If flagellate infection is confirmed, especially when concurrent oropharyngeal infection is present, per oral antiprotozoal treatment should be initiated [66]. In case exudative subspectacular infection has extended into the periocular and facial subcutaneous tissues, it might be necessary to surgically debride and flush these lesions (Figure 40), similar to what is performed during the treatment of subcutaneous abscesses or granulomas in reptiles.

Figure 40. Whenever subspectacular infection of the spectacle has extended to the periocular and/or facial region, making an incision in between the scales is often required to allow drainage and local treatment of the subcutaneous infection.

Once the exudate has been removed from the subspectacular space and infection has been eliminated, the distention of the spectacle often decreases quickly and the normal appearance of the subspectacular space and fluid restores gradually. As described for pseudobuphthalmos, the spectacle may collapse and wrinkle if chronic and severe distention occurred (Figure 41), and it is highly advisable to confirm patency of the lacrimal duct during or at the end of the treatment.

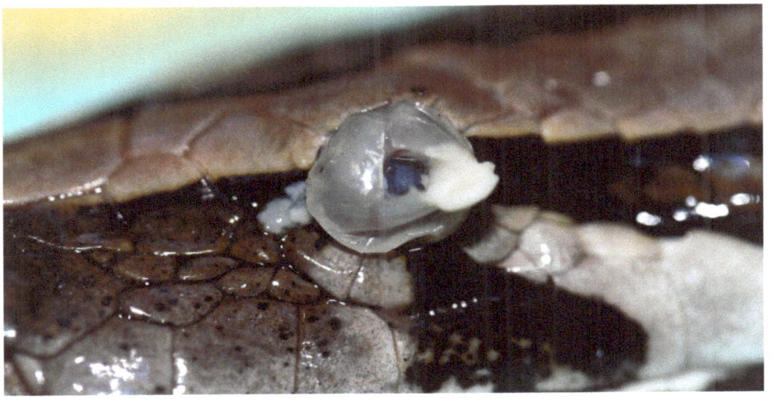

Figure 41. Collapse and wrinkling of the spectacle after performing a partial spectaculectomy and removal of the exudate in a Burmese python (*Python bivittatus*) are frequently seen in cases of chronic infection.

If persistent pseudobuphthalmos occurs, e.g., in cases of developmental disorders causing lacrimal duct obstruction or secondary to subspectacular infection, attempts can be made to recreate lacrimal drainage. The latter procedure is referred to as conjunctivoralostomy and comprises the creation of a new channel with a metal cannula or an appropriately sized needle from the fornix between the lateral maxillary teeth and edge of the oral cavity to the ventrolateral part of the spectacle [69]. The latter may be performed starting from the subspectacular space via a 30° wedge incision of the spectacle or from the fornix (Figure 42). Next, a 28-gauge silastic tube is threaded through the channel and fixed with sutures at its dorsal location and ventrally to the facial skin via a small incision in the labial scales. The tube has to be left in situ for at least 8 weeks in order to allow epithelialization around the tube. The latter is only feasible in reptiles with a considerable size, and postoperative fibrosis and occlusion of the newly created drainage duct are seen as the main complications [69].

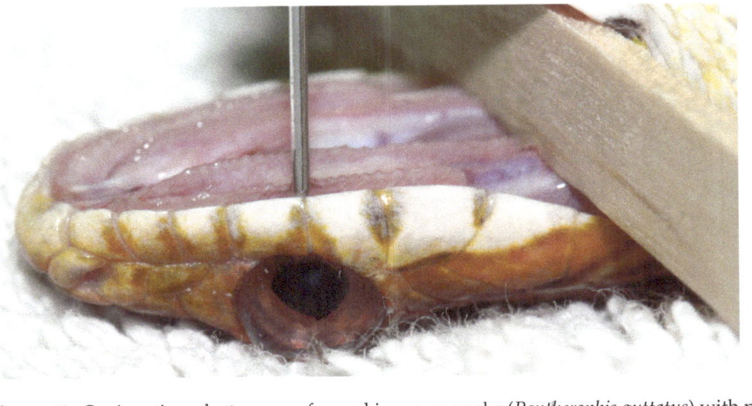

Figure 42. Conjunctivoralostomy performed in a corn snake (*Pantherophis guttatus*) with pseudobuphthalmos comprises the creation of a new channel. In this case, a metal cannula is advanced from the fornix between the lateral maxillary teeth and the facial skin to the ventral subspectacular space as the initial step of the procedure.

11. Conclusions

The reptilian spectacle is considered as the most sophisticated type of spectacle and is one of the most distinguishable traits of snakes and certain lizard species. In general, reptiles with a spectacle lack lacrimal glands. The tear film that fills the subspectacular space is produced by the Harderian gland and drains to the oral cavity via the lacrimal duct. Retention of the spectacle, spectaculitis, and injuries of the spectacle are commonly observed disorders. Especially in snakes, the complexity of the tortuous course of the lacrimal duct may vary between species and may predispose certain species towards the development of disorders of the subspectacular space, in particular pseudobuphthalmos and subspectacular infection. These conditions are caused by blockage of the lacrimal drainage system and ascending infection of the subspectacular space from the oropharyngeal cavity, respectively. A sound knowledge of the anatomy and physiology of the spectacle and its drainage system allows the establishment of an adequate diagnostic and therapeutic approach for these disorders.

Author Contributions: Conceptualization, T.H. and F.S.V.; methodology, T.H.; investigation, T.H.; writing—original draft preparation, T.H.; writing—review and editing, T.H. and F.S.V.; visualization T.H.; supervision, T.H. All authors have read and agreed to the published version of the manuscript.

Funding: This research received no external funding.

Institutional Review Board Statement: This study did not require ethical approval.

Informed Consent Statement: Not applicable.

Data Availability Statement: Not applicable.

Acknowledgments: The authors would like to express their gratitude to Frank Pasmans of the Department of Pathobiology, Pharmacology and Zoological Medicine of the Faculty of Veterinary Medicine, Ghent University for providing the photographic material used in Figure 2.

Conflicts of Interest: The authors declare no conflict of interest.

References

1. Kearney, M. Systematics of the amphisbaenia (Lepidosauria: Squamata) based on morphological evidence from recent and fossil forms. *Herpetol. Monogr.* **2003**, *17*, 1–74. [CrossRef]
2. van Doorn, K. Investigations on the Reptilian Spectacle. Ph.D. Thesis, University of Waterloo, Waterloo, ON, Canada, 2012.
3. Walls, G.L. Adaptations to media and substrates. In *The Vertebrate Eye and Its Adaptive Radiation*, 1st ed.; Walls, G.L., Ed.; Cranbrook Institute of Science: Bloomfield Hills, MI, USA, 1942; Volume 1, pp. 449–462.
4. Guerra-Fuentes, R.; Roscito, J.; Nunes, P.; Oliveira-Bastos, P.; Antoniazzi, M.; Jared, C.; Rodrigues, M. Through the looking glass: The spectacle in gymnophthalmid lizards. *Anat. Rec.* **2014**, *297*, 496–504. [CrossRef] [PubMed]
5. Underwood, G. The eye. In *Biology of the Reptilia*, 1st ed.; Gans, C., Parsons, T.S., Eds.; Academic Press: London, UK, 1970; Volume 2, pp. 1–95.
6. Bellairs, A.D.; Boyd, J.D. The lachrymal apparatus in lizards and snakes. The brille, the orbital glands, lachrymal canaliculi and origin of the lachrymal duct. *Proc. Zool. Soc. Lond.* **1947**, *117*, 81–108. [CrossRef]
7. Conrad, J.L. Phylogeny and systematics of Squamata (Reptilia) based on morphology. *Bull. Am. Mus. Nat. Hist.* **2008**, *310*, 1–182. [CrossRef]
8. Gamble, T.; Bauer, A.M.; Colli, G.R.; Greenbaum, E.; Jackman, T.R.; Vitt, L.J.; Simons, A.M. Coming to America: Multiple origins of New World geckos. *J. Evol. Biol.* **2011**, *24*, 231–244. [CrossRef]
9. Gauthier, J.A.; Kearney, M.; Maisano, J.A.; Rieppel, O.; Behlke, A.D.B. Assembling the squamate tree of life: Perspectives from the phenotype and the fossil record. *Bull. Peabody Mus. Nat. Hist.* **2012**, *53*, 3–308. [CrossRef]
10. Schwarz-Karsten, H. Über entwicklung und bau der brille bei Ophidiern und Lacertiliern und die anatomie ihrer Tränenwege. *Gegenbaurs Morphol.* **1933**, *72*, 499–540.
11. Millichamp, N.J.; Jacobson, E.R.; Wolf, E.D. Diseases of the eye and ocular adnexae in reptiles. *J. Am. Vet. Med. Assoc.* **1983**, *183*, 1205–1212.
12. O'Malley, B. General anatomy and physiology of reptiles. In *Clinical Anatomy and Physiology of Exotic Species: Structure and Function of Mammals, Birds, Reptiles and Amphibians*, 1st ed.; O'Malley, B., Ed.; Elsevier Ltd: Oxford, UK, 2005; Volume 1, pp. 17–39.
13. Johnson, G.L. Contributions to the comparative anatomy of the reptilian and the amphibian eye, chiefly based on ophthalmological examination. *Philos. Trans. R. Soc. Lond. B Biol. Sci.* **1927**, *215*, 315–353.

14. Boulenger, G.A. *Catalogue of Lizard in the British Museum (Natural History)*, 2nd ed.; Order of the Trustees of the British Museum: London, UK, 1885; Volume 2, pp. 18–568.
15. Rodrigues, M.T.; dos Santos, E.M. A new genus and species of eyelid-less and limb reduced gymnophthalmid lizard from northeastern Brazil (Squamata, Gymnophthalmidae). *Zootaxa* **2008**, *1873*, 50–60. [CrossRef]
16. Greer, A.E. On the adaptive significance of the reptilian spectacle: The evidence from scincid, teiid, and lacertid lizards. In *Advances in Herpetology and Evolutionary Biology*, 1st ed.; Rhodin, A.G.J., Miyata, K., Eds.; Museum of Comparative Zoology: Cambridge, MA, USA, 1983; Volume 1, pp. 213–221.
17. Greer, A.E. *The Biology and Evolution of Australian Lizards*, 1st ed.; Greer, A.E., Ed.; Surrey Beatty and Sons: Sydney, Australia, 1989; Volume 1, pp. 1–264.
18. Pellegrino, K.C.; Rodrigues, M.T.; Yonenaga-Yassuda, Y.; Sites, J.W.J. A molecular perspective on the evolution of microteiid lizards (Squamata, Gymnophthalmidae), and a new classification for the family. *Biol. J. Linn. Soc.* **2001**, *74*, 315–338. [CrossRef]
19. Castoe, T.A.; Doan, T.M.; Parkinson, C.L. Data partitions and complex models in Bayesian analysis: The phylogeny of gymnophthalmid lizards. *Syst. Biol.* **2004**, *53*, 448–469. [CrossRef] [PubMed]
20. Angel, F.; Rochon-Duvigneaud, A. Les divers types de paupières des sauriens et des ophidiens. *Bull. Mus. Natl. Hist. Nat.* **1941**, *13*, 517–523.
21. Rochon-Duvigneaud, A. La protection de la cornée chez les vertébrés qui rampent (serpents etpoissons anguiformes). *Ann. Ocul.* **1916**, *153*, 185–202.
22. Jonniaux, P.; Kumazawa, Y. Molecular phylogenetic and dating analyses using mitochondrial DNA squences of eyelid geckos (Squamata: Eublepharidae). *Gene.* **2008**, *407*, 105–115. [CrossRef] [PubMed]
23. Bellairs, A.D.; Underwood, G. The origin of snakes. *Biol. Rev.* **1951**, *26*, 193–237. [CrossRef]
24. Neher, E.M. The origin of the brille in *Crotalus confluentus lutosus* (Great Basin Rattlesnake) *T. Am. Ophthal. Soc.* **1935**, *33*, 535–545.
25. Bellairs, A.D. The eyelids and spectacle in geckos. *Proc. Zool. Soc. Lond.* **1948**, *118*, 420–425. [CrossRef]
26. Rehorek, S.J.; Firth, B.T.; Hutchinson, M.N. Can an orbital gland function in the vomeronasal sense? A study of the pygopodid Harderian gland. *Can. J. Zool.* **2000**, *78*, 648–654. [CrossRef]
27. Da Silva, M.O.; Heegaard, S.; Wang, T.; Nyengaard, J.R.; Bertelsen, M.F. The spectacle of the ball python (*Python regius*): A morphological description. *J. Morphol.* **2014**, *275*, 489–496. [CrossRef]
28. Da Silva, M.O.; Bertelsen, M.F.; Heegaard, S.; Garner, M.M. Spectacular manifestations of systemic diseases of the snake: A histopathological description of four cases. *J. Zoo. Aquarium. Res.* **2015**, *3*, 43–46.
29. Da Silva, M.O.; Bertelsen, M.F.; Heegaard, S.; Garner, M.M. Ophidian spectaculitis and spectacular dysecdysis: A histologic description. *Vet. Pathol.* **2015**, *52*, 1220–1226. [CrossRef] [PubMed]
30. Da Silva, M.O.; Bertelsen, M.F.; Wang, T.; Prause, J.U.; Svahn, T.; Heegaard, S. Comparative morphology of the snake spectacle using light and transmission electron microscopy. *Vet. Ophthalmol.* **2016**, *19*, 285–290. [CrossRef]
31. Millichamp, N.J. Ophthalmology of Serpentes: Snakes. In *Wild and Exotic Animal Ophthalmology: Invertebrates, Fishes, Amphibians, Reptiles, and Birds*, 1st ed.; Montiani-Ferreira, F., Moore, B.A., Ben-Shlomo, G., Eds.; Springer Nature Switzerland AG: Cham, Switzerland, 2022; Volume 1, pp. 231–271.
32. van Doorn, K.; Sivak, J.G.; Vijayan, M.M. β-Keratin composition of the specialized spectacle scale of snakes and geckos. *Can. J. Zool.* **2014**, *92*, 299–307. [CrossRef]
33. Da Silva, M.O.; Heegaard, S.; Wang, T.; Gade, J.T.; Damsgaard, C.; Bertelsen, M.F. Morphology of the snake spectacle reflects its evolutionary adaptation and development. *BMC Vet. Res.* **2017**, *13*, 258. [CrossRef] [PubMed]
34. Sivak, J.G. The role of the spectacle in the visual optics of the snake eye. *Vis. Res.* **1977**, *17*, 293–298. [CrossRef]
35. van Doorn, K.; Sivak, J.G. Blood flow dynamics in the snake spectacle. *J. Exp. Biol.* **2013**, *216*, 4190–4195. [CrossRef]
36. van Doorn, K.; Sivak, J.G. Spectral transmittance of the spectacle scale of snakes and geckos. *Contrib. Zool.* **2015**, *84*, 1–12. [CrossRef]
37. Mead, A.W. Vascularity in the reptilian spectacle. *Investig. Ophthalmol.* **1976**, *15*, 587–591.
38. Bellhorn, R.W.; Strom, A.R.; Motta, M.J.; Doval, J.; Hawkins, M.G.; Paul-Murphy, J. Snake spectacle vessel permeability to sodium fluorescein. *Vet. Ophthalmol.* **2018**, *21*, 119–124. [CrossRef]
39. Foureaux, G.; Egami, M.I.; Jared, C.; Antoniazzi, M.M.; Gutierre, R.C.; Smith, R.L. Rudimentary eyes of squamate fossorial reptiles (Amphisbaenia and Serpentes). *Anat. Rec.* **2009**, *293*, 351–357. [CrossRef]
40. Taub, A.M. Ophidian cephalic glands. *J. Morphol.* **1966**, *118*, 529–541. [CrossRef]
41. Souza, N.M.; Maggs, D.J.; Park, S.A.; Puchalski, S.M.; Reilly, C.M.; Paul-Murphy, J.; Murphy, C.J. Gross, histologic, and micro-computed tomographic anatomy of the lacrimal system of snakes. *Vet. Ophthalmol.* **2015**, *18* (Suppl. 1), 15–22. [CrossRef] [PubMed]
42. Minucci, S.; Baccari, G.C.; di Matteo, L. Histology, histochemistry, and ultrastructure of the Harderian gland of the snake *Coluber viridiflavus*. *J. Morphol.* **1992**, *211*, 207–212. [CrossRef] [PubMed]
43. Rehorek, S.J.; Legenzoff, E.J.; Carmody, K.; Smith, T.D.; Sedlmayr, J.C. Alligator tears: A reevaluation of the lacrimal apparatus of the crocodilians. *J. Morphol.* **2005**, *266*, 298–308. [CrossRef]
44. Lawton, M.P.C. Ophthalmology. In *Mader's Reptile and Amphibian Medicine and Surgery*, 3rd ed.; Divers, S.J., Stahl, S.J., Eds.; Elsevier Health Sciences: St. Louis, MO, USA, 2019; Volume 1, pp. 721–735.

45. Lawton, M.P.C. Eye. In *Mader's Reptile and Amphibian Medicine and Surgery*, 3rd ed.; Divers, S.J., Stahl, S.J., Eds.; Elsevier Health Sciences: St. Louis, MO, USA, 2019; Volume 1, pp. 1024–1027.
46. Hellebuyck, T.; Scheelings, T.F. Dysecdysis. In *Mader's Reptile and Amphibian Medicine and Surgery*, 3rd ed.; Divers, S.J., Stahl, S.J., Eds.; Elsevier Health Sciences: St. Louis, MO, USA, 2019; Volume 1, pp. 1304–1306.
47. Williams, D.L. The reptile eye. In *Ophthalmology of Exotic Pets*, 1st ed.; Williams, D.L., Ed.; John Wiley & Sons Ltd.: Chichester, UK, 2012; Volume 1, pp. 159–196.
48. Hellebuyck, T.; Pasmans, P.; Haesebrouck, F.; Martel, A. Dermatological diseases in lizards. *Vet. J.* **2012**, *193*, 38–45. [CrossRef] [PubMed]
49. Scheelings, F.S.; Hellebuyck, T. Dermatology—Skin. In *Mader's Reptile and Amphibian Medicine and Surgery*, 3rd ed.; Divers, S.J., Stahl, S.J., Eds.; Elsevier Health Sciences: St. Louis, MO, USA, 2019; Volume 1, pp. 699–711.
50. Maas, A.K.; Paul-Murphy, J.; Kumaresan-Lampman, S.; Dubielzig, R.; Murphy, C.J. Spectacle Wound Healing in the Royal Python, *Python regius*. *J. Herpetol. Med. Surg.* **2010**, *20*, 8. [CrossRef]
51. Ledbetter, E.C.; Marion, J.S.; Morrisey, J.K. Amniotic membrane grafting for traumatic complete spectaculectomy and keratomalacia in a Boelen's python (*Simalia boeleni*). *Vet. Opthalm.* **2021**, *24*, 295–300. [CrossRef]
52. Wojick, K.B.; McBride, M.P. Successful treatment of spectacle avulsion in a Brazilian Rainbow Boa (*Epicrates cenchria cenchria*) by using a porcine small intestinal submucosal graft. *J. Herpetol. Med. Surg.* **2018**, *28*, 23–28. [CrossRef]
53. Cazalot, G.; Rival, F.; Linsart, A.; Isard, P.F.; Tissier, M.; Peiffer, R.L.; Dulaurent, T. Scanning laser ophthalmoscopy and optical coherence tomography imaging of spectacular ecdysis in the corn snake (*Pantherophis guttatus*) and the California king snake (*Lampropeltis getulus californiae*). *Vet. Ophthalmol.* **2015**, *18* (Suppl. 1), 8–14. [CrossRef]
54. Tusler, C.A.; Maggs, D.J.; Kass, P.H.; Paul-Murphy, J.R.; Schwab, I.R.; Murphy, C.J. Spectral domain optical coherence tomography imaging of spectacular ecdysis in the royal python (*Python regius*). *Vet. Ophthalmol.* **2015**, *18* (Suppl. 1), 1–7. [CrossRef] [PubMed]
55. Steinmetz, A.; Neul, A.; Schmidt, V.; Pees, M. Pathology of the spectacle in a green tree python (*Morelia viridis*) as a clinical sign of an naso-orbital inflammation. *Kleintierpraxis.* **2016**, *61*, 554–559.
56. Zwart, P.; Verwer, M.A.; De Vries, G.A.; Hermanides-Nijhof, E.J.; De Vries, H.W. Fungal infection of the eyes of the snake *Epicrates chenchria maurus*: Enucleation under halothane narcosis. *J. Small. Anim. Pract.* **1973**, *14*, 773–779. [CrossRef] [PubMed]
57. McBride, M.P.; Wojick, K.B.; Georoff, T.A.; Kimbro, J.; Garner, M.M.; Wang, X.; Childress, A.L.; Wellehan, J.F. *Ophidiomyces ophiodiicola* dermatitis in eight free-ranging timber rattlesnakes (*Crotalus horridus*) from Massachusetts. *J. Zoo. Wildl. Med.* **2015**, *46*, 86–94. [CrossRef]
58. Ohkura, M.; Worley, J.J.; Hughes-Hallett, J.E.; Fisher, J.S.; Love, B.C.; Arnold, A.E.; Orbach, M.J. *Ophidiomyces ophiodiicola* on a captive black racer (*Coluber constrictor*) and a garter snake (*Thamnophis sirtalis*) in Pennsylvania. *J. Zoo. Wildl. Med.* **2016**, *47*, 341–346. [CrossRef]
59. Gardiner, D.W.; Baines, F.M.; Pandher, K. Photodermatitis and photokeratoconjunctivitis in a ball python (*Python regius*) and a blue-tongue skink (*Tiliqua* species). *J. Zoo. Wildl. Med.* **2009**, *40*, 757–766. [CrossRef]
60. Millichamp, N.J. Congenital abnormalities in snakes. In Proceedings of the Third International Colloquium of Pathology of Reptiles and Amphibians, Orlando, FL, USA, 13–15 January 1989; p. 103.
61. Ensley, P.K.; Anderson, M.P.; Bacon, J.P. Ophthalmic disorders in three snakes. *J. Zoo. Anim. Med.* **1978**, *9*, 57–59. [CrossRef]
62. Miller, W. Subspectacular abscess in a Burmese python. *Auburn. Veterin.* **1986**, *41*, 19–21.
63. Simard, J.; Marschang, R.E.; Leineweber, C.; Hellebuyck, T. Prevalence of inclusion body disease and associated comorbidity in captive collections of boid and pythonid snakes in Belgium. *PLoS ONE* **2020**, *15*, e0229667. [CrossRef]
64. Cleymaet, A.M.; Ehrhart, E.J.; Sadar, M.J.; Johnston, M.; Wotman, K.; Henriksen, M.L. Unfolding the diagnosis of subspectacular fluid opacity in a corn snake (*Pantherophis guttatus*). *Vet. Ophthalm.* **2020**, *23*, 754–759. [CrossRef]
65. Lee, S.Y.; Kim, J.W. Subspectacular abscess involved with MRSA (methicillin resistant *Staphylococcus aureus*) in a snake. *J. Veterin. Clin.* **2011**, *28*, 446–448.
66. Miller, H.A.; Freye, F.L.; Graig, T.M. Trichomonas associated with ocular and subcutaneous lesions in geckos. *Proc. Am. Assoc. Zoo. Vet.* **1994**, 102–107.
67. Hausmann, J.C.; Mans, C.; Dreyfus, J.; Reavill, D.R.; Lucio-Forster, A.; Bowman, D.D. Subspectacular Nematodiasis caused by a novel *Serpentirhabdias* species in ball pythons (*Python regius*). *J. Comp. Pathol.* **2015**, *152*, 260–264. [CrossRef]
68. Kurochkin, Y.V.; Guscov, E.P. A new nematode species from the eye of *Natrix natrix*. In *Helminths of Man, Animals and Plants and Their Control: Papers on Helminthology Presented to Academician K.I. Skryabin on His 85th Birthday*, 1st ed.; Skryabin, K.I., Ed.; Izdatelstvo: Moscow, Russia, 1963; Volume 1, pp. 183–185.
69. Millichamp, N.J.; Jacobson, E.R.; Dziezyc, J. Conjunctivoralostomy for treatment of an occluded lacrimal duct in a blood python. *J. Am. Vet. Med. Assoc.* **1986**, *189*, 1136–1138.

animals MDPI

Article

Evaluating the Physiologic Effects of Alfaxalone, Dexmedetomidine, and Midazolam Combinations in Common Blue-Tongued Skinks (*Tiliqua scincoides*)

Haerin Rhim, Ashleigh M. Godke, M. Graciela Aguilar and Mark A. Mitchell *

Department of Veterinary Clinical Sciences, School of Veterinary Medicine, Louisiana State University, Baton Rouge, LA 70803, USA; hrhim1@lsu.edu (H.R.); agodke3@lsu.edu (A.M.G.); magui32@lsu.edu (M.G.A.)
* Correspondence: mmitchell@lsu.edu

Simple Summary: Common blue-tongued skinks (*Tiliqua scincoides*) are popular pets due to their docile temper. Because of their popularity, they are routinely presented to veterinarians for examinations or procedures; however, to date, there has been limited research evaluating sedation protocols for this species. This study aimed to test different sedation combinations in these skinks: alfaxalone alone, alfaxalone with midazolam, dexmedetomidine with midazolam, and a combination of alfaxalone, dexmedetomidine, and midazolam. All four combinations provided safe sedation, but there were different physiologic responses noted. According to our trials, the combinations of all three drugs or alfaxalone with midazolam are recommended for minor procedures.

Abstract: Common blue-tongued skinks (*Tiliqua scincoides*) are popular pet reptiles; however, there has been limited research to investigate sedatives for this species. The purpose of this study was to measure the physiologic effects of four combinations of alfaxalone, dexmedetomidine, and midazolam for minor procedures such as intubation and blood collection. Eleven common blue-tongued skinks (*Tiliqua scincoides*) were used for this prospective, randomized cross-over study. The subcutaneous combinations were used as follows: 20 mg/kg alfaxalone (A), 10 mg/kg alfaxalone and 1 mg/kg midazolam (AM); 0.1 mg/kg dexmedetomidine and 1 mg/kg midazolam (DM); and 5 mg/kg alfaxalone, 0.05 mg/kg dexmedetomidine, and 0.5 mg/kg midazolam (ADM). Heart rate, respiratory rate, palpebral reflex, righting reflex, escape reflex, toe pinch withdrawal reflex, tongue flicking, and the possibility of intubation were recorded at baseline and every 5 min for 60 min. Venous blood gases were measured at baseline, full sedation, and recovery. Heart and respiratory rates decreased significantly in all groups, but the reductions were most prominent in DM and ADM. Analgesic effects, as measured by the toe pinch withdrawal reflex, were only observed in DM and ADM. Intubation was possible in all four protocols; however, it was not possible in two DM skinks. Based on these trials, ADM and AM are recommended for minor procedures in blue-tongue skinks.

Keywords: anesthesia; blood gas; blue-tongue skink; intubation; lizard; pH; PCO_2; reptile; sedation; *Tiliqua scincoides*

Citation: Rhim, H.; Godke, A.M.; Aguilar, M.G.; Mitchell, M.A. Evaluating the Physiologic Effects of Alfaxalone, Dexmedetomidine, and Midazolam Combinations in Common Blue-Tongued Skinks (*Tiliqua scincoides*). *Animals* **2024**, *14*, 2636. https://doi.org/10.3390/ani14182636

Academic Editor: Tom Hellebuyck

Received: 2 August 2024
Revised: 6 September 2024
Accepted: 9 September 2024
Published: 11 September 2024

1. Introduction

Common blue-tongued skinks (*Tiliqua scincoides*) are popular as pets and exhibit animals at zoological institutions because of their inquisitive behavior and ease of care. Because of the popularity of these lizards, veterinarians are being asked to provide medical and surgical care for these animals. However, to date, there has been a dearth of evidence-based research related to the medical and surgical management of these animals. A search of PubMed and Google Scholar using the key words "*Tiliqua*" or "blue-tongued skink" on 1 July 2024 revealed only twenty-one articles related to the medical and surgical care of these animals, with five being case reports and sixteen being clinically related research.

Limited evidence-based research can make it challenging for veterinarians to identify "best practices" for managing these animals in the captive setting.

One of the challenges a veterinarian faces when working with any new species is identifying safe, consistent sedation and anesthetic protocols to manage different diagnostic and surgical procedures. From the previously noted literature search, six papers were case reports or evidence-based articles for non-anesthetic purposes that described sedation or anesthesia in common blue tongue skinks, while one was an evidence-based research paper that used a specific anesthetic regimen on a study population of skinks [1–7]. In these seven articles, the sedation and anesthetic protocols were used to perform different procedures, including computed tomography scans, endoscopic sexing, radiography, and induction of anesthesia. The sedatives/anesthetics and doses used in these articles included propofol (9 mg/kg intravenous [IV] and 5 mg/kg IV) [1,2], medetomidine (0.2 mg/kg) and ketamine (1.1 mg/kg intramuscular [IM]) [3], alfaxalone alone (9 mg/kg IV and 10 mg/kg IM) [4,5], alfaxalone (10 mg/kg) and midazolam (0.4 mg/kg IM) [5], and midazolam (0.7 mg/kg), dexmedetomidine (0.05 mg/kg), and ketamine (5 mg/kg IM) [7]. The evidence-based study only evaluated alfaxalone, and the dose used was sufficient to obtain a loss of righting reflex for 9 min [4]. Based on these limited findings, it is important to pursue additional evidence-based research to identify safe and reliable sedation protocols for this species.

Alfaxalone is an injectable anesthetic commonly used in captive reptiles. Based on a search of the same engines and dates using the keywords alfaxalone and reptile, 42 original research studies have described the use of alfaxalone in reptiles as a solitary sedative or in combination with other sedatives [4,8–49]. Alfaxalone is a neuroactive steroid that acts on GABAA receptors as a positive allosteric modulator; it can also act as a direct agonist when used in high doses [50]. Major benefits attributed to alfaxalone are that it has minimal inhibitory effects on the cardiovascular and respiratory systems compared to other commonly used injectable anesthetics, and it has a rapid onset of action [12,51]. However, large volumes are needed in reptiles compared to other sedatives, which makes subcutaneous (SC) injection preferred over IM injection.

Dexmedetomidine, an α_2-adrenoceptor agonist, is also a widely used injectable anesthetic in veterinary medicine. Based on a review of the same methods noted previously using the keywords dexmedetomidine, medetomidine, and reptile, 46 original research studies described using dexmedetomidine (n = 24) or medetomidine (n = 23) in reptiles [10,17,26,28,31–35,43,52–87]. The α_2-adrenoceptor agonists have been found to provide sedation, muscle relaxation, and visceral analgesia, while cardiovascular and respiratory depression are the most commonly described adverse effects. Ineffective sedation has been reported in brown anole lizards (*Anolis sagrei*), Japanese grass lizards (*Takydromus tachydromoides*), and Argentine tegus (*Salvator merianae*) when dexmedetomidine was given as a single agent [32,33,70]. However, it is more common to combine α_2-adrenoceptor agonists with other sedatives/anesthetics to reduce the dosages of each drug and minimize any potential adverse effects. Another benefit of α_2-adrenoceptor agonists is that they can be reversed once a procedure is completed.

Midazolam is a commonly used benzodiazepine in veterinary medicine, and it is used both alone or in combination with other sedatives and anesthetics. The benzodiazepines enhance the effects of GABA on GABAA receptors, resulting in sedation. The most common uses for midazolam in veterinary medicine are as a premedication before surgery, as a sedative, and for seizure control [88,89]. According to the same search criteria mentioned previously and using the keywords midazolam and reptile, 36 original research studies were found describing midazolam use in reptiles [17,26,28,32,34,35,64,66,70,72,77,80,82,83,85,86,90–109]. As a sole agent, midazolam has been found to provide both successful and limited sedative effects in reptiles [32,91,93,99]; however, it is more common to combine midazolam with other sedatives or anesthetics to achieve a synergistic effect [70]. This also has been proven in pigs and rats, where the combination of α_2-adrenoceptor agonists and midazolam provided maximal pain control for a longer duration than those used alone [110,111]. A major benefit of using multiple drugs in combination is that the adverse effects of each drug can be minimized by reducing the dose of each drug [112]. Another benefit of midazolam is that it can be reversed with flumazenil.

Given the limited data available for blue-tongue skinks, further evidence-based research is needed to better ascertain the value of sedatives for this species. The purpose of this study was to determine the preferred combinations and doses of alfaxalone, dexmedetomidine, and midazolam for providing safe and reliable sedation for performing basic procedures in blue-tongue skinks. The specific hypotheses for this study were that (1) all sedation protocols would be safe and provide sufficient sedation for minor procedures such as tracheal intubation, collecting radiographic images, and blood collection from the skinks; (2) sedation would occur within 10 min after SC injection; and (3) that the combinations with dexmedetomidine would result in decreased heart rates (HR) and respiratory rates (RR).

2. Materials and Methods

A prospective, crossover study was conducted under the regulations set forth by the Louisiana State University Institutional Animal Care and Use Committee (23-077).

2.1. Animals

Eleven adult (yearling) non-sexed common blue-tongued skinks were used in the study. The skinks were obtained from a private breeder and were used to routine handling. Physical examinations were performed on the skinks prior to the study, and the animals were found to be clinically healthy with mean ± standard deviation (SD) bodyweights of 666.8 ± 123.9 g. The animals were housed in 18″ × 12″ × 36″ (45 × 31 × 91 cm) enclosures (Rubbermaid, Wooster, OH, USA) in a room with an ambient temperature and humidity of 86 °F (30 °C) and 20–30%, respectively. The skinks were exposed to a 12 h photoperiod using ambient fluorescent lighting; no ultraviolet B lighting was provided. Animals were provided ad-libitum chlorinated tap water daily and fed three times a week with 30 g of wet cat food (Wellness, Burlington, MA, USA). The skinks were fasted the day before any sedation trial. The sedation trials were performed in a room with an ambient temperature of 75 °F (23.9 °C), but the skinks were kept in an incubator (TLC-40; Brinsea, Titusville, FL, USA) set at 85 °F (29.4 °C) during trials to maintain a consistent body temperature. The skinks were only removed from the incubator for the initial injection and for measuring the physiologic parameters at baseline and every 5 min until the trial was completed; the skinks were returned to the incubator after each sampling period.

2.2. Pilot Sedation Trials

A set of pilot studies were done to determine the final doses for the primary sedation trials. Three drugs were used in four different combinations for the pilot studies: alfaxalone (Alfaxan, 10 mg/mL; Jurox Inc., Kansas City, MO, USA), dexmedetomidine (Dexmedesed, 0.5 mg/mL; Dechra Veterinary Products, Overland Park, KS, USA), and midazolam (5 mg/mL; Hikma Pharmaceuticals, Berkeley Heights, NJ, USA). Four doses of alfaxalone; two doses of alfaxalone with midazolam; two doses of dexmedetomidine with 1mg/kg midazolam; and one dose of alfaxalone, dexmedetomidine, and midazolam were evaluated (in total, 18 trials, Table 1). These doses were selected based on the authors' previous clinical experiences with skinks. All injections were given SC using one syringe in the right flank, caudal to the right forearm. Skinks were randomly selected for each pilot study using a random number generator (random.org; accessed on 14 August 2023), and two skinks were used for each dose tested. A 2 week washout was provided between each pilot study. Moreover, all doses were tested in randomized dosing order, and the individual assessing the skink during each trial was blinded to the drug and dose because a second individual obtained and administered the drugs.

Prior to starting any trial, a baseline HR, RR, palpebral reflex, righting reflex, escape reflex, toe pinch withdrawal reflex, tongue flicking, and the possibility of intubation were recorded. Heart rate was measured using a crystal Doppler (Parks Medical Electronics, Aloha, OR, USA) for 15 s and multiplying the number by 4 to determine beats/minute, while RR was measured by counting rib excursions for one full minute to limit the impact of breath-holding behavior. The palpebral reflex was evoked by touching a cotton-tipped applicator to the medial and lateral canthus of the eye and noting a blink. The righting reflex was evoked by placing the skink on its back and measuring the time required to return its body to a sternal position. The escape reflex was induced by pinching the distal third of the tail and evaluating whether the skink moved away from this stimulus. This reflex was also used to determine the suitability for collecting radiographic images using the sedation protocols. For the toe pinch withdrawal reflex, the toes of the hindlimbs were used. A padded hemostat was used to apply pressure to a digit to determine if the skink withdrew the leg as an indicator of deep pain. The tongue flicking was based on whether the tongue moved freely. For the reflexes and possibility of intubation, a 1–3 ordinal scale was used: (1) reflex was present (<1 s) or unable to intubate at all, (2) reflex was delayed (1–4 s) or unable to intubate easily but might be possible with physical restraint of the mouth, and (3) reflex was absent (>5 s) or able to intubate smoothly. The HR, RR, reflexes, and attempts at intubation were recorded every 5 min until all 5 reflexes returned to baseline (score = 1). Full sedation was determined to be when the righting and escape reflexes were both scored as 3 (absent). All skinks recovered uneventfully from the pilot studies.

Table 1. Description of the drugs, sample sizes, and doses used to determine sedation protocols for common blue-tongued skinks.

Drugs	Pilot	Primary	Alfaxalone	Dexmedetomidine	Midazolam
	N of Skinks				
A					
	2		10 mg/kg		
	2		15 mg/kg		
	2		17 mg/kg		
	2	11	20 mg/kg		
AM					
	2	11	10 mg/kg		1 mg/kg
	2		15 mg/kg		1 mg/kg
DM					
	2			0.05 mg/kg	1 mg/kg
	2	11		0.1 mg/kg	1 mg/kg
ADM					
	2	11	5 mg/kg	0.05 mg/kg	0.5 mg/kg

2.3. Primary Sedation Trials

The doses used in the pilot trials that had the least impact on HR and RR, provided sedation, and consistently allowed for intubation were selected for the final complete sedation trial evaluating all 11 skinks. The final doses selected were as follows: 20 mg/kg alfaxalone [A]; 10 mg/kg alfaxalone and 1 mg/kg midazolam [AM]; 0.1 mg/kg dexmedetomidine and 1 mg/kg midazolam [DM]; and 5 mg/kg alfaxalone, 0.05 mg/kg dexmedetomidine, and 0.5 mg/kg midazolam [ADM]. The injection site and route were the same as in the pilot trials. All 11 skinks were randomly assigned to the order in which they received each combination, and a 2 week washout was provided between trials. Two to four combinations were used on each trial day using three to six individuals, and the individuals assessing the skinks were blinded to the drug and dose. To avoid a potential bias attributed to a refractory response to the sedatives over time, the order of the sedatives used was included as an independent variable in the statistical modeling. The primary sedation trials were performed under the same environmental conditions as described for the pilot trials. Similar to the pilot study, all parameters were measured at baseline and every 5 min until the skink recovered. However, unlike the pilot study, 0.05 mg/kg flumazenil (0.1 mg/mL; Hikma Pharmaceuticals) and 0.5 mg/kg atipamezole

(Revertidine, 5 mg/mL; Modern Veterinary Therapeutics, Miami, FL, USA) were given 60 min SC after induction to skinks receiving dexmedetomidine or midazolam, respectively, if the skinks were not recovered. The time to loss of righting reflex, time to intubation, time to recovery, duration of intubation, and duration of the loss of righting reflex were also analyzed. Intubation was assessed only for feasibility and the tube was not maintained. Attempts were made every 5 min. Again, recovery was defined as the time when all reflexes returned to baseline (ordinal score = 1). Blood samples were collected from the ventral tail vein of six randomly selected skinks using a 25-gauge needle fastened to a 1 mL syringe to measure blood gases at baseline, time of full sedation, and recovery for each of the drug combinations. The same 6 skinks were sampled for each drug combination to minimize individual physiologic variation. The venous gas analysis was conducted using the VetScan i-STAT 1 (Abaxis, Union City, CA, USA) and CG4+ cartridge. The body temperature for the i-STAT was corrected to 86 °F (30 °C) to match the skink body temperature. No more than 0.15 mL of whole blood was collected for each sample, and thus, blood sampling was limited to <1% of body weight.

2.4. Statistical Analysis

The sample size for the primary trials was based on the following a priori data: an alpha = 0.05, power = 0.8, an expected difference in full sedation time between drug combinations of 8 min, and standard deviations between groups of 8 min. The sample size for the blood gas component of the study was based on an alpha = 0.05, power = 0.8, a mean difference in pH between baseline and full sedation of 0.3, and a standard deviation of 0.15. The minimum sample sizes for each of these comparisons were 10 and 5, respectively. Because our IACUC recommends a 10% buffer to minimize the likelihood of a type II error, an additional subject was added to each group to obtain the final numbers of 11 and 6 for the primary sedation trial and blood gas component, respectively. Sample size calculations were performed using MedCalc® Statistical Software version 22.006 (MedCalc Software Ltd., Ostend, Belgium; https://www.medcalc.org; accessed on 14 August 2023). The distributions of the continuous data and their residuals were assessed for normality using the Shapiro–Wilk test, q-q plots, skewness, kurtosis, and histograms. Data that met the assumptions of normality are reported by the mean ± SD, while non-normally distributed data are reported by the median (interquartile range [IQR]) values. When the residuals met the assumption of normality, linear mixed models were used to determine if there were differences in the outcome data (HR, time to loss of righting reflex, duration of intubation, duration of loss of righting reflex, and blood gases values [pH, PCO_2, HCO_3, and lactate]) by group (A, AM, DM, and ADM), time (0–60 min with 5 min intervals), and order drugs (1–4). Skink was entered into the model as the random factor, and group, time, order, and their interaction terms were entered as fixed factors. Body weight was also added to the model as a covariate. The Akaike's information criterion was used to help assess model fit. Because the residuals for RR were not normal, a generalized linear mixed model was used to determine whether the fixed factors group, time, order, and their interaction terms impacted RR. Skink served as the random variable in the model and body weight was added as a covariate. Bonferroni's multiple comparisons test was used as a post-hoc test when differences were noted. The time to intubation was analyzed using a generalized linear mixed model with group and order as fixed variables and body weight as a covariate. Dunn's multiple comparison test was used as a post-hoc test. Ordinal data, including all reflexes and the possibility of intubation, were analyzed using the duration of the loss of each reflex as a continuous outcome variable using generalized linear mixed models. Group and order were included in the model as fixed factors and body weight as a covariate. Data were analyzed using SPSS V27.0 (IBM Statistics, Armonk, NY, USA) and GraphPad Prism V9.0 (GraphPad Software, San Diego, CA, USA). A p-value < 0.05 was used to determine statistical significance.

3. Results

All skinks recovered well without any adverse effects from both the pilot and primary studies. Following SC injection, erythema was immediately noted in more than half of the trials with all drugs but had dispersed in all cases within 10 min.

3.1. Pilot Sedation Trials

Every skink that received DM was sedated for over 180 min, regardless of the doses, and thus were reversed with flumazenil and atipamezole. Within the pilot trials, times to recovery for the two skinks in each trial were as follows: 25 and 30 min for 10 mg/kg alfaxalone; 35 and 35 min for 15 mg/kg alfaxalone; 30 and 50 min for 17 mg/kg alfaxalone; 60 and 130 min for 20 mg/kg alfaxalone; 60 and 65 min for 10 mg/kg alfaxalone with 1 mg/kg midazolam; 60 and 65 min for 15 mg/kg alfaxalone with 1 mg/kg midazolam; 170 and >180 (reversed at 180 min with atipamezole) minutes for 0.05 mg/kg dexmedetomidine with 1 mg/kg midazolam; >220 (reversed at 220 min with atipamezole and flumazenil) and 180 min for 0.1 mg/kg dexmedetomidine with 1 mg/kg midazolam; and 100 and 105 min for 5 mg/kg alfaxalone, 0.05 mg/kg dexmedetomidine, and 0.5 mg/kg midazolam. All lizards receiving DM or ADM were bradycardic and bradypneic at the time all their reflexes returned to baseline, unless they were reversed. The loss of righting reflex was recorded in all skinks in the pilot study except one of the skinks receiving 10 mg/kg alfaxalone. Intubation was only possible in 100% (2/2) of the pilot animals for 20 mg/kg alfaxalone, 10 mg/kg alfaxalone and 1 mg/kg midazolam, 0.1 mg/kg dexmedetomidine with 1 mg/kg midazolam, and 5 mg/kg alfaxalone with 0.05 mg/kg dexmedetomidine and 0.5 mg/kg midazolam. Intubation was not always possible with the other drug combinations (10 mg/kg alfaxalone [0/2]; 15 mg/kg alfaxalone [1/2]; 17 mg/kg alfaxalone [1/2]; and 0.05 mg/kg dexmedetomidine with 1 mg/kg midazolam [1/2]).

3.2. Primary Sedation Trials

In the primary trials, differences in the reflexes were found between drugs and over time (Figure 1, Table 2). The times to the loss of the righting reflex were the shortest in A and AM, although this difference was not significantly different ($p = 0.06$). All skinks lost their righting reflex during the trials except a single skink in the DM group; this animal was excluded from the final calculation because the skink never achieved that outcome. There was a significant difference in the duration of the loss of the righting reflex between drugs ($F = 5.22$, $p = 0.011$), with ADM being significantly longer than A ($p = 0.01$) and DM ($p = 0.046$). The total loss of the escape reflex in all skinks was only observed in the ADM skinks, whereas in the other drug groups, it occurred in most of the skinks (A: 8/11; AM: 9/11; DM: 8/11). The duration of the loss of the escape reflex was different by drug ($F = 8.6$, $p = 0.001$). The duration of the lost escape reflex was the longest in ADM (55 [45–60] min) and was almost equal to the entire sedation period after induction. The complete loss of the reflex in all 11 skinks was only obtained in ADM. Post-hoc tests confirmed significant differences in the escape reflex between A-ADM ($p = 0.001$) and AM-ADM ($p = 0.001$).

The palpebral reflex was never lost in the skinks receiving A (0 [0–0] min) or AM (0 [0–0] min) but was lost for 50 (40–55) min and 55 (50–55) min in all skinks receiving DM and ADM, respectively. There was a significant difference in the duration of the loss by drug group ($F = 102.5$, $p < 0.001$), with pairwise differences found between all groups (all $p < 0.001$) except A-AM and DM-ADM ($p > 0.99$). The toe pinch withdrawal reflex was never lost for skinks given A (0 [0–0] min) or AM (0 [0–0] min) but was lost for 10 (0–40) min and 40 (15–50) min for DM (8/11; 73%) and ADM (9/11; 82%), respectively. Again, a significant difference in the duration of the loss by drug group was observed ($F = 14.5$, $p < 0.001$), with pairwise differences (all $p < 0.05$) between all groups except A-AM and DM-ADM (all $p > 0.99$). The duration of the loss of the tongue flicking was significantly different by drug ($F = 14.0$, $p < 0.001$). The durations of time for the lost tongue flicking for each drug were as follows: A, 25 (0–30) min; AM, 30 (15–35) min, DM, 50 (25–60) min, and ADM, 55 (50–60) min. Significant differences were noted between all groups (all $p < 0.05$), except A-AM ($p = 0.196$) and DM-ADM ($p = 0.09$).

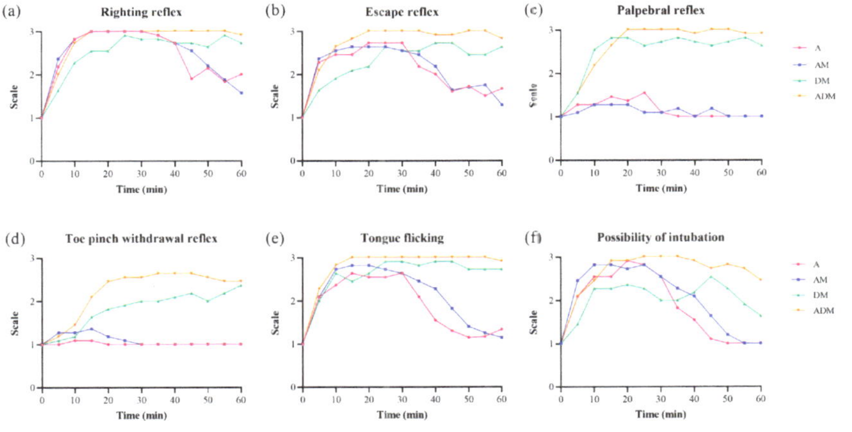

Figure 1. Frequencies for all reflexes by drugs over time for 11 common blue-tongued skinks. (**a**) The loss of righting reflex was observed in all skinks except one skink which was given dexmedetomidine-midazolam (DM). (**b**) The complete loss of the escape reflex in all 11 skinks was only achieved in alfaxalone-dexmedetomidine-midazolam (ADM). Most of the skinks receiving alfaxalone (A) or alfaxalone-midazolam (AM) maintained their (**c**) palpebral reflexes and (**d**) toe pinch withdrawal reflexes. (**e**) The complete loss of tongue flicking was found in DM and ADM, and (**f**) the possibility of intubation was highest in ADM followed by AM and A.

Table 2. Descriptive statistics for the durations of time (minutes) that reflexes were lost in blue-tongued skinks following sedation with alfaxalone (A), alfaxalone-midazolam (AM), dexmedetomidine-midazolam (DM), and alfaxalone-dexmedetomidine-midazolam (ADM). The mean ± standard deviation or median (interquartile range) are reported.

Variables	A	AM	DM	ADM
Time to loss of righting reflex	8.18 ± 3.37	8.18 ± 4.05	14.09 ± 6.64	9.09 ± 4.37
Duration of loss of righting reflex *	38.18 ± 8.86 [a]	41.82 ± 9.83	40.91 ± 18.07 [b]	55.45 ± 4.98 [ab]
Duration of loss of escape reflex *	20 (0–35) [c]	30 (15–35) [d]	35 (0–45)	55 (45–60) [cd]
Time to intubation *	10 (5–20) [e]	5 (5–10) [f]	10 (10–30) [efg]	5 (5–15) [g]
Duration of intubation *	23.18 ± 9.11 [h]	31.36 ± 10.46 [i]	28.18 ± 21.24 [j]	48.18 ± 10.06 [hij]
Time to recovery †	55 (45–60)	60 (50–60)	60	60

The significant difference ($p < 0.05$) confirmed between groups (*). As DM and ADM groups were reversed at 60 min, time to recovery was not analyzed by group (†). [a] ($p = 0.01$), [b] ($p = 0.046$), [c,d] ($p = 0.001$), [e,f,g] ($p < 0.001$), [h] ($p < 0.001$), [i] ($p = 0.002$), [j] ($p < 0.01$).

There was a significant difference in time to intubation by drug ($F = 16.9$, $p < 0.001$), with skinks receiving DM taking longer than A, AM, and ADM (all $p < 0.001$). The skinks that could not be intubated (>60 min) were again excluded from the analysis. The duration of intubation was significantly different by drug group ($F = 11.9$, $p < 0.001$), with differences noted between ADM and A ($p < 0.001$), AM ($p = 0.002$), and DM ($p < 0.001$). One skink in the A group and two in the DM could only be intubated for 5 min, and the time that was available for the two DM skinks was 45 and 50 min. Since two skinks in the DM group could not be intubated, 4/11 DM skinks only provided 0–5 min of time for intubation. The remaining 7/11 DM skinks could be intubated for more than 40 min. The duration of intubation was consecutive in A, AM, and ADM, whereas consecutive intubation was only achieved in 3/11 DM skinks. All ADM skinks provided more than 40 min of intubation, except for a single animal where only 20 min was possible.

The majority of skinks administered A and AM recovered prior to the 60 min deadline, except for a single skink given A (130 min) and two skinks given AM (reversed at 60 min). None of the skinks receiving DM or ADM recovered by the 60 min deadline, and thus, all 11 animals in both treatments received the reversals for dexmedetomidine and midazolam. All skinks recovered within 10 min of receiving the reversal agents.

Heart rate and RR significantly decreased after injection in all groups (Figure 2; Supplementary Table S1). The distribution of HR was different based on drug ($F = 654.5$, $p < 0.001$), time ($F = 42.68$, $p < 0.001$), and drug × time ($F = 5.4$, $p < 0.001$). The RR was also significantly different based on drug ($F = 129.0$, $p < 0.001$), time ($F = 37.6$, $p < 0.001$), and drug × time ($F = 2.1$, $p < 0.001$). The order the drugs were provided over the course of the trial did not affect HR or RR (all $p > 0.05$). Mean ± SD and median (IQR) baseline HR and RR were 79.7 ± 6.2 and 28.5 (20–39.8), respectively. In A and AM skinks, the lowest median HR was 56 for both groups, whereas the lowest HR in the DM and ADM groups were 24 and 28, respectively. Compared to the baseline HR values, a significant difference was observed in A for 15–45 min, AM for 15–40 min, and both DM and ADM for 10–60 min (all $p < 0.05$). The RR was also confirmed to be significantly different from baseline in A for 10–55 min (all $p < 0.05$), AM for 10–40 min (all $p < 0.05$), and DM and ADM for 5–60 min (all $p < 0.001$). The lowest median RR for the A and AM groups were 9 and 7, respectively, while the lowest RR in the DM and ADM groups was 0.

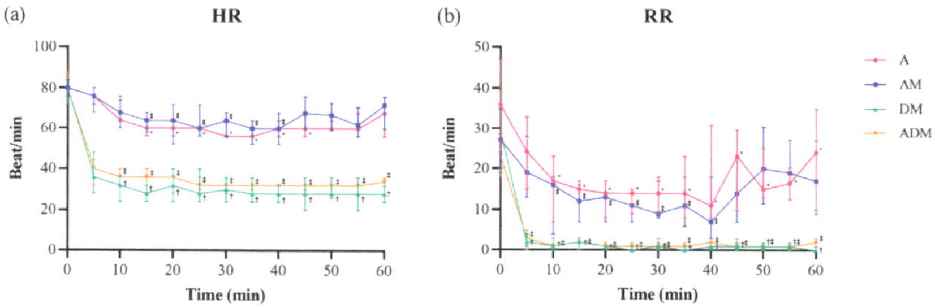

Figure 2. (**a**) Heart rate (HR) and (**b**) respiratory rate (RR) over time for each of the four drugs administered subcutaneously to 11 skinks (A: alfaxalone 20 mg/kg; AM: alfaxalone 10 mg/kg + midazolam 1 mg/kg; DM: dexmedetomidine 0.1 mg/kg + midazolam 1 mg/kg; ADM: alfaxalone 5 mg/kg + dexmedetomidine 0.05 mg/kg + midazolam 0.5 mg/kg). The median (dot) and interquartile range (bar) are shown. Both parameters showed statistical differences by drug, time, and drug × time (all $p < 0.05$). All four drugs showed significant decreases from baseline in HR and RR ($p < 0.05$). However, the duration of the shown differences was almost doubled in DM and ADM. Significantly different time points from the baseline value are marked (* A, ⁑ AM, † DM, and ‡ ADM).

There were significant differences in pH, PCO_2, and lactate over time (all $p < 0.001$; Figure 3; Supplementary Table S1). For pH and PCO_2, values at the time of full sedation were significantly lower and higher, respectively, from baseline and recovery values (all $p < 0.001$); lactate was only found to be significantly higher at full sedation compared to baseline ($p < 0.001$). The lactate was increased at recovery but was not statistically significant from baseline ($p = 0.054$). A significant difference for the interaction drug × time was only found for PCO_2 ($F = 3.2$, $p = 0.008$). When analyzing by each drug, DM showed differences between baseline and full sedation ($p = 0.023$), and ADM at full sedation demonstrated a significantly higher value from baseline and recovery (all $p < 0.001$).

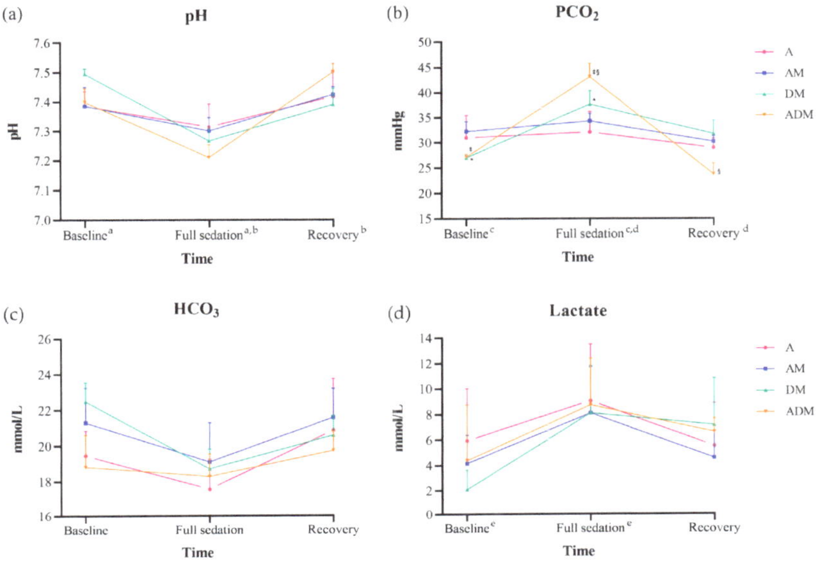

Figure 3. Blood gas results for (**a**) pH, (**b**) PCO_2, (**c**) HCO_3, and (**d**) lactate from six skinks administered alfaxalone (A), alfaxalone-midazolam (AM), dexmedetomidine-midazolam (DM), and alfaxalone-dexmedetomidine-midazolam (ADM). The data are reported by the mean (dot) and standard deviation (bar). The pH, PCO_2, and lactate showed a significant difference by time (all $p < 0.001$), and only PCO_2 was significant by drug × time ($p = 0.008$). Significant post-hoc comparisons by time are indicated by [a–e] (all $p < 0.001$). When analyzing PCO_2 by each drug, skinks receiving DM significantly increased at full sedation from baseline (*; $p = 0.023$), while skinks receiving ADM showed significant differences between full sedation and baseline and recovery (‡, §; all $p < 0.001$).

4. Discussion

The results of this study partially proved our first hypothesis that all four drug protocols would be safe and provide sufficient sedation to allow for minor procedures such as intubation, taking radiographs, and blood collection. Sedation aims to achieve central nerve system (CNS) depression, resulting in an awake but calm state in which the animal is generally uninterested in its surroundings but responsive to noxious stimulation [113]. Even though every regimen induced sedation in this species, there was a noticeable difference between these combinations, including the depth of sedation, duration of time the skinks could be intubated, and duration of sedation provided, with some cohorts not fully being sedated. From the authors' experience, it is not uncommon to observe within species variation in reptiles when using the same protocol. While we controlled temperature to minimize that impact on metabolism, a recent study implicated obesity and right-to-left cardiac shunts as reasons induction times for anesthesia can be prolonged in reptiles [114]. It is possible that differences in body fat between animals, shunting, and other physiologic differences led to some of the variability found in this study.

In the pilot study, a dose-dependent effect of alfaxalone was confirmed for the duration of sedation, which allowed us to select a dose that would provide the most complete sedation as a single-agent trial. The DM and ADM doses selected for the primary trial provided more than 100 min of sedation in the pilot trials; thus, they could prove valuable for longer minor procedures, if desired. The doses of A and AM selected for the primary trial offered appropriate time for minor procedures meant in this study, especially if reversals are not available. If AM is utilized over A alone, a 50% lower alfaxalone could be used, and the midazolam can be reversed if desired. Although the durations of sedation were adequate at less than 60 min for the desired procedures, considering that there are no reversal agents available for alfaxalone itself, DM or ADM may be chosen to provide that option. Additionally, in the two groups given dexmedetomidine, all skinks had decreased HR and RR even when reflexes returned to baseline values, so reversal is recommended when the goal for sedation is achieved. A study conducted in Argentine tegus found that a combination of 0.2 mg/kg dexmedetomidine and 1 mg/kg of midazolam provided sedation for 80–350 min [70]. The dose of dexmedetomidine used in the tegus was double the dose used in our study, and both the HR and RR were significantly decreased for 6 and 4 h post-sedation, respectively, even though the righting and escape reflexes were returning after 100 min. In our pilot trials, all skinks receiving DM and ADM regained those reflexes at around 80–90 min as well.

The second hypothesis was also partially proven because the reflexes did decrease within 5–10 min of SC injection in 75% of skinks, 81.8% when excluding the DM group. The time to loss of righting reflex and intubation was faster and smoother in the groups that received alfaxalone (A, AM, and ADM) compared to the group that did not (DM). This rapid onset of SC alfaxalone has also been noted in ball pythons (*Python regius*) and leopard geckos (*Eublepharis macularius*) [18,35]. As our findings have proved, these SC drugs can induce effective sedation in blue-tongue skinks and avoid unnecessary pain caused by the large volume being injected IM. The time to intubation in A, AM, and ADM was reasonable, and our ability to perform the intubation was considered straightforward. However, in DM-treated skinks, the time to intubation was highly variable. Among the four DM individuals that could not be intubated or experienced delayed intubation, one of the skinks also did not lose the righting reflex, and the escape reflex was retained in two individuals. These findings affirm that the ability to intubate does not necessarily agree with the loss of righting and escape reflexes. Despite successful initial intubation or the loss of other reflexes, skinks given DM still occasionally reacted to touching their mouths and closed them. This was why consecutive intubation was only achieved in three DM skinks during the primary trials. For intubation, a sensory blockade around the mouth and tongue was required; however, in A and AM skinks, a loss of pain response was not necessary for intubation, and in some DM individuals, intubation was not possible even when the tongue could not move. These results suggest that analgesia is not a prerequisite for intubation.

All or most A and AM skinks maintained their palpebral and toe pinch withdrawal reflexes, contrary to those of the DM and ADM groups. Similarly to our study, 9 mg/kg IV alfaxalone in blue-tongue skinks did not affect the animals' responses to interdigital pinching [3]. Based on these results, alfaxalone should not be used as a sole agent for sedation protocols where a painful response may be elicited. In the present study, the addition of dexmedetomidine appeared to provide a certain level of analgesia for 10 (0–40) min in the DM and 40 (15–50) min in the ADM groups. The analgesic effect of α_2-adrenoceptor agonists is provided through supraspinal and spinal mechanisms [115]. The α_1-adrenoceptor agonist can also provide analgesic effects like an α_2 agonist, but only at high doses, making them clinically restrictive [116]. Although a direct comparison has not been made with other α_2-adrenoceptor agonists, dexmedetomidine (medetomidine) has a much greater (10 times more than xylazine) α_2 selectivity, so it is preferred when pain control is required [117,118]. Unfortunately, the use of α_2 agonists as sole analgesics is not recommended because of their negative cardiovascular effects [115]. Therefore, they are commonly used in combination with ketamine or opioid analgesics to produce synergistic effects that can be achieved while reducing the doses of each drug [115]. Dexmedetomidine has also been shown to have an opioid-level analgesic effect in dogs when used in combination with ketamine [119]. Regardless, additional analgesics are recommended for blue-tongue skinks because some of the animals receiving dexmedetomidine reacted to painful stimuli. A limitation of these trials was that we only tested a deep pain response by pinching the toe; however, other forms of superficial, thermal, or electrical stimuli can also induce pain. The antinociceptive effects of a DM or ADM combination evoked by thermal, superficial, or deep pain stimuli have been observed in ball pythons, Argentine tegus, and American alligators (*Alligator mississippiensis*) but not in leopard geckos [18,35,70,81]. Only one-third of leopard geckos that received DM lost their superficial skin pinch response and none of the lizards lost their deep pain response. Future studies with blue-tongue skinks should be done to further evaluate and characterize how this species responds to painful stimuli.

Anesthesia is a state of unconsciousness characterized by controlled but reversible CNS depression and perception in which the patient is not awakened by noxious stimuli [113]. According to these criteria, we could achieve some level of anesthesia using ADM but not in A, AM, or DM. Skinks given DM appeared immobile but were somewhat sensitive to stimuli, staying unresponsive when carefully turned over but responding to exaggerated movements when flipped. They also displayed heightened sensitivity and aggressive defensive responses, especially during recovery. These behavioral changes have also been reported in Argentine tegus when using dexmedetomidine [70]. Such responses were not observed when the skinks received ADM, and it was likely through the synergism with the alfaxalone and the deeper sedation observed when all three drugs were used together that these behaviors were eliminated.

Heart rate and RR were significantly decreased in all drug groups, but the reductions were more prominent in the DM and ADM groups, indicating a profound suppressive effect of dexmedetomidine. While this finding was consistent with our third hypothesis, we did also find that HR and RR decreased in the skinks receiving the drug combinations without dexmedetomidine. A decrease in HR from alfaxalone alone has been documented in green iguanas (*Iguana iguana*) using 20 and 30 mg/kg IM and estuarine crocodiles (*Crocodylus porosus*) using 3 mg/kg IV, but conversely, no change in HR was reported in corn snakes (*Pantherophis guttatus*) given 5, 10, and 15 mg/kg SC and bearded dragons (*Pogona vitticeps*) given 15 mg/kg SC [13,14,40,47]. Since all skinks given A or AM maintained their HR > 50 beats/minute, the level of suppression was not considered a serious abnormality. The possibility that the baseline HR and RR values were falsely elevated because of manual handling and excitement should also be considered. It has been reported that even gentle handling increases the HR of the green iguanas [120]. A previous study evaluating a low IV dose of alfaxalone (9 mg/kg) in blue-tongued skinks found that it did not cause noticeable respiratory depression [3]. However, alfaxalone has been associated with decreasing the respiratory rate in other reptiles, including green iguanas, ball pythons, loggerhead sea turtles (*Caretta caretta*), corn snakes, and bearded dragons [8,13,22,25,40,47]. Compared to α_2 agonists, most of them given 20 mg/kg or less of A maintained respiration rates >6 breaths/minute, similar to the skinks in the present study. However, apnea or bradypnea has been observed in green iguanas and ball pythons given very high doses of A (30 mg/kg) [13,25]. According to one study conducted on bearded dragons comparing injection routes, IV alfaxalone caused the biggest respiratory suppression over other routes [47]. The species differences noted in these studies further reinforce the importance of closely monitoring the HR and RR of reptiles undergoing sedation or anesthesia protocols because we cannot expect all animals to respond similarly to these drugs.

Using dexmedetomidine (DM and ADM), HR decreased to <30 beats/minute and the RR plunged to <2 breaths/minute, representing a >50% decrease in HR and a >90% decrease in RR from the baseline values. Moreover, the period showing a significant difference in HR and RR compared to the baseline values in the DM and ADM groups was twice that of A and AM. This influence on the HR and RR by an α_2 agonist has been previously reported in reptiles, including ball pythons, leopard geckos, Argentine tegus, and red-footed tortoises (*Geochelone carbonaria*) [18,70,73,80]. These cardiovascular effects were found to be minimized in dogs by giving these drugs via IM or SC routes rather than IV, and at lower doses [121]. However, a significant decrease in HR and RR was still observed in the skinks even when injected SC. This could be related to dosing, as the doses commonly used in reptiles are 10 times higher than that recommended for dogs. Similarly, in humans, dexmedetomidine was found to be the least respiration-depressing anesthetic among those compared; however, the dose used in humans was 100 times lower (1 μg/kg) than that commonly used in reptiles [122]. While these skinks did experience significant reductions in their HR and RR, they did all have unremarkable recoveries and continue to be healthy more than 8 months after the completion of the study. A few reports have described successful sedation in reptiles using dexmedetomidine without alterations in the RR, including leopard geckos and yellow-bellied sliders (*Trachemys scripta scripta*) [18,62]. In those studies, the doses used for DM were similar to what we used in the skinks, while the dexmedetomidine, at twice the dose used in skinks, was combined with ketamine in the slider turtles. Again, these findings further reinforce the importance of developing species-specific sedation protocols to identify potential complications that might occur so that the clinical team managing the reptile patient can provide appropriate corrective measures (e.g., reversal) if needed.

Although the body temperature of the skinks was not measured due to the influence of ambient temperature on reptiles, it is known that sedation with DM in dogs results in a decrease in body temperature [123]. Additionally, since drug metabolism in reptiles can be significantly impacted by body temperature, it is important to maintain the reptiles at a similar environmental and, thus, body temperature in experimental studies [12]. A study of red-eared sliders found that the administration of alfaxalone at 20 °C resulted in significantly longer recovery times and lower heart rates compared to 30 °C. Even with the same dose of alfaxalone, none of the sliders could be intubated at the higher temperature with a dose of 10 mg/kg, and only 30% could be intubated with 20 mg/kg. However, at the lower temperature, 80% could be intubated with the 10 mg/kg dose and 100% with the 20 mg/kg dose. In the present study, the skinks were maintained at the same temperature and exposed to all four sedation protocols to reduce within-animal and temperature-based biases. However, if the DM or ADM protocols reduced body temperatures to lower than the environmental temperature, it could have impacted the skinks. In the future, measuring the core body temperatures of animals being exposed to these sedation protocols could be used to determine whether body temperature is impacted so that corrective measures, such as increasing the environmental temperature, can be implemented.

Given that the authors could only sample a subset of the skink for blood gases, two studies conducted on eastern copperheads (*Agkistrodon contortrix*) and eastern ratsnakes (*Pantherophis alleghaniensis*) were referenced for comparison [124]. The blue-tongued skinks in the present study had slightly higher pH values at baseline (average 7.41) compared to the eastern copperheads but similar values to the eastern ratsnakes. The lowest pH recorded in a skink at full sedation was 7.02 with a PCO_2 of 50.7 and HCO_3 of 14.8. When the skink recovered, the pH returned to the baseline range, indicating respiratory acidosis due to reduced RR. The pH significantly decreased at the point of full sedation and then increased at recovery, while there was a corresponding initial increase in PCO_2 due to the decreased RR, followed by a decrease in PCO_2 as the RR increased at recovery. Since blood was collected when full sedation was first recorded (approximately 20 min after injection), the parameters reported may not represent the peak or trough values during sedation. However, the marked decrease in pH and increase in PCO_2 indicated that these sedatives indeed impacted respiratory ventilation. Although the critical limits for these two parameters in reptiles have not been reported, this point should be emphasized in patients where decreased ventilation could be fatal.

The alterations in the blood gases aligned with the significant differences noted over time at the individual drug levels for both DM and ADM, where the most severe respiratory depression occurred. No changes in pH, PCO_2, or lactate were observed in dogs anesthetized with isoflurane during continuous infusion with dexmedetomidine [125]. However, it is again worth noting that the dose used was 10 times lower than the single dose administered in reptiles. Additionally, the anesthesia in dogs was maintained using isoflurane with tracheal intubation, which provided adequate oxygen even though they did not provide intermittent positive pressure ventilation (IPPV). In our study, a single injection at a lower concentration did not achieve sufficient sedation, indicating the need for further research on the effects of continuous low-dose infusion on respiration and blood gases. A decrease in blood pH has also been described in loggerhead sea turtles given

10 mg/kg alfaxalone IV, and the decreased pH was attributed to a marked decrease in RR and thus increase in PCO_2 [22]. The turtles were given IPPV, but it was not clear whether the resolution of hypoxemia was due to manual ventilation or to the return of spontaneous breathing. One study conducted in rattlesnakes anesthetized with propofol demonstrated that IPPV significantly affected blood gas parameters [126]. Snakes that received fewer mechanical ventilation compared to baseline had significantly lower pH and PaO_2 and higher $PaCO_2$ than subjects that received more ventilation. Based on these findings, it is recommended to compensate for the physiological changes identified in our study through appropriate IPPV when using these sedation protocols.

Lactate is produced as part of glycolysis in an anaerobic environment and is cleared through the liver and kidneys [127]. It is known that excessive excitement, such as escaping from a predator, can rapidly increase oxygen demand and, thus, lactate in reptiles. This has been reported in crocodiles and gopher tortoises, and in both species, individuals that were physically stunned, so that there was no persistent stress, or that were less stressed due to restricted movement by trapping, showed a smaller increase in lactate compared to those in groups that were stressed [128,129]. In our study, the skinks showed significant increases in lactate at full sedation but subsequently decreased by recovery. The increase was most likely attributed to the hypoxia caused by the decreased RR and stress caused by blood collection. The skink with the lowest pH also had the highest lactate concentration. This alteration in lactate likely reflected the reduction in oxygen and the need to buffer the respiratory acidosis. A study in rattlesnakes similarly found that lactate increased and then decreased after anesthesia; however, in that study, there was no significant difference in lactate concentrations across levels of mechanical ventilation, which made analysis of respiration influences difficult [126]. These results further reinforce the value of measuring blood gases in reptile patients to gain a better understanding of the physiologic consequences of sedatives or anesthetics in these animals.

There were several potential limitations with this study. Sample size is traditionally a limitation in these types of clinical trials. However, to increase the power associated with the study, a complete crossover study was done to reduce within-skink variability between sedation protocols. Moreover, since we were able to reject the null hypotheses for the primary hypotheses we tested, the sample size was sufficient to limit the impact of a type II error. For the blood gas analysis, we collected venous blood gas samples. This is commonly done for reptiles because of the challenges in obtaining arterial samples without invasive methods (e.g., surgical cut-down). While arterial blood gases are preferred, venous blood gases are less of a concern for reptiles, especially three-chambered heart species, because of the inherent mixing of blood during systole. In our case, the venous blood gases still provided an opportunity to characterize trends over time and between sedation protocols.

5. Conclusions

Overall, blue-tongued skinks were sufficiently sedated to perform an examination, collect blood, and be positioned for radiographs using all four sedation protocols. Intubation was also achieved using all four sedation protocols; however, with DM (dexmedetomidine-midazolam), 4/11 animals could not be intubated or intubated over 45 min. The authors found that ADM (alfaxalone-dexmedetomidine-midazolam) provided the best overall sedation for performing these minor procedures, followed by AM (alfaxalone-midazolam). These findings further reinforce the importance of using multiple drugs in combination at lower doses to achieve a synergistic effect and minimize side effects.

Supplementary Materials: The following supporting information can be downloaded at: https://www.mdpi.com/article/10.3390/ani14182636/s1, Table S1: Descriptive statistics for heart rate, respiratory rate, and blood gas parameters (pH, PCO2, HCO3, and lactate) measured in 11 common blue-tongued skinks provided four different combinations of sedatives.

Author Contributions: Conceptualization, H.R. and M.A.M.; methodology, H.R., A.M.G., M.G.A. and M.A.M.; software, H.R. and M.A.M.; validation, H.R. and M.A.M.; formal analysis, H.R. and M.A.M.;

investigation, H.R. and M.A.M.; resources, M.A.M.; data curation, H.R., A.M.G. and M.G.A.; writing—original draft preparation, H.R.; writing—review and editing, H.R., A.M.G., M.G.A. and M.A.M.; visualization, H.R.; supervision, M.A.M.; project administration, M.A.M.; funding acquisition, M.A.M. All authors have read and agreed to the published version of the manuscript.

Funding: This research was funded by Fluker Farms (Port Allen, LA, USA).

Institutional Review Board Statement: The animal study protocol was approved by the Institutional Animal Care and Use Committee of Louisiana State University (23-077, 7/3/23).

Informed Consent Statement: Not applicable.

Data Availability Statement: The original datasets presented in the study are included in the article/Supplementary Materials, further inquiries can be directed to the corresponding author.

Acknowledgments: The authors wish to thank the LSU Division of Laboratory Animal Medicine caretakers for caring for these animals.

Conflicts of Interest: The authors declare no conflicts of interest.

References

1. Scheelings, T.F. Surgical management of maxillary and mandibular fractures in an eastern bluetongue skink, *Tiliqua scincoides scincoides*. *J. Herpetol. Med. Surg.* **2007**, *17*, 136–140. [CrossRef]
2. Köchli, B.; Schmid, N.; Hatt, J.-M.; Dennler, M.; Steinmetz, H.W. Nonsurgical treatment of a bilateral mandibular fracture in a blue-tongued skink. *Exot. DVM* **2008**, *10*, 25–28.
3. Scheelings, T.F.; Baker, R.T.; Hammersley, G.; Hollis, K.; Elton, I.; Holz, P. A preliminary investigation into the chemical restraint with alfaxalone of selected Australian squamate species. *J. Herpetol. Med. Surg.* **2011**, *21*, 63–67. [CrossRef]
4. Kehoe, S.P.; Guzman, D.S.-M.; Sokoloff, A.M.; Grosset, C.; Weber, E.S., III; Murphy, B.; Culp, W.T.N. Partial glossectomy in a blue-tongued skink (*Tiliqua scincoides*) with lingual squamous cell carcinoma. *J. Herpetol. Med. Surg.* **2016**, *26*, 36–41. [CrossRef]
5. Vergneau-Grosset, C.; Carmel, É.N.; Raulic, J.; Tucoulet, J.; Summa, N.; Langlois, I.; Benoit, J.-M. Vitamin d toxicosis in a blue-tongued skink (*Tiliqua scincoides*) presented with epistaxis and tongue discoloration. *J. Herpetol. Med. Surg.* **2021**, *30*, 224–231. [CrossRef]
6. McKenzie, A.; Li, T.; Doneley, B. A comparison of two techniques to identify the sex of the eastern blue-tongue skink (*Tiliqua scincoides scincoides*). *Aust. Vet. J.* **2022**, *100*, 407–413. [CrossRef]
7. Vetere, A.; Di Girolamo, N.; Porter, I.; Tollefson, C.; Di Ianni, F.; Nardini, G. Sex identification in juvenile and adult Indonesian blue-tongued skinks (*Tiliqua gigas*) through cystoscopy and accuracy of contrast radiography. *J. Am. Vet. Med. Assoc.* **2023**, *261*, 1–8. [CrossRef] [PubMed]
8. Bertelsen, M.F.; Sauer, C.D. Alfaxalone anaesthesia in the green iguana (*Iguana iguana*). *Vet. Anaesth. Analg.* **2011**, *38*, 461–466. [CrossRef]
9. Knotek, Z.; Hrda, A.; Kley, N.; Knotkova, Z. Alfaxalone anaesthesia in veiled chameleon (*Chamaeleo calyptratus*). In Proceedings of the 18th Annual Conference ARAV, Seattle, WA, USA, 10–12 April 2011; pp. 179–181.
10. Wenger, S.; Wyss, F.; Peterson, S.; Gull, J.; Hatt, J.M. The use of alfaxalone for induction of anaesthesia in selected reptile species: A preliminary clinical investigation. In Proceedings of the Diseases of Zoo and Wild Animals, Vienna, Austria, 8–11 May 2013; p. 24.
11. Hansen, L.L.; Bertelsen, M.F. Assessment of the effects of intramuscular administration of alfaxalone with and without medetomidine in horsfield's tortoises (*Agrionemys horsfieldii*). *Vet. Anaesth. Analg.* **2013**, *40*, e68–e75. [CrossRef]
12. Kischinovsky, M.; Duse, A.; Wang, T.; Bertelsen, M.F. Intramuscular administration of alfaxalone in red-eared sliders (*Trachemys scripta elegans*)—Effects of dose and body temperature. *Vet. Anaesth. Analg.* **2013**, *40*, 13–20. [CrossRef]
13. Knotek, Z.; Hrdá, A.; Knotková, Z.; Trnková, Š.; Babák, V. Alfaxalone anaesthesia in the green iguana (*Iguana iguana*). *Acta Vet. Brno* **2013**, *82*, 109–114. [CrossRef]
14. Olsson, A.; Phalen, D.; Dart, C. Preliminary studies of alfaxalone for intravenous immobilization of juvenile captive estuarine crocodiles (*Crocodylus porosus*) and Australian freshwater crocodiles (*Crocodylus johnstoni*) at optimal and selected sub–optimal thermal zones. *Vet. Anaesth. Analg.* **2013**, *40*, 494–502. [CrossRef] [PubMed]
15. Scheelings, T.F. Use of intravenous and intramuscular alfaxalone in macquarie river turtles (*Emydura macquarii*). *J. Herpetol. Med. Surg.* **2013**, *23*, 91–94. [CrossRef]
16. Shepard, M.K.; Divers, S.; Braun, C.; Hofmeister, E.H. Pharmacodynamics of alfaxalone after single-dose intramuscular administration in red-eared sliders (*Trachemys scripta elegans*): A comparison of two different doses at two different ambient temperatures. *Vet. Anaesth. Analg.* **2013**, *40*, 590–598. [CrossRef] [PubMed]
17. Knotek, Z. Alfaxalone as an induction agent for anaesthesia in terrapins and tortoises. *Vet. Rec.* **2014**, *175*, 327. [CrossRef]
18. Doss, G.A.; Fink, D.M.; Sladky, K.K.; Mans, C. Comparison of subcutaneous dexmedetomidine–midazolam versus alfaxalone–midazolam sedation in leopard geckos (*Eublepharis macularius*). *Vet. Anaesth. Analg.* **2017**, *44*, 1175–1183. [CrossRef]
19. Knotek, Z. Induction to inhalation anaesthesia in agamid lizards with alfaxalone. *Veterinární Medicína* **2017**, *62*, 41–43. [CrossRef]

20. Morici, M. Intravenous Alfaxalone Anaesthesia in Two Squamate Species: *Eublepharis macularius* and *Morelia spilota cheynei*. Ph.D. Thesis, University of Messina, Messina, Italy, 2017.
21. Perrin, K.L.; Bertelsen, M.F. Intravenous alfaxalone and propofol anesthesia in the bearded dragon (*Pogona vitticeps*). *J. Herpetol. Med. Surg.* **2017**, *27*, 123–126. [CrossRef]
22. Phillips, B.E.; Posner, L.P.; Lewbart, G.A.; Christiansen, E.F.; Harms, C.A. Effects of alfaxalone administered intravenously to healthy yearling loggerhead sea turtles (*Caretta caretta*) at three different doses. *J. Am. Vet. Med. Assoc.* **2017**, *250*, 909–917. [CrossRef]
23. Heuvel, M.v.d. Repeated Measurement Comparison of Different Protocols for Anesthesia and Analgesia Consisting of Alfaxalone, Meloxicam, and Butorphanol or Tramadol IM in Leopard Geckos (*Eublepharis macularius*). Master's Thesis, University Utrecht, Utrecht, The Netherlands, 2017.
24. Kleinschmidt, L.M.; Hanley, C.S.; Sahrmann, J.M.; Padilla, L.R. Randomized controlled trial comparing the effects of alfaxalone and ketamine hydrochloride in the Haitian giant galliwasp (*Celestus warreni*). *J. Zoo Wildl. Med.* **2018**, *49*, 283–290. [CrossRef]
25. James, L.E.; Williams, C.J.A.; Bertelsen, M.F.; Wang, T. Anaesthetic induction with alfaxalone in the ball python (*Python regius*): Dose response and effect of injection site. *Vet. Anaesth. Analg.* **2018**, *45*, 329–337. [CrossRef] [PubMed]
26. Morici, M.; Di Giuseppe, M.; Spadola, F.; Oliveri, M.; Knotkova, Z.; Knotek, Z. Intravenous alfaxalone anaesthesia in leopard geckos (*Eublepharis macularius*). *J. Exot. Pet Med.* **2018**, *27*, 11–14. [CrossRef]
27. Sadar, M.J.; Ambros, B. Use of alfaxalone or midazolam–dexmedetomidine–ketamine for implantation of radiotransmitters in bullsnakes (*Pituophis catenifer sayi*). *J. Herpetol. Med. Surg.* **2018**, *28*, 93–98. [CrossRef]
28. Yaw, T.J.; Mans, C.; Johnson, S.M.; Doss, G.A.; Sladky, K.K. Effect of injection site on alfaxalone-induced sedation in ball pythons (*python regius*). *J. Small Anim. Pract.* **2018**, *59*, 747–751. [CrossRef] [PubMed]
29. Bardi, E.; Di Cesare, F.; D'Urso, E.; Gioeni, D.; Rabbogliatti, V.; Romussi, S. Total intramuscular multimodal anesthesia in pond sliders (*Trachemys scripta*) undergoing endoscopic gonadectomy. In Proceedings of the 4th International Conference on Avian Herpetological and Exotic Mammal Medicine, London, UK, 28 April–2 May 2019.
30. Ferreira, T.H.; Mans, C.; Di Girolamo, N. Evaluation of the sedative and physiological effects of intramuscular lidocaine in bearded dragons (*Pogona vitticeps*) sedated with alfaxalone. *Vet. Anaesth. Analg.* **2019**, *46*, 496–500. [CrossRef]
31. Strahl-Heldreth, D.E.; Clark-Price, S.C.; Keating, S.C.J.; Escalante, G.C.; Graham, L.F.; Chinnadurai, S.K.; Schaeffer, D.J. Effect of intracoelomic administration of alfaxalone on the righting reflex and tactile stimulus response of common garter snakes (*Thamnophis sirtalis*). *Am. J. Vet. Res.* **2019**, *80*, 144–151. [CrossRef]
32. Chen, K.; Keating, S.; Strahl-Heldreth, D.; Clark-Price, S. Effects of intracoelomic alfaxalone–dexmedetomidine on righting reflex in common garter snakes (*Thamnophis sirtalis*): Preliminary data. *Vet. Anaesth. Analg.* **2020**, *47*, 793–796. [CrossRef]
33. Nara, T.; Kondoh, D.; Yanagawa, M. Comparison of anesthetic protocols indicates Japanese grass lizards (*Takydromus tachydromoides*) are insensitive to medetomidine. *Res. Bull. Obihiro Univ.* **2020**, *41*, 1–5.
34. Rasys, A.M.; Divers, S.J.; Lauderdale, J.D.; Menke, D.B. A systematic study of injectable anesthetic agents in the brown anole lizard (*Anolis sagrei*). *Lab. Anim.* **2020**, *54*, 281–294. [CrossRef]
35. Yaw, T.J.; Mans, C.; Johnson, S.; Bunke, L.; Doss, G.A.; Sladky, K.K. Evaluation of subcutaneous administration of alfaxalone-midazolam and dexmedetomidine-midazolam for sedation of ball pythons (*Python regius*). *J. Am. Vet. Med. Assoc.* **2020**, *256*, 573–579. [CrossRef]
36. Aymen, J.; Queiroz-Williams, P.; Hampton, C.C.E.; Cremer, J.; Liu, C.-C.; Nevarez, J.G. Comparison of ketamine–dexmedetomidine–midazolam versus alfaxalone–dexmedetomidine–midazolam administered intravenously to American alligators (*Alligator mississippiensis*). *J. Herpetol. Med. Surg.* **2021**, *31*, 132–140. [CrossRef]
37. Bertelsen, M.F.; Buchanan, R.; Jensen, H.M.; Leite, C.A.C.; Abe, A.S.; Wang, T. Pharmacodynamics of propofol and alfaxalone in rattlesnakes (*Crotalus durissus*). *Comp. Biochem. Physiol. Part A Mol. Integr. Physiol.* **2021**, *256*, 110935. [CrossRef] [PubMed]
38. Morici, M.; Lubian, E.; Costa, G.L.; Spadola, F. Difference between cranial and caudal intravenous alfaxalone administration in yellow-bellied sliders (*Trachemys scripta scripta*). *Acta Vet. Eurasia* **2021**, *47*, 88–92. [CrossRef]
39. Prinz, C. Intramuscular vs Intravenous Administration of Alfaxalon for Induction in General Anaesthesia in *Geochelone platynota* and *Astrochelys radiata*. University of Veterinary Medicine Vienna, Wien, Austria, 2021. Available online: https://phaidra.vetmeduni.ac.at/open/o:1490 (accessed on 8 September 2024).
40. Rockwell, K.; Boykin, K.; Padlo, J.; Ford, C.; Aschebrock, S.; Mitchell, M. Evaluating the efficacy of alfaxalone in corn snakes (*Pantherophis guttatus*). *Vet. Anaesth. Analg.* **2021**, *48*, 364–371. [CrossRef]
41. Webb, J.K.; Keller, K.A.; Chinnadurai, S.K.; Kadotani, S.; Allender, M.C.; Fries, R. Optimizing the pharmacodynamics and evaluating cardiogenic effects of the injectable anesthetic alfaxalone in prairie rattlesnakes (*Crotalus viridis*). *J. Zoo Wildl. Med.* **2021**, *52*, 1105–1112. [CrossRef]
42. Cardinali, M. Immobilization of a juvenile Komodo dragon using alfaxalone subcutaneously. *Vet. Anaesth. Analg.* **2022**, *49*, 521–523. [CrossRef]
43. Freitag, F.A.V.; Barboza, T.K.; Dutton, C.; Buck, R.K. Alfaxalone for anesthesia of a giant snake. *Vet. Anaesth. Analg.* **2022**, *49*, 147–148. [CrossRef]
44. Gantner, L.; Portier, K.; Quintard, B. Comparison of intramuscular alfaxalone with medetomidine-ketamine for inducing anaesthesia in *Trachemys scripta* spp. Undergoing sterilization. *Vet. Anaesth. Analg.* **2023**, *50*, 421–429. [CrossRef] [PubMed]

45. Morici, M.; Spadola, F.; Oliveri, M.; Lubian, E.; Knotek, Z. Anaesthetic induction with alfaxalone in jungle carpet python (*Morelia spilota cheynei*). *Vet. Arh.* **2023**, *93*, 129–136. [CrossRef]
46. Shippy, S.; Allgood, H.; Messenger, K.; Hernandez, J.A.; Gatson, B.; Martin de Bustamante, M.G.; Alexander, A.B.; Wellehan, J.F.X.; Johnson, A. Pharmacokinetics and pharmacodynamics of intramuscular alfaxalone in central bearded dragons (*Pogona vitticeps*): Effect of injection site. *Vet. Anaesth. Analg.* **2023**, *50*, 280–288. [CrossRef]
47. Webb, J.K.; Keller, K.A.; Chinnadurai, S.K.; Kadotani, S.; Allender, M.C.; Fries, R. Use of alfaxalone in bearded dragons (*Pogona vitticeps*): Optimizing pharmacodynamics and evaluating cardiogenic effects via echocardiography. *J. Am. Vet. Med. Assoc.* **2023**, *261*, 126–131. [CrossRef]
48. Suganthan, H.; Stefano, D.D.; Buck, L.T. Alfaxalone is an effective anesthetic for the electrophysiological study of anoxia-tolerance mechanisms in western painted turtle pyramidal neurons. *PLoS ONE* **2024**, *19*, e0298065. [CrossRef] [PubMed]
49. Zec, S.; Mitchell, M.A.; Rockwell, K.; Lindemann, D. Evaluating the anesthetic and physiologic effects of intramuscular and intravenous alfaxalone in eastern mud turtles (*Kinosternon subrubrum*). *Animals* **2024**, *14*, 460. [CrossRef] [PubMed]
50. Lambert, J.J.; Belelli, D.; Peden, D.R.; Vardy, A.W.; Peters, J.A. Neurosteroid modulation of GABAa receptors. *Prog. Neurobiol.* **2003**, *71*, 67–80. [CrossRef] [PubMed]
51. Jones, K.L. Therapeutic review: Alfaxalone. *J. Exot. Pet Med.* **2012**, *21*, 347–353. [CrossRef]
52. Lock, B.A.; Heard, D.J.; Dennis, P. Preliminary evaluation of medetomidine/ketamine combinations for immobilization and reversal with atipamezole in three tortoise species. *Bull. Assoc. Reptil. Amphib. Vet.* **1998**, *8*, 6–11. [CrossRef]
53. Sleeman, J.M.; Gaynor, J. Sedative and cardiopulmonary effects of medetomidine and reversal with atipamezole in desert tortoises (*Gopherus agassizii*). *J. Zoo Wildl. Med.* **2000**, *31*, 28–35.
54. Greer, L.L.; Jenne, K.J.; Diggs, H.E. Medetomidine-ketamine anesthesia in red-eared slider turtles (*Trachemys scripta elegans*). *J. Am. Assoc. Lab. Anim. Sci.* **2001**, *40*, 8–11.
55. Chittick, E.J.; Stamper, M.A.; Beasley, J.F.; Lewbart, G.A.; Horne, W.A. Medetomidine, ketamine, and sevoflurane for anesthesia of injured loggerhead sea turtles: 13 cases (1996–2000). *J. Am. Vet. Med. Assoc.* **2002**, *221*, 1019–1025. [CrossRef]
56. Dennis, P.M.; Heard, D.J. Cardiopulmonary effects of a medetomidine-ketamine combination administered intravenously in gopher tortoises. *J. Am. Vet. Med. Assoc.* **2002**, *220*, 1516–1519. [CrossRef]
57. Terrell, G.H.-J.; Jeff, C.H.K.; Heaton-Jones, D.L. Evaluation of medetomidine–ketamine anesthesia with atipamezole reversal in American alligators (*Alligator mississippiensis*). *J. Zoo Wildl. Med.* **2002**, *33*, 36–44. [CrossRef]
58. Young, J.S.; Min-Su, K.; Young, K.S.; Kang-Moon, S.; Tchi-Chou, N. Anesthetic effects of medetomidine-tiletamine/zolazepam combination in green iguanas (*Iguana iguana*). *J. Vet. Clin.* **2005**, *22*, 194–197.
59. Tchi-Chou, N. The reverse effects of atipamezole on medetomidine-tiletamine/zolazepam combination anesthesia in the green iguana (*Iguana iguana*). *J. Vet. Clin.* **2006**, *23*, 18–21.
60. Harms, C.A.; Eckert, S.A.; Kubis, S.A.; Campbell, M.; Levenson, D.H.; Crognale, M.A. Field anaesthesia of leatherback sea turtles (*Dermochelys coriacea*). *Vet. Rec.* **2007**, *161*, 15–21. [CrossRef] [PubMed]
61. Olsson, A.; Phalen, D. Preliminary studies of chemical immobilization of captive juvenile estuarine (*Crocodylus porosus*) and Australian freshwater (*C. Johnstoni*) crocodiles with medetomidine and reversal with atipamezole. *Vet. Anaesth. Analg.* **2012**, *39*, 345–356. [CrossRef]
62. Schnellbacher, R.W.; Hernandez, S.M.; Tuberville, T.D.; Mayer, J.; Alhamhoom, Y.; Arnold, R.D. The efficacy of intranasal administration of dexmedetomidine and ketamine to yellow-bellied sliders (*Trachemys scripta scripta*). *J. Herpetol. Med. Surg.* **2012**, *22*, 91–98. [CrossRef]
63. Olsson, A.; Phalen, D. The effects of decreased body temperature on the onset, duration and action of medetomidine and its antagonist atipamezole in juvenile farmed estuarine crocodiles (*Crocodylus porosus*). *Vet. Anaesth. Analg.* **2013**, *40*, 272–279. [CrossRef] [PubMed]
64. Emery, L.; Parsons, G.; Gerhardt, L.; Schumacher, J.; Souza, M. Sedative effects of intranasal midazolam and dexmedetomidine in 2 species of tortoises (*Chelonoidis carbonaria* and *Geochelone platynota*). *J. Exot. Pet Med.* **2014**, *23*, 380–383. [CrossRef]
65. McGuire, J.L.; Hernandez, S.M.; Smith, L.L.; Yabsley, M.J. Safety and utility of an anesthetic protocol for the collection of biological samples from gopher tortoises. *Wildl. Soc. Bull.* **2014**, *38*, 43–50. [CrossRef]
66. Nardini, G.; Silvetti, S.; Magnelli, I.; Girolamo, N.d.; Bielli, M. Medetomidine-ketamine-midazolam and buthorphanol (MKMB) as intramuscolar injectable combination for anesthesia in loggerhead sea turtles (*Caretta caretta*). *Veterinaria* **2014**, *28*, 27–31.
67. Morici, M.; Interlandi, C.; Costa, G.L.; Di Giuseppe, M.; Spadola, F. Sedation with intracloacal administration of dexmedetomidine and ketamine in yellow-bellied sliders (*Trachemys scripta scripta*). *J. Exot. Pet Med.* **2017**, *26*, 188–191. [CrossRef]
68. Stegmann, G.F.; Franklin, C.; Wang, T.; Axelsson, M.; Williams, C.J.A. Long-term surgical anaesthesia with isoflurane in human habituated Nile crocodiles. *J. S. Afr. Vet. Assoc.* **2017**, *88*, 1–6. [CrossRef]
69. Barrillot, B.; Roux, J.; Arthaud, S.; Averty, L.; Clair, A.; Herrel, A.; Libourel, P.-A. Intramuscular administration of ketamine-medetomidine assures stable anaesthesia needed for long-term surgery in the Argentine tegu *Salvator merianae*. *J. Zoo Wildl. Med.* **2018**, *49*, 291–296. [CrossRef]
70. Bisetto, S.P.; Melo, C.F.; Carregaro, A.B. Evaluation of sedative and antinociceptive effects of dexmedetomidine, midazolam and dexmedetomidine–midazolam in tegus (*Salvator merianae*). *Vet. Anaesth. Analg.* **2018**, *45*, 320–328. [CrossRef] [PubMed]
71. Bochmann, M.; Wenger, S.; Hatt, J.-M. Preliminary clinical comparison of anesthesia with ketamine/medetomidine and s-ketamine/medetomidine in *Testudo* spp. *J. Herpetol. Med. Surg.* **2018**, *28*, 40–46. [CrossRef]

72. Budden, L.; Doss, G.A.; Clyde, V.L.; Mans, C. Retrospective evaluation of sedation in 16 lizard species with dexmedetomidine-midazolam with or without ketamine. *J. Herpetol. Med. Surg.* **2018**, *28*, 47–50. [CrossRef]
73. Bunke, L.G.; Sladky, K.K.; Johnson, S.M. Antinociceptive efficacy and respiratory effects of dexmedetomidine in ball pythons (*Python regius*). *Am. J. Vet. Res.* **2018**, *79*, 718–726. [CrossRef]
74. Cermakova, E.; Ceplecha, V.; Knotek, Z. Efficacy of two methods of intranasal administration of anaesthetic drugs in red-eared terrapins (*Trachemys scripta elegans*). *Veterinární Medicína* **2018**, *63*, 87–93. [CrossRef]
75. Fink, D.M.; Doss, G.A.; Sladky, K.K.; Mans, C. Effect of injection site on dexmedetomidine-ketamine induced sedation in leopard geckos (*Eublepharis macularius*). *J. Am. Vet. Med. Assoc.* **2018**, *253*, 1146–1150. [CrossRef]
76. Monticelli, P.; Ronaldson, H.L.; Hutchinson, J.R.; Cuff, A.R.; d'Ovidio, D.; Adami, C. Medetomidine–ketamine–sevoflurane anaesthesia in juvenile Nile crocodiles (*Crocodylus niloticus*) undergoing experimental surgery. *Vet. Anaesth. Analg.* **2019**, *46*, 84–89. [CrossRef]
77. Turcu, M.R.; Pavel, R.; Degan, A.; GÎRdan, G.; Micsa, C.; Ovidiu, R.; IoniȚĂ, L. The use of two different anesthetic protocols for ovariectomy in *Trachemys scripta elegans*. *Sci. Work. Ser. C Vet. Med.* **2020**, *66*, 57–61.
78. Scheelings, T.F.; Gatto, C.; Reina, R.D. Anaesthesia of hatchling green sea turtles (*Chelonia mydas*) with intramuscular ketamine-medetomidine-tramadol. *Aust. Vet. J.* **2020**, *98*, 511–516. [CrossRef] [PubMed]
79. Yaw, T.J.; Mans, C.; Martinelli, L.; Sladky, K.K. Comparison of subcutaneous administration of alfaxalone–midazolam–dexmedetomidine with ketamine–midazolam–dexmedetomidine for chemical restraint in juvenile blue poison dart frogs (*Dendrobates tinctorius azureus*). *J. Zoo Wildl. Med.* **2020**, *50*, 868–873. [CrossRef] [PubMed]
80. Eshar, D.; Rooney, T.A.; Gardhouse, S.; Beaufrère, H. Evaluation of the effects of a dexmedetomidine-midazolam-ketamine combination administered intramuscularly to captive red-footed tortoises (*Chelonoidis carbonaria*). *Am. J. Vet. Res.* **2021**, *82*, 858–864. [CrossRef]
81. Karklus, A.A.; Sladky, K.K.; Johnson, S.M. Respiratory and antinociceptive effects of dexmedetomidine and doxapram in ball pythons (*Python regius*). *Am. J. Vet. Res.* **2021**, *82*, 11–21. [CrossRef] [PubMed]
82. Rooney, T.A.; Eshar, D.; Gardhouse, S.; Beaufrère, H. Evaluation of a dexmedetomidine–midazolam–ketamine combination administered intramuscularly in captive ornate box turtles (*Terrapene ornata ornata*). *Vet. Anaesth. Analg.* **2021**, *48*, 914–921. [CrossRef]
83. Turner, R.C.; Gatson, B.J.; Hernandez, J.A.; Alexander, A.B.; Aitken-Palmer, C.; Vigani, A.; Heard, D.J. Sedation and anesthesia of Galapagos (*Chelonoidis nigra*), Aldabra (*Aldabrachelys gigantea*), and African spurred tortoises (*Centrochelys sulcata*): A retrospective review (2009–2019). *Animals* **2021**, *11*, 2920. [CrossRef]
84. Emmel, E.S.; Rivera, S.; Cabrera, F.; Blake, S.; Deem, S.L. Field anesthesia and gonadal morphology of immature western Santa Cruz tortoises (*Chelonoidis porteri*). *J. Zoo Wildl. Med.* **2021**, *51*, 848–855.
85. Heniff, A.C.; Petritz, O.A.; Carpenter, R.G.; Lewbart, G.A.; Balko, J.A. Anesthetic efficacy of dexmedetomidine-ketamine in eastern box turtles (*Terrapene carolina carolina*) is enhanced with the addition of midazolam and when administered in the forelimb versus the hindlimb. *Am. J. Vet. Res.* **2023**, *1*, 1–9. [CrossRef]
86. Masi, M.; Vetere, A.; Casalini, J.; Corsi, F.; Di Ianni, F.; Nardini, G. Comparison of subcutaneous versus intramuscular dexmedetomidine–midazolam–ketamine–morphine (DMKM) mixture as chemical restraint for endoscopic sex determination in Aldabra giant tortoises (*Aldabrachelys gigantea*). *Animals* **2023**, *13*, 3626. [CrossRef]
87. LutvİKadİC, I.; MaksimoviĆ, A. A comparison of anesthesia induction by two different administration routes and doses of ketamine and medetomidine in red-eared sliders (*Trachemys scripta elegans*). *Ank. Üniversitesi Vet. Fakültesi Derg.* **2024**, *71*, 231–237. [CrossRef]
88. Schwartz, M.; Muñana, K.R.; Nettifee-Osborne, J.A.; Messenger, K.M.; Papich, M.G. The pharmacokinetics of midazolam after intravenous, intramuscular, and rectal administration in healthy dogs. *J. Vet. Pharmacol. Ther.* **2013**, *36*, 471–477. [CrossRef] [PubMed]
89. Le Chevallier, D.; Slingsby, L.; Murrell, J. Use of midazolam in combination with medetomidine for premedication in healthy dogs. *Vet. Anaesth. Analg.* **2019**, *46*, 74–78. [CrossRef]
90. Bienzle, D.; Boyd, C.J. Sedative effects of ketamine and midazolam in snapping turtles (*Chelydra serpentina*). *J. Zoo Wildl. Med.* **1992**, *23*, 201–204.
91. Harvey-Clark, C. Midazolam fails to sedate painted turtles, *Chrysemys picta*. *Bull. Assoc. Reptil. Amphib. Vet.* **1993**, *3*, 7–8. [CrossRef]
92. Holz, P.; Holz, R.M. Evaluation of ketamine, ketamine/xylazine, and ketamine/midazolam anesthesia in red-eared sliders (*Trachemys scripta elegans*). *J. Zoo Wildl. Med.* **1994**, *25*, 531–537.
93. Oppenheim, Y.C.; Moon, P.F. Sedative effects of midazolam in red-eared slider turtles (*Trachemys scripta elegans*). *J. Zoo Wildl. Med.* **1995**, *26*, 409–413.
94. Santos, A.L.Q.; Gomes, D.O.; Lima, C.A.d.P.; Nascimento, L.R.; Menezes, L.T.; Kaminishi, Á.P.S. Anesthesia of turtle *Trachemys dorbigni* (duméril e bibron, 1835)-testudine: Emydidae with the combination of midazolam and propofol. *Pubvet* **2011**, *5*, 15. [CrossRef]
95. Simone, S.B.S.d.; Santos, A.L.Q. Effects of the combination midazolam, fentanyl and ketamine in boas *Boa constrictor* linnaeus, 1758 (*squamata boidae*). *Pubvet* **2011**, *5*, ref. 35.

96. Alves-Júnior, J.R.F.; Bosso, A.C.S.; Andrade, M.B.; Werther, K.; Santos, A.L.Q. Association of midazolam with ketamine in giant Amazon river turtles *podocnemis expansa* breed in captivity. *Acta Cir. Bras.* **2012**, *27*, 144–147. [CrossRef]
97. Santos, A.L.Q.; Oliveira, S.R.P.d.; Kaminishi, Á.P.S.; Andrade, M.B.; Menezes, L.T.; Souza, R.R.d.; Ferreira, C.H.; Nascimento, L.R.; Moraes, F.M.d. Evaluation of the use of the combination of propofol and midazolam in the chemical restraint and anesthesia of the turtle-de-beard *Phrynops geoffroanus* schweigger, 1812 (testudines, chelidae). *Pubvet* **2012**, *6*, ref. 21.
98. Olsson, A.; Phalen, D. Comparison of biochemical stress indicators in juvenile captive estuarine crocodiles (*Crocodylus porosus*) following physical restraint or chemical restraint by midazolam injection. *J. Wildl. Dis.* **2013**, *49*, 560–567. [CrossRef] [PubMed]
99. Arnett-Chinn, E.R.; Hadfield, C.A.; Clayton, L.A. Review of intramuscular midazolam for sedation in reptiles at the national aquarium, Baltimore. *J. Herpetol. Med. Surg.* **2016**, *26*, 59–63. [CrossRef]
100. Lopes, I.G.; Armelin, V.A.; da Silva Braga, V.H.; Florindo, L.H. The influence of midazolam on heart rate arises from cardiac autonomic tones alterations in Burmese pythons, *Python molurus*. *Auton. Neurosci.* **2017**, *208*, 103–112. [CrossRef] [PubMed]
101. Simone, S.B.S.D.; Hirano, L.Q.L.; Santos, A.L.Q. Effects of midazolam in different doses in redtail boa *Boa constrictor* linnaeus, 1758 (squamata: Boidae). *Ciência Anim. Bras.* **2017**, *18*, e22230.
102. Miller, L.J.; Fetterer, D.P.; Garza, N.L.; Lackemeyer, M.G.; Donnelly, G.C.; Steffens, J.T.; Van Tongeren, S.A.; Fiallos, J.O.; Moore, J.L.; Marko, S.T. A fixed moderate-dose combination of tiletamine+ zolazepam outperforms midazolam in induction of short-term immobilization of ball pythons (*Python regius*). *PLoS ONE* **2018**, *13*, e0199339. [CrossRef]
103. B Larouche, C. The Use of Midazolam, Isoflurane, and Nitrous oxide for Sedation and Anesthesia of Ball Pythons (*Python regius*). Ph.D. Thesis, University of Guelph, Guelph, ON, Canada, 2019.
104. Bressan, T.F.; Sobreira, T.; Carregaro, A.B. Use of rodent sedation tests to evaluate midazolam and flumazenil in green iguanas (*Iguana iguana*). *J. Am. Assoc. Lab. Anim. Sci.* **2019**, *58*, 810–816. [CrossRef]
105. Larouche, C.B.; Beaufrère, H.; Mosley, C.; Nemeth, N.M.; Dutton, C. Evaluation of the effects of midazolam and flumazenil in the ball python (*Python regius*). *J. Zoo Wildl. Med.* **2019**, *50*, 579–588. [CrossRef]
106. Larouche, C.B.; Johnson, R.; Beaudry, F.; Mosley, C.; Gu, Y.; Zaman, K.A.; Beaufrère, H.; Dutton, C. Pharmacokinetics of midazolam and its major metabolite 1-hydroxymidazolam in the ball python (*Python regius*) after intracardiac and intramuscular administrations. *J. Vet. Pharmacol. Ther.* **2019**, *42*, 722–731. [CrossRef]
107. Larouche, C.B.; Mosley, C.; Beaufrère, H.; Dutton, C. Effects of midazolam and nitrous oxide on the minimum anesthetic concentration of isoflurane in the ball python (*Python regius*). *Vet. Anaesth. Analg.* **2019**, *46*, 807–814. [CrossRef]
108. González, V.R.; Britez, V.C.; Bazán, Y.; Maldonado, A.E.; Vetter, R.; Fiore, F. Effect of ketamine-midazolam/tiletamine-zolazepam protocols on physiological parameters in *Chelonoidis carbonaria* turtles subjected to routine exploratory procedures. *Rev. Investig. Vet. Peru (RIVEP)* **2022**, *33*, ref. 19.
109. Hirano, L.Q.L.; de Oliveira, A.L.R.; de Barros, R.F.; Veloso, D.F.M.C.; Lima, E.M.; Santos, A.L.Q.; Moreno, J.C.D. Pharmacokinetics and pharmacodynamics of dextroketamine alone or combined with midazolam in *Caiman crocodilus*. *J. Vet. Pharmacol. Ther.* **2024**, *47*, 427–436. [CrossRef] [PubMed]
110. Nishimura, R.; Kim, H.-y.; Matsunaga, S.; Hayashi, K.; Sasaki, N.; Tamura, H.; Takeuchi, A. Sedative effect induced by a combination of medetomidine and midazolam in pigs. *J. Vet. Med. Sci.* **1993**, *55*, 717–722. [CrossRef] [PubMed]
111. Boehm, C.A.; Carney, E.L.; Tallarida, R.J.; Wilson, R.P. Midazolam enhances the analgesic properties of dexmedetomidine in the rat. *Vet. Anaesth. Analg.* **2010**, *37*, 550–556. [CrossRef]
112. Sánchez, A.; Belda, E.; Escobar, M.; Agut, A.; Soler, M.; Laredo, F.G. Effects of altering the sequence of midazolam and propofol during co-induction of anaesthesia. *Vet. Anaesth. Analg.* **2013**, *40*, 359–366. [CrossRef] [PubMed]
113. Tranquilli, W.J.; Grimm, K.A. Introduction: Use, definitions, history, concepts, classification, and considerations for anesthesia and analgesia. In *Veterinary Anesthesia and Analgesia*, 5th ed.; Grimm, K.A., Lamont, L.A., Tranquilli, W.J., Greene, S.A., Robertson, S.A., Eds.; WILEY Blackwell: Hoboken, NJ, USA, 2015; pp. 1–10.
114. Kristensen, L.; Malte, C.L.; Malte, H.; Wang, T.; Williams, C.J.A. Obesity prolongs induction times in reptiles. *Comp. Biochem. Physiol. Part A Mol. Integr. Physiol.* **2022**, *271*, 111255. [CrossRef]
115. Valverde, A.; Skelding, A.M. Alternatives to opioid analgesia in small animal anesthesia: Alpha-2 agonists. *Vet. Clin. Small Anim. Pract.* **2019**, *49*, 1013–1027. [CrossRef]
116. Yaksh, T.L. Pharmacology of spinal adrenergic systems which modulate spinal nociceptive processing. *Pharmacol. Biochem. Behav.* **1985**, *22*, 845–858. [CrossRef]
117. Gil, D.W.; Cheevers, C.V.; Kedzie, K.M.; Manlapaz, C.A.; Rao, S.; Tang, E.; Donello, J.E. A-1-adrenergic receptor agonist activity of clinical α-adrenergic receptor agonists interferes with α-2-mediated analgesia. *J. Am. Soc. Anesthesiol.* **2009**, *110*, 401–407. [CrossRef]
118. Virtanen, R.; Savola, J.-M.; Saano, V.; Nyman, L. Characterization of the selectivity, specificity and potency of medetomidine as an α2-adrenoceptor agonist. *Eur. J. Pharmacol.* **1988**, *150*, 9–14.
119. Lovell, S.; Simon, B.; Boudreau, E.C.; Mankin, J.; Jeffery, N. Randomized clinical trial comparing outcomes after fentanyl or ketamine-dexmedetomidine analgesia in thoracolumbar spinal surgery in dogs. *J. Vet. Intern. Med.* **2022**, *36*, 1742–1751. [CrossRef] [PubMed]
120. Cabanac, A.; Cabanac, M. Heart rate response to gentle handling of frog and lizard. *Behav. Process.* **2000**, *52*, 89–95. [CrossRef] [PubMed]
121. Klide, A.M.; Calderwood, H.W.; Soma, L.R. Cardiopulmonary effects of xylazine in dogs. *Am. J. Vet. Res.* **1975**, *36*, 931–935.

122. Giovannitti, J.A., Jr.; Thoms, S.M.; Crawford, J.J. Alpha-2 adrenergic receptor agonists: A review of current clinical applications. *Anesth. Prog.* **2015**, *62*, 31–39. [CrossRef]
123. Cardoso, C.G.; Marques, D.R.C.; da Silva, T.H.M.; de Mattos-Junior, E. Cardiorespiratory, sedative and antinociceptive effects of dexmedetomidine alone or in combination with methadone, morphine or tramadol in dogs. *Vet. Anaesth. Analg.* **2014**, *41*, 636–643. [CrossRef]
124. Cerreta, A.J.; Cannizzo, S.A.; Smith, D.C.; Minter, L.J. Venous hematology, biochemistry, and blood gas analysis of free-ranging eastern copperheads (*Agkistrodon contortrix*) and eastern ratsnakes (*Pantherophis alleghaniensis*). *PLoS ONE* **2020**, *15*, e0229102. [CrossRef] [PubMed]
125. Congdon, J.M.; Marquez, M.; Niyom, S.; Boscan, P. Cardiovascular, respiratory, electrolyte and acid–base balance during continuous dexmedetomidine infusion in anesthetized dogs. *Vet. Anaesth. Analg.* **2013**, *40*, 464–471. [CrossRef]
126. Bertelsen, M.F.; Buchanan, R.; Jensen, H.M.; Leite, C.A.C.; Abe, A.S.; Nielsen, S.S.; Wang, T. Assessing the influence of mechanical ventilation on blood gases and blood pressure in rattlesnakes. *Vet. Anaesth. Analg.* **2015**, *42*, 386–393. [CrossRef]
127. Allen, S.E.; Holm, J.L. Lactate: Physiology and clinical utility. *J. Vet. Emerg. Crit. Care* **2008**, *18*, 123–132. [CrossRef]
128. Franklin, C.E.; Davis, B.M.; Peucker, S.K.J.; Stephenson, H.; Mayer, R.; Whittier, J.; Lever, J.; Grigg, G.C. Comparison of stress induced by manual restraint and immobilisation in the estuarine crocodile, *Crocodylus porosus*. *J. Exp. Zool. Part A Comp. Exp. Biol.* **2003**, *298A*, 86–92. [CrossRef]
129. Goessling, J.M.; Mendonça, M.T. Physiological responses of gopher tortoises (*Gopherus polyphemus*) to trapping. *Conserv. Physiol.* **2021**, *9*, coab003. [CrossRef] [PubMed]

animals

MDPI

Article

Comparison of Two Intravenous Propofol Doses after Jugular Administration for Short Non-Surgical Procedures in Red-Eared Sliders (*Trachemys scripta elegans*)

Lucia Victoria Bel [1,*], Paolo Selleri [2], Carmen Maria Turcu [1], Constantin Cerbu [3], Ioana Adriana Matei [4], Marco Masi [2] and Iulia Melega [1,*]

[1] Department of Surgery and ICU, Faculty of Veterinary Medicine, University of Agricultural Sciences and Veterinary Medicine Cluj-Napoca, Calea Manastur No. 3-5, 400372 Cluj-Napoca, Romania; smaria-carmen.turcu@usamvcluj.ro

[2] Clinica per Animali Esotici, Veterinary Specialists Centre, 00154 Rome, Italy; paolsell@gmail.com (P.S.); marco8912@hotmail.it (M.M.)

[3] Department of Infectious Diseases, Faculty of Veterinary Medicine, University of Agricultural Sciences and Veterinary Medicine Cluj-Napoca, Calea Manastur No. 3-5, 400372 Cluj-Napoca, Romania; contantin.cerbu@usamvcluj.ro

[4] Department of Microbiology, Immunology and Epidemiology, Faculty of Veterinary Medicine, University of Agricultural Sciences and Veterinary Medicine Cluj-Napoca, Calea Manastur No. 3-5, 400372 Cluj-Napoca, Romania; ioana.matei@usamvcluj.ro

* Correspondence: lucia.bel@usamvcluj.ro (L.V.B.); iulia.melega@usamvcluj.ro (I.M.); Tel.: +40-742321306 (L.V.B.)

Simple Summary: Anesthesia in red-eared sliders is necessary both for surgical procedures and for imaging techniques. Propofol is a nonbarbiturate anesthetic agent used for induction in many species and should be administered intravenously. Jugular intravenous cannulas provide safe access with the least lymph contamination for anesthetic administration. In this study, red-eared sliders were anesthetized with 5 mg/kg and 10 mg/kg propofol administered in the jugular veins. Our results indicate that the 10 mg/kg dose is efficient for inducing anesthesia for short non-painful procedures, whereas the 5 mg/kg dose did not prove to be enough for anesthetic induction.

Abstract: This study compares the effects of two different doses of propofol administered intravenously (IV), in the jugular vein, to red-eared sliders (*Trachemys scripta elegans*). In this crossover study, 5 or 10 mg/kg propofol was administered to six *Trachemys scripta elegans* after cannulation of the jugular vein. Each turtle received each dose, G1 (5 mg/kg IV) and G2 (10 mg/kg IV), after a 7-day washout period. The parameters evaluated were heart rate, palpebral reflex, cloacal reflex, muscle relaxation, ease of handling, sensitivity to anterior and posterior pinch stimuli, and possibility of intubation. Additionally, respiratory rate was measured when possible, and the times from propofol administration to full recovery and from intubation to extubation were recorded. None of the turtles in G1 could be intubated, and this dose provided little relaxation and ease of handling, with a duration of effect until full recovery of 12.16 ± 8.32 (SD) min for this group. In G2, five out of the six turtles could be intubated, and the duration of effect was 32.33 ± 5.85 (SD) min. Heart rates were influenced by manipulation for catheter placement. There were statistically significant differences (p value ≤ 0.05) between the two groups in muscle relaxation degree, handling, cloacal reflex, and possibility of intubation. The 5 mg/kg propofol dose was not sufficient to induce anesthesia, even when administered in the jugular vein, in red-eared sliders. A dose of 10 mg/kg IV or higher should be used.

Keywords: propofol; anesthesia; *Trachemys scripta elegans*; turtle; jugular vein

Citation: Bel, L.V.; Selleri, P.; Turcu, C.M.; Cerbu, C.; Matei, I.A.; Masi, M.; Melega, I. Comparison of Two Intravenous Propofol Doses after Jugular Administration for Short Non-Surgical Procedures in Red-Eared Sliders (*Trachemys scripta elegans*). *Animals* **2024**, *14*, 1847. https://doi.org/10.3390/ani14131847

Academic Editor: Patricia V. Turner

Received: 29 April 2024
Revised: 4 June 2024
Accepted: 17 June 2024
Published: 21 June 2024

1. Introduction

In reptiles, the onset of anesthesia from inhalatory anesthetics delivered by mask can be prolonged and, in some cases, like turtles, very difficult to achieve. This is due to their anatomical and physiological characteristics, most importantly, the dive reflex. Moreover, freshwater turtles, such as *Trachemys* spp., exhibit large cardiac shunts, owing to the incomplete anatomical division of the ventricle [1]. Due to these shunts, systemic venous blood recirculates into systemic arteries and lowers arterial oxygen partial pressure, slowing the uptake of inhaled anesthesia [2–4]. As a consequence, there is a clinical demand for viable injectable anesthetics in turtles.

Propofol (2,6-diisopropylphenol) is a nonbarbiturate anesthetic, used extensively for anesthetic induction in amphibians and reptiles [5,6], that is eliminated through hepatic metabolism to glucuronide metabolites and renal excretion [7]. It provides rapid induction and has been used in lizards [8], snakes [9], and chelonians [10–12]. The intramuscular (IM) administration of propofol is not recommended, both because it has been shown to cause inflammation and necrosis in different species [13] and because it is ineffective for producing sedation or anesthesia. Several studies report the use of propofol in reptiles, with a rapid induction time and short recovery, with dosages ranging from 5 to 20 mg/kg [6,8,10,14–18]. Some studies even advise the possibility of intracoelomic [19] or intraosseous [20] administration in red-eared sliders. Nevertheless, the recommended route of administration for propofol in all species is intravenous [15]. We hypothesized that the administered effective dose could be lower due to jugular vein administration, and thus decided on the two doses, 5 and 10 mg/kg.

Venous access is very important for many medical-based procedures, specifically medication administration. It can be challenging in many reptilian species, but mostly in Chelonians, due to the inability to visualize veins and the difficulty in restraint and handling. This is primarily because they can retract their head and limbs into their shell. The most used venipuncture sites in turtles are the subcarapaceal or supravertebral vessel, the dorsal coccygeal vein, the brachial vein, the occipital plexus, and the external jugular vein [15]. The last is the blood sampling site of choice in turtles and tortoises and should be used when possible [21,22]. This recommendation is based on the fact that it carries the least risk of lymph contamination compared to the subcarapaceal plexus or occipital sinus.

In this context, the aim of this study was to compare the sedative and muscle relaxant effects of two dose rates of propofol (5 and 10 mg/kg) in red-eared sliders, administered intravenously via the jugular vein. To our knowledge, this is the first study assessing the effects of propofol administered via this route in *Trachemys scripta elegans*.

2. Materials and Methods

2.1. Turtles

Six healthy adult female red-eared sliders (*Trachemys scripta elegans*) weighing 1.37 ± 0.593 (SD) kg were used in this study. All turtles were considered clinically healthy based on physical examination and history.

Turtles were anesthetized to perform CT scans with or without a contrast substance, as a method of screening for reproductive disorders. They were privately owned animals, living outside in a pond, from a single owner, who had signed their informed consent. This study was performed in compliance with directive 2010/63/EU as well as the applicable items of the REFLECT statement.

Prior to this study, the animals were acclimatized one day before and were kept in the same environment. Their housing consisted of plastic water tubs, equipped with an area for basking under artificial lighting and an area for retreat, with the temperature of the water and basking spot being in the POTZ—preferred optimum temperature zone—of the species (20 °C to 25 °C in the water and 25 °C to 45 °C for basking). The turtles were provided with UV B lighting and fed a commercial diet with extra green, s every second day. For the purpose of this study and to limit possible complications, all turtles were kept for 10 days and then returned to the owner's pond.

2.2. Study Design

An uncontrolled crossover design was used, meaning that each individual received both dosages, 5 mg/kg (G1) propofol (PropoVet Multidose, 10 mg/mL, Zoetis, Copenhagen, Denmark) and 10 mg/kg propofol (G2), IV, in one of the jugular veins, with a 7-day washout period. On each of the days of the experiments, the turtles were weighed, and an IV cannula was inserted in either the right or the left jugular vein, without any prior sedation. After local antisepsis with chlorhexidine and alcohol, one examiner would insert either a 24 G or a 26 G IV catheter, depending on the size of the animal.

The heads of the turtles were slightly directed ventrally, and the jugular vein was identified. This vessel lies very superficially, running from the tympanic membrane at the jaw angle to the base of the neck, and compression at the level of the coelomic inlet is useful for better visualization. Once the cannulation was successful, 1 mL/kg IV sterile water was used to flush and confirm the correct placement, and propofol was administered in a bolus, in about 10 s. Turtles were then placed back in ventral recumbency, and various measurements were taken every 5 min (heart rate, palpebral reflex, cloacal reflex, muscle relaxation, ease of handling, sensitivity to anterior and posterior pinch stimuli, possibility of intubation). Respiratory rate was measured when possible; when animals were ventilated, this was performed with room air using a small Ambu bag, at 8 breaths/minute. Heart rate (HR) was obtained by Doppler ultrasound (Ultrasound Blood Flow Detector MD4, Sonomed, Warsaw, Poland) after placing the probe at the left thoracic inlet and directing it towards the heart. Palpebral and cloacal reflexes were evaluated by touching the lateral and medial canthus of the eye with a cotton-tipped applicator and the cloacal opening, respectively, and were assigned a score of '0' when absent and '1' when present.

Using the three-point scale of Santos et al. [12], muscle relaxation, ease of handling and sensitivity to pinch stimuli were evaluated. When the turtle kept its head up and retracted, muscle relaxation score was considered to be '0'; when the head, legs, and tail retained a mild degree of muscle tone, the muscle relaxation score was considered to be '1'; and when the head, tail, and legs remained extended, the muscle relaxation score was '2'. In the case of handling, the same principles were applied: '0' when there was difficulty in flexing and extending the head, legs, and tail and in opening the mouth manually; '1' when there was mild resistance to manipulation of the head, legs, and tail, or to opening the animal's mouth; and '2' when no resistance to manipulation of the head, legs, and tail, or to opening the animal's mouth, was identified. The score for sensitivity to toe pinch stimulus was considered to be '0' when the withdrawal response after a pinch using tongue forceps performed on both anterior and posterior limbs was accompanied by movement of the head or another limb, and '1' when the absence of a response to the toe pinch or responses suggestive of a spinal reflex (limb withdrawal unaccompanied by limb or head movement) was seen. Intubation was graded '0' when possible and '1' when impossible.

The induction of anesthesia was defined as the time point at which turtles achieved both maximal muscle relaxation and ease of handling. 'Recovery time' was considered to be when there was a full return of baseline scores for palpebral reflex, cloacal reflex, muscle relaxation, and ease of handling. The times from propofol administration and full recovery and the time from intubation to extubation were also recorded. To ensure blinding and eliminate assessment bias, assessors were not informed beforehand about the dose administered to the turtles.

The statistical analysis was performed in EpiInfoTM7 (CDC). The means and standard deviation of heart rate were calculated and compared between the two dosage groups using both an ANOVA Parametric Test for Inequality of Population Means and a Kruskal–Wallis test for two groups. For the other variables (muscle relaxation, handling, and sensitivity to pinching degrees), intubation grading and cloacal and palpebral reflex presence frequencies and 95% confidence intervals were calculated, and frequencies were compared using chi-square, and a p value < 0.05 was considered to be significant.

3. Results

No complications were observed throughout the study and all the blind jugular cannulations were successful (Figure 1). The CT images were unremarkable, showing follicles in different stages of development.

Figure 1. Right jugular vein cannulation with a 24 G IV catheter in a female turtle.

The median heart rate was 48 (25–75% range: 45–60) for G1 and 46 (25–75% range: 42–60) for G2, considering all times. The median for each time measurement can be seen in Figure 2. A lower heart rate mean was observed at times T10 for the group receiving a 5 mg dosage and at T25 for the group receiving 10 mg/kg. When compared, the differences between the means were not statistically significant ($p > 0.05$) whether considering measurements at all times or for each time measurement.

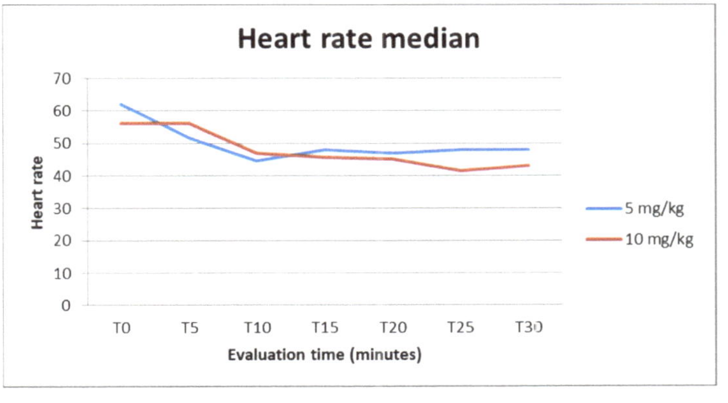

Figure 2. Evolution of median heart rate for both G1 and G2 during the 30-minute evaluation time.

In the case of muscle relaxation degree (0, 1, 2), considering measurements at all times, among the two groups, we observed a statistically significant difference (x^2 = 10.7985, df = 2, p = 0.0045) (Figure 3). Between T20 and T30, the prevalence of degrees of muscle relaxation was similar among the two groups (Supplementary File S1).

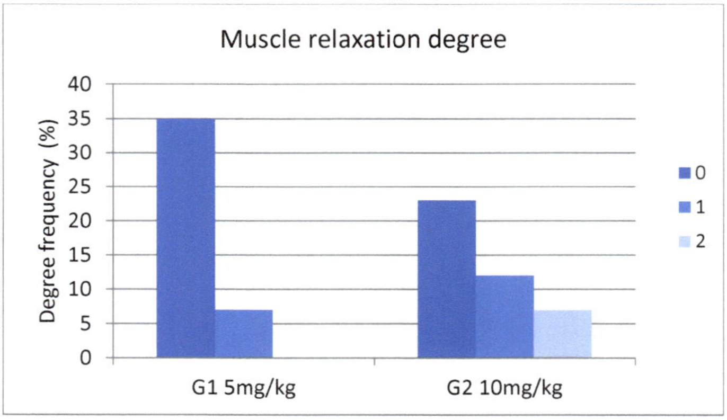

Figure 3. Muscle relaxation degree for G1 (**left**) and G2 (**right**) during the evaluation of the protocol, where 0 = no muscle relaxation, 1 = mild muscle relaxation, and 2 = fully relaxed.

Similarly, in the case of handling degree (0, 1, 2), considering measurements at all times, a statistically significant difference was observed among the two groups (x^2 = 7.785, df = 2, p = 0.0203) (Figure 4). Between T20 and T30 the prevalence of degrees of muscle relaxation was similar among the two groups (Supplementary File S1).

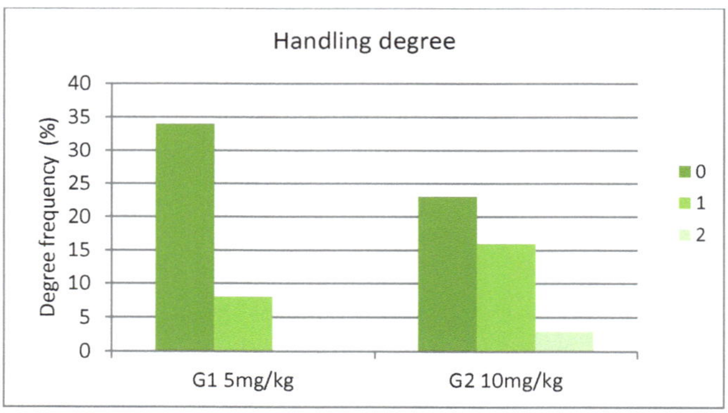

Figure 4. Handling degree for G1 (**left**) and G2 (**right**) during the evaluation of the protocol, where 0 = reactivity to handling, 1 = mild response to handling, and 2 = no response when handled.

None of the turtles from G1, receiving the 5 mg/kg jugular vein IV propofol dose, could be intubated (Supplementary File S1). On the other hand, when receiving the 10 mg/kg dose, five of the turtles could be intubated (Supplementary File S1) and stayed intubated for 8.6 ± 4.15 (SD) min.

Intubation was possible in four cases (66.67%, 95% CI: 22.28–95.67) at T0 and three cases (50%, 95% CI: 11.81–88.19) at T10 (two of the turtles were still intubated after 10 min), all in the group with 10 mg dosages (Supplementary File S1). The difference in intubation success was statistically significant between the dosage groups when considering measurements at all times (x^2 = 7.6364, df = 1, p = 0.002979).

The cloacal reflex was lost in four cases (66.67%, 95% CI: 22.28–95.67) at T0 and 1 case (16.67%, 95% CI: 0.42–64.12) at both T5 and T10, both in group G2 (Supplementary File S1), with a significant difference between the groups (x^2 = 6.4615, df = 1, p = 0.00645267).

The palpebral reflex was lost in one case (16.67%, 95% CI: 0.42–64.12) at T0 in the group with 10 mg/kg dosages (Supplementary File S1). The difference was not statistically significant between the dosage groups when taking into consideration measurements at all times ($x^2 = 1.012$, df = 1, $p = 0.25$).

The duration from T0 (administration of propofol) to intubation was 3.2 ± 2.68 (SD) min for G2, and the duration of effect until full recovery was 12.16 ± 8.32 (SD) min for G1 and 32.33 ± 5.85 (SD) for G2.

In our study, we were not able to evaluate respiration rate throughout the different recorded times, but we could observe apnea in all individuals that could be intubated. Nevertheless, once signs of spontaneous breathing could be observed, extubation followed. We then observed an increase in the respiratory rate, which was also observed in the G2 group, probably because of manipulation for evaluation of the different parameters.

4. Discussion

Reptiles, and turtles, in particular, possess a renal portal system. This is a ring of vessels around the kidney, including the cranial portal vein and caudal portal vein, with blood flowing from the caudal area through the coccygeal and iliac veins, that then continues into the afferent renal portal vein, which transfers blood to the kidneys [23,24]. Blood perfusion tubules then leave the kidney through the efferent portal vein, which joins the post-cava vein. This system seems to be involved in kidney perfusion during dehydration, avoiding ischemic damage by ensuring adequate perfusion [25,26]. A system of valves located between the abdominal and the femoral veins either directs the blood flow, influenced by hydration status, directly to the kidneys or bypasses them and directs the flow directly to the circulatory system. This is why the common recommendation is to administer drugs in the cranial part of the body, to avoid this renal portal system, even though there is some controversy surrounding this subject. Some studies have shown that the administration of a combination of the same dosage of dexmedetomidine and ketamine in the cranial or caudal quarter of the body showed a strong difference in sedation power [27,28], and on the contrary, some have proven that the clinical relevance of the renal portal system is minimal and might lead to interactions only in dehydrated animals [29,30].

The anesthetic effects of propofol were evaluated in different reptile species. In *Trachemys scripta* spp., Di Giuseppe et al. [31] evaluated the administration of 5 mg/kg of propofol through the occipital venous sinus (cranial part of the body), the subcarapacial venous sinus (cranial part of the body), and the coccygeal vein (caudal part of the body). When administering the substance in the occipital venous sinus, they observed a full anesthetic plain in all turtles, but this was not the case for the other two groups. They hypothesize that in the case of the subcarapaceal plexus, a higher dosage might be required, 10 or 20 mg/kg as shown by Ziolo et al. [11] and that the poor effect of coccygeal administration is due to the renal portal system. Morici et al. [32] evaluated the difference in response when administering another anesthetic, alfaxalone, at two different sites intravenously in the cranial area: the cervical dorsal sinus and the caudal area, through the coccygeal vein. Their results are consistent with other studies regarding the influence of the renal portal system, proving once again that cranial substance administration will result in a better effect at a lower dosage [32].

Our study is the first to evaluate intravenous propofol administration in the jugular vein in this species. We chose 5 and 10 mg/kg dosages, to identify the lowest dosage that could be used to permit endotracheal intubation in this species, and the jugular vein as it is one of the safest sites for intravenous administration. The location of this vein in turtles is superficial under the skin and can be approachable with or without chemical restraint, depending on the cooperation of the patient. The vein runs on both sides of the neck, in a caudodorsal direction from the dorsal aspect of the tympanum to the coelomic cavity. For better visibility, the cervical folds can be tightened, or in some species, an otoscope can be inserted in the mouth of the animal [22]. The vein can be used for phlebocentesis or catheter placement and substance administration. Once a catheter is in place, it can then also be used for anesthesia induction or supplementation. The 5 mg/kg dose did not prove as effective as the 10 mg/kg dose, and even though the jugular vein is in the cranial part of the body and a very safe route for IV administration (since it is not lymph-contaminated and no shunts will be involved in the excretion of the substances administered via this vein), we did not obtain similar results to Di Giuseppe et al. [31] with occipital venous sinus administration. In their case, the mean induction time was within 1.8 ± 30.8 min (with one terrapin that did not lose the mandibular tone) and the mean tracheal tube insertion time was within 3.89 ± 2.45 min [31]. We obtained a very high standard deviation concerning the induction time; thus, we assume that with the 5 mg/kg propofol dose, there is more of an individual response and that is why our study showed a more stable response at a higher dose, that of 10 mg/kg.

Another reason for such different results might be the influence of the health status of the animals or the environmental temperature. In one study evaluating the administration of alfaxalone in *Trachemys scripta* spp. at two different temperatures, colder temperatures were shown to affect anesthesia and time of recovery [33]. The room temperature in our case was 24 °C, turtles were kept at 26 °C for acclimatization, and cloacal temperature was not evaluated since it is not as relevant compared to its relevance in mammals. Di Giuseppe's turtles were kept at a temperature 1 °C higher compared to 3.2 ± 2.68 min in the current study. Using a 10 mg/kg dose, Ziolo et al. [11] obtained a mean induction time of 1.7 ± 2.4 min, compared to our 3.2 ± 2.68 min, when administering it in the subcarapaceal plexus. We can only assume that the substance distribution in the blood flow is faster when administered in this area.

The choice of the subcarapaceal sinus as a site for substance administration is mostly related to its ease of access in unsedated animals, with minimal manipulation [34], but is not recommended in clinical practice due to reports of accidental submeningeal injection and clinical abnormalities [35,36]. Kristensen et al. [14] reported in a study the use of atropine and propofol at 15 mg/kg, administered in the subcarpaceal sinus, as described by Ziolo and Bertelsen in 2009 [11]. In their study, repeated dosing of propofol was needed for intubation, with the average propofol dose required being 17.8 mg/kg. Moreover, in one of the turtles (out of eight), induction with propofol was unsuccessful. This might be due to the possible extravasation of the substance when this route is used, a complication best described by Rockwell et al. [37], making the use of the jugular or brachial veins the safest route of IV administration for anesthetic substances when intravenous access is needed. One of the few studies to use propofol administration in the jugular vein in other turtles, Cicarelli et al. [38], used a dose ranging from 5 to 8 mg/kg, for intravenous anesthesia in a marine turtle species, *Caretta caretta*. These dosages allowed a sufficient degree of anesthesia for the minimally invasive procedures they performed. Also, the procedures took around 10–15 min. In the same study, bronchoscopy was performed, and during this time, total intravenous anesthesia with propofol was used. Marine turtles have a higher capacity to maintain apnea than semi-aquatic ones, and none of these turtles were intubated. We would assume that if intubation was required, the propofol dose should have been higher. At the same time, these turtles were a very different species to the semiaquatic one in our study and were not clinically healthy, since all the procedures were performed for diagnostic and treatment purposes for pulmonary disease, hence the lower dose necessary for induction in these individuals.

Concerning the effects of IV propofol on the cardiovascular system, previously, a stable level and even a decrease in HR have been reported in most reptiles [8,11,12,39–41] except in the case of Bertelsen et al. [39], which is consistent with our study and those on dogs and cats [42–44]. In most individuals, we observed an increase in heart rate, independent of the propofol dose that was used, after administration, even though our T0 was considered to be the moment of propofol administration. We believe that this is due to stress caused by manipulation and cannot be compared to a 'baseline' value which would have also been very high, which is consistent with the proof of tachycardia as a consequence of physical restraint [45,46]. Also, respiratory depression has been recorded in various species as a concern when using propofol, like fish [47], amphibians [48,49], reptiles [10], avian [50], and mammalian [51] species. In our case, we observed apnea in all individuals that could be intubated.

Prolonged recovery is a common post-anesthetic complication cited by veterinarians practicing reptile medicine [52]. Compared to other studies [11], we consider the average half-hour recovery time in our study to be efficient for nonsurgical procedures, such as radiology, ultrasound, and CT scanning. The use of other anesthetic substances, like inhalatory anesthesia, sedatives, or opioids, when performing anesthesia for surgical procedures might prolong recovery time, and more research is advised to evaluate more complex anesthetic protocols useful for more complex procedures in red-eared sliders. As there are no analgesic properties to propofol, we recommend pairing it with proven analgesic drugs in turtles undergoing invasive or potentially painful procedures. We would also recommend a thorough pre-anesthetic physical examination and, if needed, at least a protein analysis. There have also been reports of death [22] at high propofol dosages, because propofol is highly protein-bound in vivo, meaning that hypoproteinemia will result in a higher unbound fraction and thus effective dose, possibly leading to an overdose. What our study showed was that propofol administered IV in the jugular vein at 10 mg/kg may be used in red-eared sliders to produce a suitable degree of muscle relaxation for diagnostic or small medical procedures and was enough to intubate five out of six turtles. We have previously used the same dose in practice but injected it IV in the coccygeal vein (personal communication), but this was not enough for intubation. Jugular IV canulation, despite being 100% successful in our case, might be challenging in unsedated turtles, so it could be of future interest to evaluate the use of sedatives like benzodiazepines before propofol administration [53].

Limitations of the Study

1. Absence of Control Group: This study lacks a control group. This absence is justified by the lack of a consensus regarding a safe anesthetic protocol in turtles, making it difficult, if not impossible, to establish a positive control group.
2. Small Sample Size: This study includes a small number of turtles due to the limited availability of individuals. As a result, the experiment was conducted as a small-scale crossover trial to limit the variability in the within-sliders comparisons.
3. Generalization Limitation: Because of the small sample size and the absence of a control group, the results of this study cannot be generalized widely. However, they are considered an important advancement toward establishing a specific propofol dose for general use in turtle anesthesia.

These points suggest that while this study has limitations, it represents progress in the field and lays the groundwork for future research.

5. Conclusions

The results of this small-scale study show that propofol at 10 mg/kg administered intravenously in the jugular vein provides a fast onset of anesthesia and a short recovery time. Jugular vein cannulation, although slightly challenging, is the safest method of propofol administration, and it should be used whenever possible. The 5 mg/kg IV dose has little effectiveness, but it might be of interest to re-evaluate it in conjunction with the administration of sedatives.

Supplementary Materials: The following supporting information can be downloaded at: https://www.mdpi.com/article/10.3390/ani14131847/s1. The data that support the findings of this study are available from the corresponding author, L.V.B., upon reasonable request.

Author Contributions: Conceptualization, L.V.B.; data curation, L.V.B. and C.M.T.; investigation, L.V.B. and M.M.; software and statistical analyses I.A.M. and C.C.; supervision, P.S.; writing—original draft preparation, L.V.B.; writing—review and editing, I.M., C.C., and P.S. All authors have read and agreed to the published version of the manuscript.

Funding: This research received no external funding.

Institutional Review Board Statement: The animal study protocol was approved by the Institutional Ethics Committee of USAMV Cluj Napoca (260b/04.06.2021).

Informed Consent Statement: The owner of the turtles consented in writing to all procedures.

Data Availability Statement: The data presented in this study are available on request from the corresponding author. The data are not publicly available due to the General Data Protection Regulation.

Acknowledgments: The authors would like to thank the owner of the turtles and the personnel from Clinica per Animali Esotici for their support.

Conflicts of Interest: The authors declare no conflicts of interest.

References

1. Hicks, J.W. Cardiac shunting in reptiles: Mechanisms, regulation and physiological functions. In *Biology of the Reptilia*; Morphology G: The Visceral Organs; Gans, C., Gaunt, A.S., Eds.; Society for the Study of Amphibians and Reptiles: Ohio, OH, USA, 1998; Volume 19, pp. 425–483.
2. Eger, E.I.; Severinghaus, J.W. Effect of uneven pulmonary distribution of blood and gas on induction with inhalation anesthetics. *Anesthesiology* **1964**, *25*, 620–626. [CrossRef] [PubMed]
3. Tanner, G.E.; Angers, D.G.; Barash, P.G.; Mulla, A.; Miller, P.L.; Rothstein, P. Effect of left-to-right, mixed left-to-right, and right-to-left shunts on inhalational anesthetic induction in children: A computer model. *Anesth. Analg.* **1985**, *64*, 101–107. [CrossRef] [PubMed]
4. Williams, C.J.A.; Malte, C.L.; Malte, H.; Bertelsen, M.F.; Wang, T. Ectothermy and cardiac shunts profoundly slow the equilibration of inhaled anesthetics in a multi-compartment model. *Sci. Rep.* **2020**, *10*, 17157. [CrossRef] [PubMed]
5. Longley, L.A. Amphibian anesthesia. In *Anesthesia of Exotic Pets*; Saunders Elsevier: Edinburgh, UK, 2008; pp. 245–260.
6. Bertelsen, M.F.; Buchanan, R.; Jensen, H.M.; Leite, C.A.; Abe, A.S.; Nielsen, S.S.; Wang, T. Assessing the influence of mechanical ventilation on blood gases and blood pressure in rattlesnakes. *Vet. Anaesth. Analg.* **2015**, *42*, 386–393. [CrossRef] [PubMed]
7. Digger, T.; Viira, D.J. Anaesthesia and surgical pain relief—The ideal general anesthetic agent. *Hosp. Pharmacist.* **2003**, *10*, 432–440.
8. Bennet, R.A.; Schumacher, J.; Hedjazi Haring, K.; Newell, S.M. Cardiopulmonary and anesthetic effects of propofol administered intraosseously to green iguana. *J. Am. Vet. Med. Ass.* **1998**, *212*, 93–98. [CrossRef]
9. Anderson, N.L.; Wack, R.F.; Calloway, L.; Hetherington, T. Cardiopulmonary effects and efficacy of propofol as an anesthetic agent in brown tree snakes (*Boiga irregularis*). *Bull. Assoc. Reptil. Amphib. Vet.* **1999**, *9*, 9–15. [CrossRef]
10. MacLean, R.A.; Harms, C.A.; Braun-McNeill, J. Propofol anesthesia in loggerhead (*Caretta caretta*) sea turtles. *J. Wildl. Dis.* **2008**, *44*, 143–150. [CrossRef] [PubMed]
11. Ziolo, M.S.; Bertelsen, F. Effects of propofol administered via the supravertebral sinus in red-eared sliders. *J. Am. Vet. Med. Assoc.* **2009**, *234*, 390–393. [CrossRef]
12. Olsson, A.; Phalen, D.; Dart, C. Preliminary studies of alfaxalone for intravenous immobilization of juvenile captive estuarine crocodiles (*Crocodylus porosus*) and Australian freshwater crocodiles (*Crocodylus johnstoni*) at optimal and selected sub-optimal thermal zones. *Vet. Anesth. Analg.* **2013**, *405*, 494–502. [CrossRef]
13. McKune, C.M.; Brosnan, R.J.; Dark, M.J.; Haldorson, G.J. Safety and efficacy of intramuscular propofol administration in rats. *Vet. Anesth. Analg.* **2008**, *35*, 495–500. [CrossRef] [PubMed]
14. Kristensen, L.; Zardo, J.Q.; Hansen, S.M.; Bertelsen, M.F.; Alstrup, A.K.; Wang, T.; Williams, C.J. Effect of atropine and propofol on the minimum anaesthetic concentration of isoflurane in the freshwater turtle *Trachemys scripta* (yellow-bellied slider). *Vet. Anesth. Analg.* **2023**, *50*, 180–187. [CrossRef] [PubMed]
15. Schumacher, J.; Yelen, T. Anesthesia and Analgesia. In *Reptile Medicine and Surgery*, 2nd ed.; Mader, D.R., Ed.; Elsevier Saunders: Philadelphia, PA, USA, 2006; pp. 442–452.
16. Santos, A.L.; Bosso, A.C.; Alves Júnior, J.R.; Brito, F.M.M.; Pachally, J.R.; Ávila Junior, R.H. Pharmacological restraint of captivity giant amazonian turtle *Podocnemis expansa* (testudines, podocnemididae) with xylazine and propofol. *Acta Cir. Bras.* **2008**, *23*, 270–273. [CrossRef] [PubMed]
17. McFadden, M.S.; Bennett, R.A.; Reavill, D.R.; Ragetly, G.R.; Clark-Price, S.C. Clinical and histologic effects of intracardiac administration of propofol for induction of anesthesia in ball pythons (*Python regius*). *J. Am. Vet. Med. Assoc.* **2011**, *239*, 803–807. [CrossRef] [PubMed]
18. Alves-Júnior, J.R.; Bosso, A.C.; Andrade, M.B.; Jayme, V.D.S.; Werther, K.; Santos, A.L.Q. Association of acepromazine with propofol in giant Amazon turtles *Podocnemis expansa* reared in captivity. *Acta Cir. Bras.* **2012**, *27*, 552–556. [CrossRef] [PubMed]
19. Schroeder, C.A.; Johnson, R.A. The efficacy of intracoelomic fospropofol in red-eared sliders. *J. Zoo. Wildl. Med.* **2013**, *44*, 941. [CrossRef] [PubMed]
20. Fonda, D. Intraosseous anesthesia with propofol in red-eared sliders (*Trachemys scripta elegans*). *J. Vet. Anesth.* **1999**, *26*, 46–47.
21. Mans, C. Venipuncture techniques in chelonian species. *Lab Anim.* **2008**, *37*, 303–304. [CrossRef] [PubMed]
22. Bel, L.V.; Selleri, P. Use of a light source to help identify the jugular vein in chelonians. *J. Small. Anim. Pract.* **2018**, *59*, 197. [CrossRef]

23. Murray, M.M. Cardiopulmonary anatomy and physiology. In *Reptile Medicine and Surgery*, 2nd ed.; Mader, D.R., Ed.; Elsevier Saunders: Philadelphia, PA, USA, 2006; pp. 124–128.
24. O'Malley, B. (Ed.) Tortoises and Turtles. In *Clinical Anatomy and Physiology of Exotic Species*. Elsevier Saunders: Philadelphia, PA, USA, 2005; pp. 41–56.
25. Holz, P.; Barker, I.K.; Crawshaw, G.J.; Dobson, H. The anatomy and perfusion of the renal portal system in the red-eared slider (*Trachemys scripta elegans*). *J. Zoo Wild. Med.* **1997**, *28*, 378–385.
26. Holz, P.; Raidal, S. Comparative renal anatomy of exotic species. *Vet. Clin. N. Am. Exot. Anim. Pract.* **2006**, *9*, 1–11. [CrossRef] [PubMed]
27. Lahner, L.; Mans, C.; Sladky, K.K. Comparison of anesthetic induction and recovery times after intramuscular, subcutaneous or intranasal dexmedetomidine-ketamine administration in red-eared slider turtles (*Trachemys scripta elegans*). In Proceedings of the AAZV Conference, Kansas City, MO, USA, 22–28 October 2011.
28. Sladky, K.K.; Mans, C. Clinical anesthesia in reptiles. *J. Exot. Pet. Med.* **2012**, *21*, 17–31. [CrossRef]
29. Holz, P.; Barker, I.K.; Burger, J.P.; Crawshaw, G.J.; Conlon, P.D. The effect of the renal portal system on pharmacokinetic parameters in the red-eared slider (*Trachemys scripta elegans*). *J. Zoo Wildl. Med.* **1997**, *28*, 386–393. [PubMed]
30. Giorgi, M.; Salvadori, M.; De Vito, V.; Owen, H.; Demontis, M.P.; Varoni, M.V. Pharmacokinetic/ Pharmacodynamic assessments of 10 mg/kg tramadol intramuscular injection in yellow-bellied slider turtles (*Trachemys scripta* Scripta). *J. Vet. Pharmacol. Ther.* **2015**, *38*, 488–496. [CrossRef] [PubMed]
31. Di Giuseppe, M.D.; Faraci, L.; Luparello, M. Preliminary survey on the influence of renal portal system during propofol anesthesia in yellow-bellied turtle (*Trachemys scripta scripta*). *Vet. Med. Allied Sci.* **2018**, *2*, 1–3.
32. Morici, M.; Lubian, E.; Costa, G.L.; Spadola, F. Difference between cranial and caudal intravenous alfaxalone administration in yellow-bellied sliders (*Trachemys scripta scripta*). *Acta Vet. Eurasia* **2021**, *47*, 88–92. [CrossRef]
33. Shepard, M.K.; Divers, S.; Braun, C.; Hofmeister, E.H. Pharmacodynamics of alfaxalone after single-dose intramuscular administration in red-eared sliders (*Trachemys scripta elegans*): A comparison of two different doses at two different ambient temperatures. *Vet. Anesth. Analg.* **2013**, *40*, 590–598. [CrossRef] [PubMed]
34. Hernandez-Divers, S.M.; Hernandez-Divers, S.J.; Wyneken, J. Angiographic, anatomic and clinical technique descriptions of a subcarapacial venipuncture site for chelonians. *J. Herpetol. Med. Surg.* **2002**, *12*, 32–37. [CrossRef]
35. Innis, C.; DeVoe, R.; Mylniczenko, N.A. Call for additional study of the safety of subcarapacial venipuncture in chelonians. In Proceedings of the ARAV, Seventh Annual Conference, South Padre Island, TX, USA, 23–39 October 2010; pp. 8–10.
36. Quesada, R.J.; Aitken-Palmer, C.; Conley, K.; Heard, D.J. Accidental submeningeal injection of propofol in gopher tortoises (*Gopherus polyphemus*). *Vet. Rec.* **2010**, *167*, 494–495. [CrossRef]
37. Rockwell, K.; Rademacher, N.; Osborn, M.L.; Nevarez, J.G. Extravasation of contrast media after subcarapacial vessel injection in three chelonian species. *J. Zoo Wildl. Med.* **2022**, *53*, 402–411. [CrossRef]
38. Ciccarelli, S.; Valastro, C.; Di Bello, A.; Paci, S.; Caprio, F.; Corrente, M.L.; Trotta, A.; Franchini, D. Diagnosis and treatment of pulmonary disease in sea turtles (*Caretta caretta*). *Animals* **2020**, *10*, 1355. [CrossRef] [PubMed]
39. Bertelsen, M.F. Anaesthesia and analgesia. In *BSAVA Manual of Reptiles*; Girling, S.J., Raiti, P., Eds.; British Small Animal Veterinary Association: Gloucester, UK, 2019; pp. 200–209.
40. Kischinovsky, M.; Duse, A.; Wang, T.; Bertelsen, M.F. Intramuscular administration of alfaxalone in red-eared sliders (*Trachemys scripta elegans*) effects of dose and body temperature. *Vet. Anesth. Analg.* **2013**, *40*, 13–20. [CrossRef] [PubMed]
41. Perpiñán, D. Reptile anesthesia and analgesia. *Companion Anim.* **2018**, *23*, 236–243. [CrossRef]
42. Muir, W.W., 3rd; Gadawski, J.E. Respiratory depression and apnea induced by propofol in dogs. *Am. J. Vet. Res.* **1998**, *59*, 157–161. [CrossRef] [PubMed]
43. Suarez, M.A.; Dzikiti, B.T.; Stegmann, F.G.; Hartman, M. Comparison of alfaxalone and propofol administered as total intravenous anaesthesia for ovariohysterectomy in dogs. *Vet. Anaesth. Analg.* **2012**, *3*, 36–244. [CrossRef] [PubMed]
44. Maney, J.K.; Shepard, M.K.; Braun, C.; Cremer, J.; Hofmeister, E.H. A comparison of cardiopulmonary and anesthetic effects of an induction dose of alfaxalone or propofol in dogs. *Vet. Anaesth. Analg.* **2013**, *40*, 237–244. [CrossRef] [PubMed]
45. Cabanac, A.; Cabanac, M. Heart rate response to gentle handling of frog and lizard. *Behav. Process.* **2000**, *52*, 89–95. [CrossRef] [PubMed]
46. Perrin, K.L.; Bertelsen, M.F. Intravenous alfaxalone and propofol anesthesia in the bearded dragon (*Pogona vitticeps*). *J. Herpetol. Med. Surg.* **2017**, *27*, 123–126. [CrossRef]
47. Fleming, G.J.; Heard, D.J.; Floyd, R.F.; Riggs, A. Evaluation of propofol and medetomidine-ketamine for short-term immobilization of Gulf of Mexico sturgeon (*Acipenser oxyrinchus* de Soti). *J. Zoo Wild. Med.* **2003**, *34*, 153–158. [CrossRef]
48. Mitchell, M.A.; Riggs, S.M.; Singleton, C.B.; Diaz-Figueroa, O.; Hale, L.K. Evaluating the clinical and cardiopulmonary effects of clove oil and propofol in tiger salamanders (*Ambystoma tigrinum*). *J. Exot. Pet. Med.* **2008**, *18*, 50–56. [CrossRef]
49. Wojick, K.B.; Langan, J.N.; Mitchell, M.A. Evaluation of MS-222 (*Tricaine methanesulfonate*) and propofol as anesthetic agents in Sonoran desert toads (*Bufo alvarius*). *J. Herp. Med. Surg.* **2010**, *20*, 79–83. [CrossRef]
50. Machin, K.L.; Caulkett, N.A. Cardiopulmonary effects of propofol infusion in canvasback ducks (*Aythya valisineria*). *J. Avian Med. Surg.* **1999**, *13*, 167–172.
51. Read, M.R. Evaluation of the use of anesthesia and analgesia in reptiles. *J. Am. Vet. Med Assoc.* **2004**, *224*, 547–552. [CrossRef] [PubMed]

52. Bertelsen, M.F.; Buchanan, R.; Jensen, H.M.; Leite, C.A.; Abe, A.S.; Wang, T. Pharmacodynamics of propofol and alfaxalone in rattlesnakes (*Crotalus durissus*). *Comp. Biochem. Physiol. A Mol. Integr. Physiol.* **2021**, *256*, 110935. [CrossRef]
53. Oppenheim, Y.C.; Moon, P.F. Sedative effects of midazolam in red-eared slider turtles (*Trachemys scripta elegans*). *J. Zoo Wildl. Med.* **1995**, *26*, 409–413.

animals

MDPI

Article

Comparison of Subcutaneous versus Intramuscular Dexmedetomidine–Midazolam–Ketamine–Morphine (DMKM) Mixture as Chemical Restraint for Endoscopic Sex Determination in Aldabra Giant Tortoises (*Aldabrachelys gigantea*)

Marco Masi [1], Alessandro Vetere [2,*], Jacopo Casalini [3], Flavia Corsi [4], Francesco Di Ianni [2] and Giordano Nardini [3]

1 Centro Veterinario Specialistico, Via Sandro Giovannini 51/53, 00137 Roma, Italy; apophismasi@gmail.com
2 Department of Veterinary Science, University of Parma, 43126 Parma, Italy
3 Clinica Veterinaria Modena Sud, Piazza Dei Tintori 1, 41057 Spilamberto, Italy
4 Institute of Molecular Biotechnology of the Austrian Academy of Sciences (IMBA), Vienna BioCenter (VBC), Dr. Bohr-Gasse 3, 1030 Vienna, Austria
* Correspondence: alessandro.vetere@unipr.it

Simple Summary: Anaesthesia in chelonians poses a challenge due to their resistance, making injectable anaesthesia highly valuable for this order of reptiles. In this study, we evaluated the effects of a dexmedetomidine–midazolam–ketamine–morphine (DMKM) combination delivered subcutaneously (SC) or intramuscularly (IM) in twenty-one Aldabra giant tortoises (*Aldabrachelys gigantea*) as a chemical restraint for a minimally invasive procedure for celioscopic sex identification. The intramuscular administration of a DMKM combination seems to be a good option to be used as a safe, reliable, short-lasting anaesthesia for minimally invasive procedures such as celioscopy.

Citation: Masi, M.; Vetere, A.; Casalini, J.; Corsi, F.; Di Ianni, F.; Nardini, G. Comparison of Subcutaneous versus Intramuscular Dexmedetomidine–Midazolam–Ketamine–Morphine (DMKM) Mixture as Chemical Restraint for Endoscopic Sex Determination in Aldabra Giant Tortoises (*Aldabrachelys gigantea*). *Animals* **2023**, *13*, 3626. https://doi.org/10.3390/ani13233626

Academic Editor: Tom Hellebuyck

Received: 8 October 2023
Revised: 20 November 2023
Accepted: 21 November 2023
Published: 23 November 2023

Abstract: Sex identification through coelioscopy is a minimally invasive surgical technique used to determine the sex of chelonians by directly visualizing their internal reproductive organs. An adequate anaesthesiologic plan is essential to guarantee patient immobilization and proper analgesia during the entire surgical procedure. In this study, we evaluated the effects of a combination of dexmedetomidine (0.05 mg/kg), midazolam (1 mg/kg), ketamine (8 mg/kg), and morphine (1 mg/kg) (DMKM) randomly delivered intramuscularly (IM) or subcutaneously (SC) in twenty-one Aldabra giant tortoise (*Aldabrachelys gigantea*) into the right antebrachium for celioscopic sex identification. Heart rate (HR), respiratory rate (RR), and body temperature (BT) were measured, along with the skeletal muscle tone of the thoracic and pelvic limbs, neck retraction reflex, palpebral reflex, and jaw tone every 15 min. The anaesthesiologic plan was considered to be adequate at the loss of the thoracic and pelvic limb retraction reflexes. After a 45 min interval, if the anaesthetic plan was deemed insufficient for the celioscopic procedure, a 5 mg/kg dose of propofol was administered intravenously into the subcarapacial venous plexus. At the end of the procedure, atipamezole (0.5 mg/kg) and flumazenil (0.05 mg/kg) were administered intramuscularly into the left antebrachium as reversal agents. Both HR and RR decreased from baseline to both 15 and 30 min. Due to the persistence of thoracic and pelvic limb retraction reflexes 45 min after DMKM administration, 6/11 (55%) cases in the SC group required the additional administration of propofol, in contrast to only 1/10 (10%) cases in the IM group ($p = 0.05$). The recovery times were comparable between the successfully induced animals in the IM and SC groups. In this study, the intramuscular administration of a DMKM combination quickly produced chemical restraint, suitable for celioscopic sex determination.

Keywords: DMKM; *Aldabrachelys gigantea*; anaesthesia; chelonians; reptiles

1. Introduction

Aldabrachelys gigantea, commonly known as the Aldabra giant tortoise, is a critically endangered species of tortoise native to the Aldabra Atoll in the Seychelles [1–3]. Currently, ongoing conservation measures are crucial to ensure the long-term survival of this species [4,5]. Securing the preservation of these tortoise species heavily depends on the implementation of effective captive breeding programs. Therefore, it is imperative for these programs to adhere to optimal management practices, given the significance of population density and the sex ratio of the specimens [5–7]. Immature Aldabra giant tortoises are not sexually dimorphic, reaching sexual maturity between the ages of 20 and 25 years, although this can sometimes take longer [8]. In clinical practice, the sex of immature chelonians can be easily identified endoscopically [9]. Celioscopic sexing is a minivasive technique that involves making a small incision in the body wall of the anaesthetized chelonian in the prefemoral fossa for coelomic access. A rigid endoscope is then inserted and used to directly identify the gonads and reproductive structures for accurate sex identification [9]. Administering anaesthesia for coelioscopy in chelonians is a critical aspect to ensure the safety and well-being of the animal during the procedure. Anaesthesia allows the chelonian to be relaxed and unconscious, preventing pain and discomfort during the coelioscopic examination [9–12]. Injectable anaesthetics are commonly used for chelonians undergoing coelioscopy [11]. Some commonly used anaesthetic agents for chelonians include injectable opioids, dissociative anaesthetics, and injectable or inhalant induction agents [10,12]. These medications are typically administered intramuscularly or intravenously to induce and maintain anaesthesia [10]. Because reptiles have a notoriously prolonged recovery, it is important to choose a suitable combination of drugs that should have wide safety margins, should be short-acting, and should be reversible where possible, promoting more rapid recovery after anaesthesia [10,12–14]. Dexmedetomidine, an a2-adrenergic agonist, provides sedation, muscle relaxation properties, and analgesia in many species, reptiles included [10]. Dexmedetomidine is frequently used in conjunction with either ketamine or benzodiazepines to minimize the required drug dosage [10–12]. Midazolam is a short-acting, water-soluble benzodiazepine with sedative, anticonvulsant, amnestic, anxiolytic, and skeletal muscle relaxant properties, with minimal effects on the cardiopulmonary system [15]. It is efficiently absorbed when administered intramuscularly (IM) or subcutaneously (SC), exhibiting limited impact on cardiovascular and pulmonary functions. Its effects can be reversed with flumazenil [15,16]. When utilized as the sole agent, midazolam frequently offers merely mild or unreliable sedation in reptiles [10]. Its use in combination with opioids and other injectable agents (e.g., ketamine, medetomidine, and dexmedetomidine) in a Galapagos tortoise (*Chelonoidis nigra*) and three African spurred tortoises (*Geochelone sulcata*) has been reported [16,17]. Ketamine is a dissociative anaesthetic and N-methyl-D-aspartate receptor antagonist. In snapping turtles (*Chelydra serpentina*), the coadministration of ketamine and midazolam produced more profound effects than either drug alone [18]. Similarly, in loggerhead sea turtles (*Caretta caretta*), ketamine and medetomidine alone achieved less sedative effects than a combination of the two, highlighting the benefits of multimodal mechanisms of action [19]. Morphine is a mu opioid receptor agonist that offers strong pain-relieving properties in a variety of veterinary species [10,12,20]. Substantial evidence suggests that morphine could prove to be a viable analgesic option for certain reptile species, chelonians included [20,21]. In clinical doses, it is typically associated with minimal respiratory side effects [20]. Nevertheless, in certain species, higher dosages can lead to pronounced respiratory depression [21,22]. The onset of action following morphine administration can be notably extended (2–8 h), and the duration of its effects may exhibit significant variability across different species [21,22]. In a recent study involving 10 hatchling green sea turtles (*Chelonia mydas*), researchers found that the intramuscular administration of ketamine–medetomidine–tramadol was a reliable and safe anaesthesia protocol. The turtles displayed no response to any of the potentially painful stimuli, indicating the effectiveness of the anaesthesia. However, it should be noted that anaesthesia did result in apnoea throughout the procedure [23]. In another case report of a leopard tortoise (*Stigmochelys pardalis*), a subcutaneous coadministration of medetomidine–midazolam–ketamine produced 30 min of deep sedation, which was sufficient to allow endoscopy-guided ectopic egg removal from the urinary bladder [24]. Propofol is an intravenous anaesthetic agent used extensively in reptilian anaesthesia because of its rapid onset of the anaesthetic effect and short recovery times. This expedites endotracheal intubation, which can then be followed by the maintenance of anaesthesia through the use of an inhalation agent [14]. However, rapid administration of propofol can lead to respiratory depression [25]. Propofol requires intravascular or intraosseous administration, which can be challenging in chelonians. Intrathecal propofol injection is reported in chelonians and can lead to various complications, such as fore and limb paresis, coma, and spinal necrosis [10]. The objective of this study was to investigate the hypothesis that combining dexmedetomidine, midazolam, ketamine, and morphine (DMKM) would result in efficient anaesthesia for short and minimally invasive procedures, such as celioscopic sex identification, without requiring additional propofol as an induction agent. This combination was aimed at reducing the dosage of each drug while taking advantage of their multimodal mechanisms of action. Additionally, we evaluated two different routes of administration: subcutaneous (SC) and intramuscular (IM).

2. Materials and Methods

2.1. Animals

Ethical approval for the study was given by the University of Parma (PROT. N. 04/CESA/2023). The owner gave informed consent to allow participation of their animals in the study.

Twenty-one captive-bred, clinically healthy Aldabra giant tortoises (*Aldabrachelys gigantea*) (AGTs) with a median body weight of 2090 g (IQR: 1390–4480 g; range: 890–7190 g) and aged between three and five years were included in the study. Exclusion criteria included any history or visible signs of systemic illness (e.g., signs of infection, external wounds, inflammation, trauma, or neoplasia) or discomfort. No animals were excluded in this study. All animals were divided individually into different plastic enclosures and identified with progressive identification numbers. The animals were kept under controlled environmental conditions, with a room temperature of 27 °C and a 12 h:12 h light/dark cycle [26], and housed in a plastic tank of 3 × 2 × 1 m. They were fed a diet consisting of hay and fresh greens and were given free access to water. Food was withheld 4 days before the procedure to decrease the size of the gastroenteric tract, enhancing endoscopic visibility of the reproductive tract. Every box had a heat bulb as a light source (100 W) and a heat light bulb for basking (31–34 °C). Each animal was weighed, and accurate physical examinations were performed. The animals were soaked daily in a few cm of warm water so that they could rehydrate and defecate. Exclusion criteria included any history of systemic illness or abnormalities, including dyspnoea, evidence of infection, traumatic injuries, and abnormal posture.

2.2. Study Design

A mixture of dexmedetomidine (0.05 mg/kg; Dexdomitor 0.5 mg/mL, Vétoquinol Italia S.r.l., Bertinoro, FC, Italy), midazolam (1 mg/kg; Midazolam Hameln ha 5 mg/mL, Hameln Pharmaceuticals Gloucester Business Park, Gloucester, UK), ketamine (8 mg/kg; Lobotor 100 mg/mL, Acme S.r.l., Via Fabrizio De Andrè 8, Milano, Italy), and morphine (1 mg/kg; Morfina Cloridrato Monico 10 mg/mL, Monico s.p.a, Via Ponte di Pietra 7, Venezia, Italy) was delivered to each tortoise in the right antebrachium using a syringe equipped with a 21 gauge × 16 mm needle (IV Needle; Kays spa, Via Francesco Crispi 26, Osimo, Italy) (Figure 1).

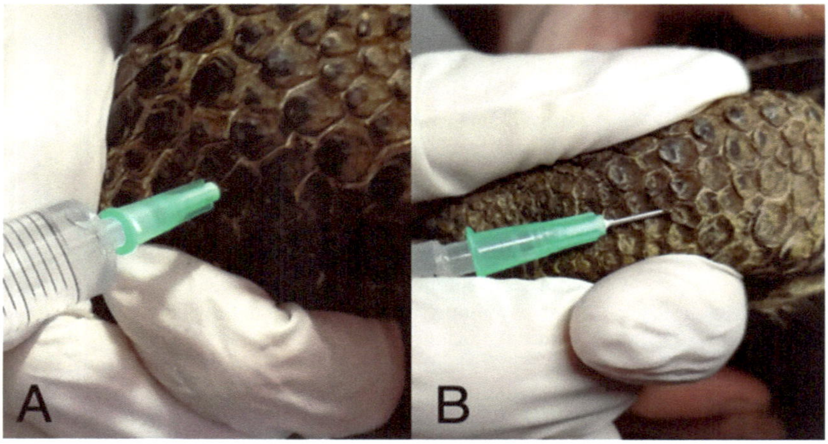

Figure 1. In the IM route (**A**), the needle was inserted for the entire length perpendicularly in the long axis of the right antebrachium, while in the SC route (**B**), the needle was inserted subcutaneously for 2–3 mm parallel to the long axis of the antebrachium.

Prior to injections, the skin was scrubbed using 2% chlorhexidine solution (Clorexinal 2%, Nuova Farmec, Settimo di Pescantina, VR, Italy). The route of administration (subcutaneously (SC) or intramuscularly (IM)) was randomly chosen for each tortoise before coelioscopy via simple randomization using randomizer software (www.randomizer.org, accessed on 20 January 2023). Before the injection, aspiration was performed to ensure whether a blood vessel had been accidentally punctured. In the case of accidental blood aspiration, the drug mixture would not be delivered, and the procedure would be repeated. Following drug administration, mild pressure was applied to the injection site using the thumb for a duration of 10 s to minimize the potential for drug leakage. However, unquantifiable drug leakage was noted in both groups of animals (Figure 2).

Figure 2. A small amount of drug mixture leaked from the injection site in both the IM and SC groups (white asterisk).

The time of drug administration was designated as time 0 (T0). The baseline heart rate (HR), respiratory rate (RR), and body temperature (BT) were recorded for 60 s prior to drug administration at T0 and then every 15 min until complete recovery of the animal, for a maximum period of 2 h from initial drug injection. HR was measured using an ultrasonic Doppler flow device (Doppler VET-BP, 8 mhz, Alcyon Italia, Via Nicotera 29, Rome, Italy) directed towards the heart in the cervicobrachial acoustic window (Figure 3).

Figure 3. The ultrasonic Doppler probe was positioned in the left cervicobrachial acoustic window.

RR was monitored by observation of skin movement in the region between the neck and shoulder or the prefemoral fossa. BT was monitored using a digital thermometer with an iron external probe (Terracheck, TFA, Herp Italia, Via Giuseppe Verdi 2, Concamarise, Italy) placed inside the cloaca. The following were evaluated every 15 min: palpebral reflex, thoracic and pelvic limb withdrawal reflexes (applying pressure with haemostat forceps), jaw tone, and ease of manual neck extension. Evaluations were performed until complete recovery and for a maximum period of 2 h from drug administration. All reflexes were denoted in a binary way as present (slowed and barely present included) or absent (complete absence of reflexes). The absence of the thoracic and pelvic limb retraction reflexes was considered an adequate anaesthetic plane for the celioscopic procedure. After 45 min, if the anaesthetic plan was considered not adequate for the celioscopic procedure (persistence of the withdrawal reflex or muscle tone of the thoracic and pelvic limbs), a 5 mg/kg propofol dose was administered intravenously in the subcarapacial venous plexus. Tortoises that exhibited apnea (18/21, 85.71%) were intubated with a non-cuffed, silicone endotracheal tube (Foschi s.r.l., Via Livornese est No. 269, Perignano, Italy) of a size appropriate for their glottis diameter and manually ventilated with a bag valve mask (one breath every 30 s) until resumption of spontaneous ventilation. Coelioscopy was performed with the animal placed in right lateral recumbency. The left prefemoral fossa was disinfected with 2.0% chlorhexidine digluconate (Clorexinal 2%, Nuova Farmec, Via Walther Fleming 7, Settimo, VR, Italy). A 0.5 mm skin incision was performed using a scalpel blade No. 11 (Mealli srl, Borgo Santi Apostoli, Firenze, Italy). The transverse and oblique muscles were bluntly dissected pulling the haemostatic forceps in a cranio-dorsal direction, reaching the coelom. A 2.7 mm, 30° viewing rigid endoscope (Storz Telepack TP100 EN, Karl Storz Endoscopia Italia S.r.l., Rome, Italy) housed within a 4.8 mm operative sheath was used to detect and identify the gonads [9]. At the end of the procedure, atipamezole (0.5 mg/kg; Antisedan 5 mg/mL, Vétoquinol Italia S.r.l.) and flumazenil (0.05 mg/kg; Anexate 0.5 mg/5 mL, AVAS Pharmaceuticals S.r.l., Milano, Italy) were administered intramuscularly in the left antebrachium as reversal agents. The recovery time was measured from the time of reversal agent administration. A tortoise was considered to have fully recovered from sedation or anaesthesia when it exhibited all assessed reflexes or responses, when the presence of palpebral reflex, thoracic, and pelvic limb withdrawal reflexes was indicated (either reduced or intact, as opposed to absent), and when the animal could maintain its head above ground and initiate spontaneous movement. During recovery, the tortoises were placed individually in a plastic tank of 3 × 2 × 1 m with a disposable absorbent pad as a substrate. Environmental conditions were the same as before the endoscopic procedure. Thirty-six hours after the end of the procedures, all of the animals ate and were returned to the owner.

A timeline diagram of the whole procedure is represented in Figure 4.

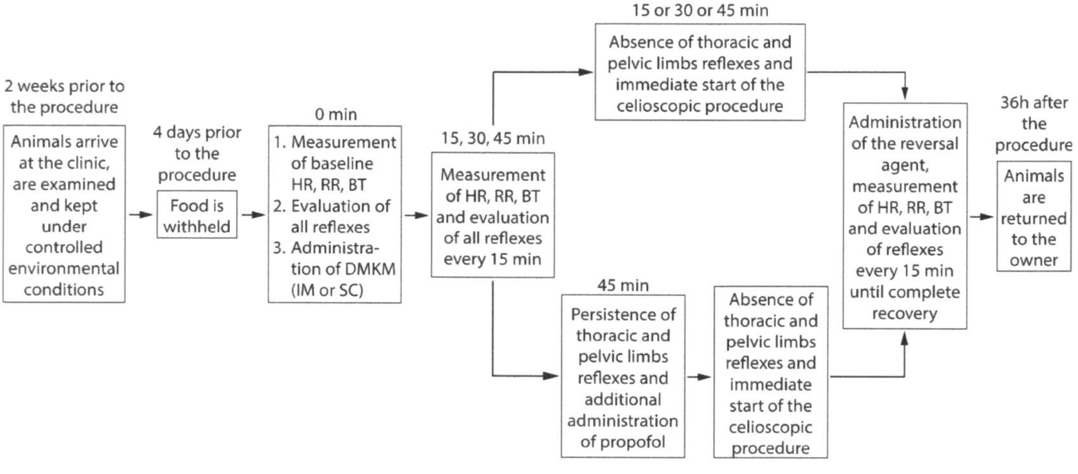

Figure 4. Timeline diagram of the procedure.

2.3. Statistical Analysis

Statistical analysis was performed using R software (version 3.6.3). The data are reported herein as the median, interquartile range (IQR = 25th–75th percentiles), and range. Time to loss and time to recovery of reflexes are reported as the grouped median, grouped interquartile range (IQR = 25th–75th percentiles), and range, to take into account the fact that timepoints were taken at discrete intervals of 15 min, and are compared using the Wilcoxon signed-rank test. Heart rate (HR), respiratory rate (RR), and body temperature (BT) were rank transformed and compared between groups and timepoints using pairwise t tests. The route (SC vs. IM) of DMKM administration (independent variable) was evaluated as a predictor for the loss of each reflex and the need for additional propofol administration (dependent variables) in logistic regression models. No threshold on the p-value was used to assess statistical significance, since there is increasing evidence that this practice should be avoided. Instead, we still report p-values in some cases together with effect sizes (odds ratios (ORs)) and 95% confidence intervals (CIs) which should always be considered all together along with other factors (e.g., related prior evidence, plausibility of mechanism, study design, and data quality, etc.) to interpret the meaningfulness of the findings in a critical manner [27–29].

3. Results

Out of the 11 tortoises, 7 (64%) belonging to the IM group were identified as male, and 4 (36%) tortoises belonging to the SC group were identified as female. The intramuscular (IM) and subcutaneous (SC) groups consisted of 10 and 11 animals, respectively. No difference was observed in body weight (g) between the two groups (SC: median = 1970, IQR = 1330–4310, range = 890–7190; IM: median = 2655, IQR = 1728–4190, range = 1370–5760; $p = 0.5$). HR and RR decreased significantly over time in both groups in a comparable manner before the administration of the antagonist to any animal (Figure 5A,B). No differences were observed between the BTs of the two groups (Figure 5C).

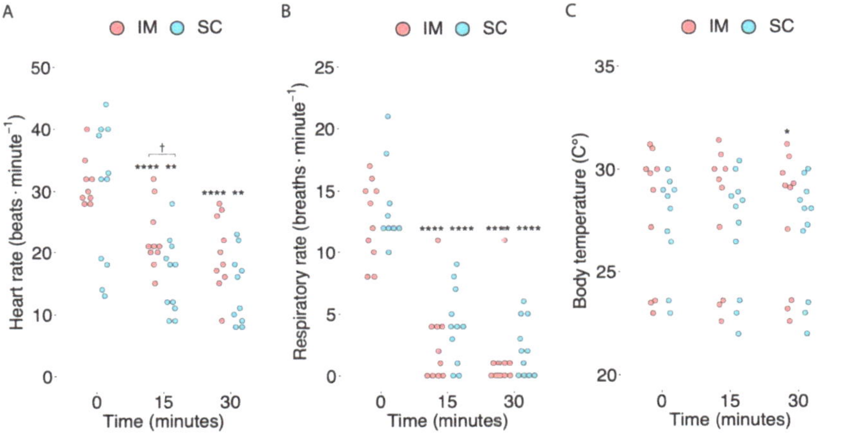

Figure 5. Distribution of the (**A**) heart rate, (**B**) respiratory rate, and (**C**) body temperature over time. For each timepoint, the values for the intramuscular (IM) ($n = 10$) and subcutaneous (SC) ($n = 11$) groups are represented in pink (on the left) and light blue (on the right), respectively. Asterisks (*) indicate the order of magnitude of the p-value between the baseline (0 min) and later timepoints within the same treatment. Daggers (†) indicate the order of magnitude of the p-value between different routes of administration at the same timepoint. An absence of symbols indicates a p-value > 0.05. * $p < 0.05$, ** $p < 0.01$, **** $p < 0.0001$, † $0.01 < p < 0.05$.

In the IM group, 9/10 (90%) animals lost the thoracic and pelvic limb withdrawal reflexes within 45 min after DMKM administration, in contrast to only 5/11 (45%) animals in the SC group (Figure 6A). Hence, due to the persistence of thoracic and pelvic limb withdrawal reflexes 45 min after DMKM administration, the remaining 6/11 (55%) animals in the SC group required the additional administration of propofol to allow the start of the celioscopic procedure after 45 min, in contrast to only 1/10 (10%) animals in the IM group (Figure 6B). These differences between the two groups suggest that, compared to SC administration, the IM administration is 10.8 times (OR = 10.8, 95% CI = 1.33–237.05, $p = 0.05$) more likely to result in loss of the thoracic and pelvic limb reflexes within 45 min from the DMKM administration, allowing one to perform the celioscopic procedure in this timeframe without the additional need for propofol.

Within 45 min after DMKM administration prior to propofol injection, 8/10 (80%) IM and 5/11 (45%) SC cases also lost the palpebral reflex, 8/10 (80%) IM and 3/11 (27%) SC cases lost neck muscle tone, and 5/10 (50%) IM and 3/11 (27%) SC cases lost jaw tone (Figure 6A). For 9/10 (90%) of the IM and 5/11 (45%) of the SC cases for which DMKM was effective within 45 min and did not need to be induced with propofol, there were no remarkable differences in the time to loss of reflexes (Table 1).

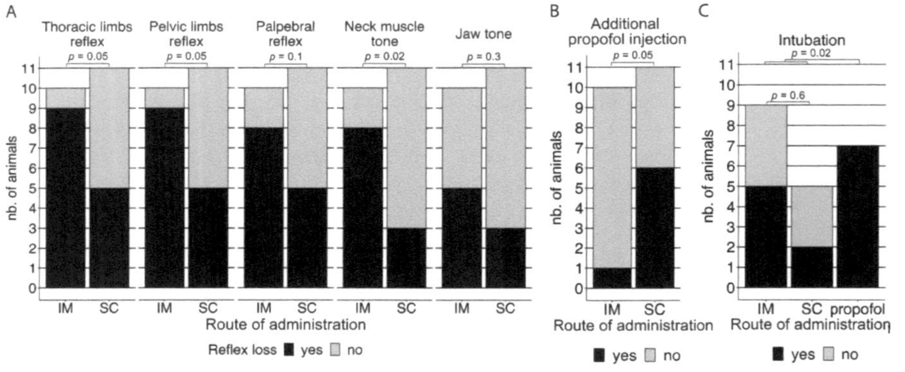

Figure 6. (**A**) Proportion of animals in the intramuscular (IM) (n = 10) and subcutaneous (SC) (n = 11) groups that lost the thoracic and pelvic limb withdrawal reflexes, palpebral reflex, neck muscle, and jaw tone within 45 min after DMKM administration prior to the decision on additional propofol injection. (**B**) Proportion of animals in the intramuscular (IM) (n = 10) and subcutaneous (SC) (n = 11) groups that had to be induced with propofol 45 min after the initial DMKM administration. (**C**) Proportion of animals for which intubation was necessary and successful in the intramuscular (IM) (n = 9) and subcutaneous (SC) (n = 5) groups successfully induced with DMKM after 45 min and the propofol (n = 7) group.

Table 1. Time to loss of reflexes only including subjects successfully induced with DMKM within 45 min (9 IM, 5 SC). The median (IQR = 25th–75th percentiles) was calculated only including the subset of animals within the ones successfully induced with DMKM which lost the stated reflex within 45 min after DMKM administration.

Parameter	IM DMKM	SC DMKM	*p*-Value
Time to loss of the thoracic limb reflex for positive cases (minutes)	16.9 (8.4–25.3) (n = 9/9)	12.5 (6.3–20.6) (n = 5/5)	0.6
Time to loss of the pelvic limb reflex for positive cases (minutes)	16.9 (8.4–25.3) (n = 9/9)	17.5 (9.4–23.8) (n = 5/5)	1
Time to loss of the palpebral reflex for positive cases (minutes)	18.8 (10–26.25) (n = 8/9)	18.8 (9.4–28.1) (n = 5/5)	1
Time to loss of neck muscle tone for positive cases (minutes)	30 (15.0–37.5) (n = 8/9)	7.5 (3.8–21.2) (n = 3/5)	0.06
Time to loss of jaw tone for positive cases (minutes)	22.5 (9.4–36.6) (n = 5/9)	18.8 (11.3–24.4) (n = 3/5)	0.8

In 1/1 (100%) of the IM and 5/6 (83%) of the SC cases successively induced with propofol, all of the measured reflexes were lost within 15 min, allowing the start of the procedure after 4 min from the time of propofol administration for the only IM case and between 4 and 14 min from the time of propofol administration for the 5/6 SC cases. In the remaining 1/6 (17%) SC cases successively induced with propofol, a delayed loss of all reflexes was observed, except for pelvic limb retraction, which was abolished within 15 min, resulting in the procedure being started 24 min after propofol administration.

Among the animals that had to be induced with propofol, 7/7 (100%) also showed apnoea, and the absence of jaw tone allowed successful intubation in all of them (Figure 6C). Among the animals successfully induced with DMKM, 12/14 (86%) of them (8 IM, 4 SC) showed apnoea, but in contrast to the group induced with propofol, only 8/14 (57%) of them (5 IM, 3 SC) lost jaw tone (Figure 6A), and intubation was successful in 7/14 (50%) (5 IM, 2 SC) of the cases (Figure 6C).

No remarkable differences were observed in the average recovery time of the individual reflexes between the IM and SC groups (Table 2). All animals reached a complete recovery of all reflexes within 60 min. Twelve hours after the coelioscopy, all of the animals were active and did not show any abnormal behaviour. No procedure-related complications were noted.

Table 2. Time to resumption of reflexes after loss. Medians (IQR = 25th–75th percentiles) were calculated only including the subset of animals that were successfully induced with DMKM and lost the respective reflexes until the end of the procedure, discarding those that had already started to partially recover the respective reflexes during the administration of the antagonist.

Parameter	IM DMKM	SC DMKM	*p*-Value
Time to resumption of the thoracic limb reflex (minutes)	8.8 (4.4–13.1) (*n* = 7)	22.5 (11.3–33.8) (*n* = 3)	0.2
Time to resumption of the pelvic limb reflex (minutes)	7.5 (3.8–11.3) (*n* = 8)	11.25 (5.6–18.8) (*n* = 3)	0.2
Time to resumption of the palpebral reflex (minutes)	10.5 (5.3–16.9) (*n* =7)	9.4 (4.7–14.1) (*n* = 5)	0.7
Time to resumption of neck muscle tone (minutes)	10.5 (5.3–16.9) (*n* = 7)	11.3 (5.6–18.8) (*n* = 3)	0.9
Time to resumption of jaw tone (minutes)	9.4 (4.7–14.1) (*n* = 5)	7.5 (3.8–11.6) (*n* = 2)	0.8
Time to complete resumption of all reflexes (minutes)	13.1 (6.6–24.4) (*n* = 8)	12.5 (6.3–26.3) (*n* = 5)	1

4. Discussion

The following databases (Google Scholar, PubMed) were searched with the following keywords: "DMKM, anaesthesia, *Aldabrachelys gigantea*, SC, IM, dexmedetomidine, ketamine, midazolam, morphine"; one textbook was consulted [30]. No studies about DMKM as an anaesthetic protocol in this species were found in these searches. In this study, we administered a DMKM combination IM or SC in young and not sexually dimorphic Aldabra giant tortoises. In a study by Lahner et al. [31], the induction times following subcutaneous administration of ketamine–dexmedetomidine to red-eared slider turtles (*Trachemys scripta elegans*) were notably longer and exhibited greater variability than those following intramuscular injection. Instead, the level of anaesthesia attained within the initial 45 min displayed no distinction between intramuscular and subcutaneous injection [32]. In another study on leopard geckos, the administration of alfaxalone–midazolam via SC injection resulted in the first effects occurring within 5 to 10 min [33]. A comparison between routes of administration of drugs in reptiles has not been well studied, and this is typically dependent on species and related to anatomical variability [33]. Historically, IM administration has been the favoured method for delivering anaesthetic and analgesic drugs to reptiles. SC administration has generally been disregarded due to the belief that its lower vascularity might result in extended and less consistent onset times, metabolism, and elimination when contrasted with the IM route [34]. The contemporary literature persists in discussing the potential for variable absorption with SC drug administration, even though current research demonstrates that anaesthetic and analgesic effects achieved through SC drug administration in reptiles are both prompt and consistent. Moreover, no noteworthy disparity in anaesthetic depth has been observed when contrasting SC and IM induction methods [10,32,33]. Additionally, in reptiles, the advantages associated with SC administration, when contrasted with IM administration, encompass the consistent accessibility of subcutaneous space across different species, the capacity to administer large volumes, and the opportunity to utilize a range of subcutaneous sites [10]. However, in this study, the subcutaneous space in the antebrachium was chosen as the injection site of the SC group instead of the region set in the front half of the body, between the neck and the forelimb, where the cutis is more distensible. This choice was made because clinically healthy tortoises tend to strongly retract their limbs into the shell, exposing only a small portion of the cutis of the dorsal surface of the forelimb. Every attempt to force the forelimbs out would have been either traumatic for the animal or unsuccessful. In every animal belonging to both the IM and SC groups, a small, unquantifiable amount of DMKM leaked from the injection site. This was supposedly related to the poor distensible skin and the well-developed antebrachial muscles, which are typical features of large terrestrial chelonians [34], in relation to the volume of the injected drug mixture. In a recent study evaluating the effect of the IM dexmedetomidine–midazolam–ketamine (DMK) combination in red-footed tortoises, the median time to maximum drug effect was reached in approximately 35 min [35]. In the literature, as there are no available studies about the effects of DMKM in Aldabra giant tortoises, a limit of 45 min was chosen as a cut-off for the time to loss of reflexes, and in the case of persistent thoracic and pelvic limb withdrawal reflexes, the animal would be induced with IV propofol. In this study, we found that the number of animals for which the anaesthesiologic plan was adequate for the celioscopic procedure within 45 min of DMKM administration was notably higher in the IM (9/10 (90%) cases) group than in the SC group (5/11 (45%) cases). For the subset of cases successfully induced with DMKM either SC or IM, the time to loss of reflexes was comparable between the two groups of animals. The signs used to assess the level of sedation in reptiles are fairly consistent. The loss of withdrawal reflexes and muscle tone is expected to follow the following sequence: pectoral limbs, pelvic limbs and neck, tail and cloaca, and finally jaw tone [36]. In this case, only 8/14 (57%) animals (5 IM, 3 SC) that were successfully induced through IM and SC DMKM lost jaw tone, and in only 7/14 (50%) (5 IM, 2 SC) of these animals, intubation was successful. During induction, the jaw reflex is often lost last, indicating a deeper state of narcosis and allowing intubation and manual ventilation [37]. Additionally, among the animals successfully induced with IM DMKM, one tortoise did not lose the neck retraction reflex, and another did not lose the palpebral reflex, indicating a lighter plane of anaesthesia. HR and RR decreased significantly over time in both groups before the start of the celioscopic procedure, as also reported in the study by Eshar et al. [35] using IM DMK administration in captive red-footed tortoises (*Chelodinis carbonaria*). In dogs and cats, the pharmacokinetics of anaesthetic drugs are influenced by the route of administration [35]. Specifically, the subcutaneous route does not always allow for complete and rapid uptake of the drug mixture. Interaction with the α-2B receptors of the precapillary sphincter in peripheral vascular beds by α-2 receptor agonists could lead to a reduction in peripheral drug absorption [36–38]. This mechanism may have contributed to the reduction in systemic absorption of MKM when mixed with dexmedetomidine in the subcutaneous group, with a consequently less sedative effect.

Propofol is commonly used alone as an induction agent in chelonians at 10 to 20 mg/kg [10] or combined with other injectable agents at a lower dosage [39,40]. A dosage of 5 mg/kg was chosen as a DMKM mixture that was previously administered, avoiding possible complications such as prolonged recovery after the coelioscopy or a dose-dependent depressant respiratory function [41,42]. In fact, all of the animals that needed propofol also lost jaw tone and were intubated, compared to 8/14 (5 IM 3 SC) tortoises in the IM and SC groups which lost jaw tone, allowing 7 of them (5 IM, 2 SC) to be intubated. Apnoea occurs in reptiles in the surgical and deep plane of anaesthesia; if they are breathing spontaneously, the anaesthetic plane should be considered to be light [10]. However, even though measuring pain and analgesia is challenging in reptiles [42], as no movements or changes in the HR were noted during the surgical procedures, the analgesia was considered adequate. Moreover, no change in the normal behaviour was noted 12 h postoperatively and 36 h after the coelioscopy, and all of the animals ate, displaying no visible signs of discomfort or lameness [42]. No swelling in the injection site was noted in any of the animals at 12 and 36 h after the coelioscopic procedure. Coelioscopy is considered a minimally invasive surgical procedure [9]. However, multimodal anaesthesia was chosen to provide a good analgesic level throughout the entire surgical procedure. In fact, a DMK combination provided sufficient sedation or light anaesthesia, allowing the execution of various clinical procedures, including minimally invasive endoscopic examinations, blood sample collection, and imaging [18,24,27,43,44]. The authors of prior studies focusing on the anaesthesia of red-footed tortoises used higher doses of dexmedetomidine–midazolam and ketamine than those used in the present DMKM protocol [35,44] and did not include morphine in the drug mixture. In this case, morphine was chosen as an additional pain relief agent, as this has been demonstrated to be effective in several reptilian species [18]. Both opioids and a2-adrenoceptor agonists can induce respiratory depression in reptiles [10,12,45–47]. The occurrence of apnoea after the administration of medetomidine and dexmedetomidine is reported in chelonians [13,35,46]. In this study, 12/14 tortoises belonging to the IM and SC groups (8 IM, 4 SC) had apnoea and needed intubation and manual ventilation. The additional use of morphine in the DMK mixture should cause more respiratory depression than using a2-adrenoceptor agonists alone. For this reason, breathing assessment should always be performed and the patient should be intubated and ventilated when needed. However, the recovery was smooth and rapid after the reversal administration, and no other apparent adverse effects were noted. This investigation was a clinical study performed during sex identification in 21 giant Aldabra tortoises. To detect possible complications, follow-up studies including several individuals are needed. Moreover, an experimental setting should be designed to evaluate the cardiocirculatory and analgesic effects of DMKM mixture in terms of its pharmacokinetic and pharmacodynamic properties in this species.

5. Conclusions

In this study, a DMKM combination delivered IM resulted in deep sedation or light anaesthesia lasting for the duration of the celioscopic procedure and a smooth and rapid recovery after reversal administration. As more than half of the animals belonging to the SC group needed an additional intravenous administration of propofol to reach the surgical plane, this route of administration should be discouraged.

Author Contributions: Conceptualization, M.M., A.V., J.C. and G.N.; methodology, M.M., G.N. and J.C.; software, F.C.; data curation, F.C.; writing—original draft preparation, M.M., A.V. and F.C.; writing—review and editing, M.M. and A.V.; supervision, F.D.I.; project administration, F.D.I. All authors have read and agreed to the published version of the manuscript.

Funding: This research received no external funding.

Institutional Review Board Statement: Ethical approval for the study was given by the University of Parma (PROT. N. 04/CESA/2023).

Informed Consent Statement: Informed consent was obtained from all subjects involved in the study.

Data Availability Statement: Data is contained within the article.

Conflicts of Interest: The authors declare no conflict of interest.

References

1. Bour, R. *Recherches Sur Des Animaux Doublement Disparus: Les Tortues Gé Antes Subfossiles De Madagascar*; Ecole Pratique Des Hautes Études: Montpellier, France, 1994; p. 253.
2. Bunbury, N.; von Brandis, R.; Currie, J.C.; van de Crommenacker, J.; Accouche, W.; Birch, D.; Chong-Seng, L.; Doak, N.; Haupt, P.; Haverson, P.; et al. Late stage dynamics of a successful feral goat eradication from the UNESCO world heritage site of aldabra atoll, seychelles. *Biol. Invasions* **2018**, *20*, 1735–1747. [CrossRef]
3. Gerlach, J.; Rocamora, G.; Gane, J.; Jolliffe, K.; Vanherck, L. Giant tortoise distribution and abundance in the seychelles islands: Past, present, and future. *Chelonian Conserv. Biol.* **2013**, *12*, 70–83. [CrossRef]
4. Gerlach, J. *Giant Tortoises of the Indian Ocean: The Genus Dipsochelys Inhabiting the Seychelles Islands and the Extinct Giants of Madagascar and the Mascarenes*; Edition Chimaira: Frankfurt, Germany, 2004; p. 207.
5. Griffiths, O.; Andre, A.; Meunier, A. Tortoise breeding and 're-wilding'on Rodrigues Island. *Chelonian Res. Monogr.* **2013**, *6*, 178–182. [CrossRef]
6. Gerlach, J. Development of distinct morphotypes in captive seychelles–aldabra giant tortoises. *Chelonian Conserv. Biol.* **2011**, *10*, 102–112. [CrossRef]

7. Kuchling, G.; Griffiths, O. Endoscopic imaging of gonads, sex ratios, and occurrence of intersexes in juvenile captive-bred aldabra giant tortoises. *Chelonian Conserv. Biol.* **2012**, *11*, 91–96. [CrossRef]
8. Swingland, I.R. Reproductive effort and life history strategy of the aldabran giant tortoise. *Nature* **1977**, *269*, 402–404. [CrossRef]
9. Hernandez-Divers, S.J.; Stahl, S.J.; Farrell, R. An endoscopic method for identifying sex of hatchling Chinese box turtles and comparison of general versus local anesthesia for coelioscopy. *J. Am. Vet. Med. Assoc.* **2009**, *234*, 800–804. [CrossRef]
10. Sladky, K.K.; Mans, C. Clinical anesthesia in reptiles. *J. Exot. Pet Med.* **2012**, *21*, 17–31. [CrossRef]
11. Innis, C.J.; Hernandez-Divers, S.; Martinez-Jimenez, D. Coelioscopic-assisted prefemoral oophorectomy in chelonians. *J. Am. Vet. Med. Assoc.* **2007**, *230*, 1049–1052. [CrossRef]
12. Read, M.R. Evaluation of the use of anesthesia and analgesia in reptiles. *J. Am. Vet. Med. Assoc.* **2004**, *224*, 547–552. [CrossRef]
13. Sleeman, J.M.; Gaynor, J. Sedative and cardiopulmonary effects of medetomidine and reversal with atipamezole in desert tortoises (Gopherus agassizii). *J. Zoo Wildl. Med.* **2000**, *31*, 28–35. [CrossRef] [PubMed]
14. Vigani, A. Chelonia (tortoises, turtles, and terrapins). *Zoo Anim. Wildl. Immobil. Anesth.* **2014**, *25*, 365–387. [CrossRef]
15. Dundee, J.W.; Halliday, N.J.; Harper, K.W.; Brogden, R.N. Midazolam. A review of its pharmacological properties and therapeutic use. *Drugs* **1984**, *28*, 519–543. [CrossRef] [PubMed]
16. Aitken-Palmer, C.; Heard, D.; Jacobson, E.; Hall, N.; Thieman, K.; Ellison, G. *Clinical Management of Cloacal Prolapse in an Adult Galapagos Tortoise (Geochelone nigra)*; Veterinary Medical Center, University of Florida: Gainesville, FL, USA, 2010; p. 47.
17. Mans, C.; Sladky, K.K. Endoscopically guided removal of cloacal calculi in three African spurred tortoises (Geochelone sulcata). *J. Am. Vet. Med. Assoc.* **2012**, *240*, 869–875. [CrossRef] [PubMed]
18. Bienzle, D.; Boyd, C.J. Sedative effects of ketamine and midazolam in snapping turtles (Chelydra serpentina). *J. Zoo Wildl. Med.* **1992**, *23*, 201–204.
19. Chittick, E.J.; Stamper, M.A.; Beasley, J.F.; Lewbart, G.A.; Horne, W.A. Medetomidine, ketamine, and sevoflurane for anesthesia of injured loggerhead sea turtles: 13 cases (1996–2000). *J. Am. Vet. Med. Assoc.* **2002**, *221*, 1019–1025. [CrossRef] [PubMed]
20. Sladky, K.K.; Kinney, M.E.; Johnson, S.M. Analgesic efficacy of butorphanol and morphine in bearded dragons and corn snakes. *JAMA* **2008**, *233*, 267. [CrossRef]
21. Sladky, K.K.; Miletic, V.; Paul-Murphy, J.; Kinney, M.E.; Dallwig, R.K.; Johnson, S.M. Analgesic efficacy and respiratory effects of butorphanol and morphine in turtles. *J. Am. Vet. Med. Assoc.* **2007**, *230*, 1356–1362. [CrossRef]
22. Mosley, C. Pain and nociception in reptiles. *Vet. Clin. Exot. Anim. Pract.* **2011**, *14*, 45–60. [CrossRef]
23. Scheelings, T.F.; Gatto, C.; Reina, R.D. Anaesthesia of hatchling green sea turtles (Chelonia mydas) with intramuscular ketamine-medetomidine-tramadol. *Aust. Vet. J.* **2020**, *98*, 511–516. [CrossRef]
24. Mans, C.; Foster, J.D. Endoscopy-guided ectopic egg removal from the urinary bladder in a leopard tortoise (Stigmochelys pardalis). *Can. Vet. J.* **2014**, *55*, 569–572. [PubMed]
25. Schumacher, J.; Yelen, T. *Anesthesia and Analgesia*; Reptile Medicine and Surgery: London, UK, 2006.
26. Falcón, W.; Baxter, R.P.; Furrer, S.; Bauert, M.; Hatt, J.M.; Schaepman-Strub, G.; Ozgul, A.; Bunbury, N.; Clauss, M.; Hansen, D.M. Patterns of activity and body temperature of Aldabra giant tortoises in relation to environmental temperature. *Ecol. Evol.* **2018**, *8*, 2108–2121. [CrossRef] [PubMed]
27. White, N.M.; Balasubramaniam, T.; Nayak, R.; Barnett, A.G. An observational analysis of the trope "A *p*-value of <0.05 was considered statistically significant" and other cut-and-paste statistical methods. *PLoS ONE* **2022**, *17*, e0264360. [PubMed]
28. Mc Shane, B.B.; Gal, D.; Gelman, A.; Robert, C.; Tackett, J.L. Abandon statistical significance. *Am. Stat.* **2019**, *73*, 235–245. [CrossRef]
29. Hurlbert, S.H.; Levine, R.A.; Utts, J. Coup de grâce for a tough old bull: "Statistically significant" expires. *Am. Stat.* **2019**, *73*, 352–357. [CrossRef]
30. Divers, S.J.; Stahl, S.J. *Mader's Reptile and Amphibian Medicine and Surgery*; Elsevier Health Sciences: London, UK, 2018; p. 1537.
31. Lahner, L.; Mans, C.; Sladky, K. Comparison of Anesthetic Induction and Recovery Times After Intramuscular, Subcutaneous or Intranasal Dexmedetomidine-Ketamine Administration in Red-Eared Slider Turtles (Trachemys Scripta Elegans). In Proceedings of the Conference American Association of Zoo Veterinarians, Kansas City, MO, USA, 27 October 2011.
32. Doss, G.A.; Fink, D.M.; Sladky, K.K.; Mans, C. Comparison of subcutaneous dexmedetomidine-midazolam versus alfaxalone-midazolam sedation in leopard geckos (Eublepharis macularius). *Vet. Anaesth. Analg.* **2017**, *44*, 1175–1183. [CrossRef] [PubMed]
33. Hawkins, S.J.; Cox, S.; Yaw, T.J.; Sladky, K. Pharmacokinetics of subcutaneously administered hydromorphone in bearded dragons (Pogona vitticeps) and red-eared slider turtles (Trachemys scripta elegans). *Vet. Anaesth. Analg.* **2019**, *46*, 352–359. [CrossRef]
34. Abdala, V.; Manzano, A.S.; Herrel, A. The distal forelimb musculature in aquatic and terrestrial turtles: Phylogeny or environmental constraints? *J. Anat.* **2008**, *213*, 159–172. [CrossRef]
35. Eshar, D.; Rooney, T.A.; Gardhouse, S.; Beaufrère, H. Evaluation of the effects of a dexmedetomidine-midazolam-ketamine combination administered intramuscularly to captive red-footed tortoises (Chelonoidis carbonaria). *Am. J. Vet. Res.* **2021**, *82*, 858–864. [CrossRef]
36. Berry, S.H. Injectable anesthetics. In *Veterinary Anesthesia and Analgesia: The Fifth Edition of Lumb and Jones*; Wiley: Hoboken, NJ, USA, 2015; pp. 277–296.
37. Murrell, J.C.; Hellebrekers, L.J. Medetomidine and dexme-detomidine: A review of cardiovascular effects and antinociceptive properties in the dog. *Vet. Anaesth. Analg.* **2005**, 32, 117–127. [CrossRef]

38. Kallio-Kujala, I.J.; Raekallio, M.R.; Honkavaara, J.; Bennett, R.C.; Turunen, H.; Scheinin, M.; Vainio, O. Peripheral alpha-2-adrenoreceptor antagonism affects the absorption of intramuscularly coadministered drugs. *Vet. Anaesth. Analg.* **2018**, *45*, 405–413. [CrossRef] [PubMed]
39. Porters, N.; De Rooster, H.; Bosmans, T.; Baert, K.; Cherlet, M.; Croubels, S.; Polis, I. Pharmacokinetics of oral transmucosal and in-tramuscular dexmedetomidine combined with buprenorphine in cats. *J. Vet. Pharmacol. Ther.* **2014**, *38*, 203–208. [CrossRef] [PubMed]
40. Schumacher, J. Chelonians (turtles, tortoises, and terrapins). In *Zoo Animal and Wildlife Immobilization and Anaesthesia*; Blackwell Publishing: Hoboken, NJ, USA, 2007; pp. 259–266.
41. Girling, S.J.; Raiti, P. *BSAVA Manual of Reptiles*, 3rd ed.; British Small Animal Veterinary Association: Quedgeley, UK, 2019.
42. Rodney, W.; Molly, S. Section 6 Anesthesia. Chapter 48 Sedation. In *Mader's Reptile and Amphibian Medicine and Surgery*; Divers, S.J., Stahl, S.J., Eds.; Elsevier Health Sciences: London, UK, 2018; p. 1537.
43. Divers, S.J. Endoscopic sex identification in chelonians and birds (psittacines, passerines, and raptors). *Vet. Clin. N. Am. Exot. Anim. Pract.* **2015**, *18*, 541–554. [CrossRef] [PubMed]
44. Meireles, Y.S.; Shinike, F.S.; Matte, D.R.; Morgado, T.O.; Kempe, G.V.; Corrêa, S.H.R.; Souza, R.L.d.; Néspoli, P.B. Ultrasound characterization of the coelomic cavity organs of the red-footed tortoise (Chelonoidis carbonaria). *Ciênc. Rural* **2016**, *46*, 1811–1817. [CrossRef]
45. Scarabelli, S.; Di Girolamo, N. Chelonian sedation and anesthesia. *Vet. Clin. Exot. Anim. Pract.* **2022**, *25*, 49–72. [CrossRef] [PubMed]
46. Hansen, L.L.; Bertelsen, M.F. Assessment of the effects of intramuscular administration of alfaxalone with and without medetomidine in Horsfield's tortoises (Agrionemys horsfieldii). *Vet. Anaesth. Analg.* **2013**, *40*, e68–e75. [CrossRef] [PubMed]
47. Karklus, A.A.; Sladky, K.K.; Johnson, S.M. Respiratory and antinociceptive effects of dexmedetomidine and doxapram in ball pythons (Python regius). *Am. J. Vet. Res.* **2021**, *82*, 11–21. [CrossRef]

animals

Brief Report

Pancuronium Bromide for Chemical Immobilization of Adult Nile Crocodiles (*Crocodylus niloticus*): A Field Study

Lionel Schilliger [1,*], Chawki Najjar [2], Clément Paillusseau [1], Camille François [1], Frédéric Gandar [3], Hela Boughdiri [4] and Marc Gansuana [4]

[1] Argos SpéNac Veterinary Clinic, 100 Bvd de la Tour Maubourg, 75007 Paris, France; paillusseauclement@gmail.com (C.P.); mille.francois21240@gmail.com (C.F.)
[2] Veterinary Clinic of El Hidhab, 1st Salaheddine Ayoubi, Hidhab 2082, Tunisia; chawkinajjar.dv@gmail.com
[3] Alvetia Veterinary Clinic, 149 Route de Guentrange, 57100 Thionville, France; fredericg.alvetia@gmail.com
[4] Djerba Explore Park, Djerba Midoun 4116, Tunisia; hela.boughdiri@gmail.com (H.B.); marc.gansuana@gmail.com (M.G.)
* Correspondence: dr.l.schilliger@club-internet.fr; Tel.: +33-1-88-61-68-31

Simple Summary: Due to their significant size and aggressiveness, the capture of adult crocodiles carries significant risk, both in terms of stress and injuries to themselves and to operators. Neuromuscular blocking agents act by inducing flaccid muscle paralysis, thereby reducing the physical and chemical risks associated with transportation and anesthesia. Pancuronium bromide and its antagonist, neostigmine methylsulfate, have been successfully used in juvenile and subadult saltwater crocodiles (*Crocodylus porosus*), but their applications in larger animals (body weight > 230 kg or total length > 3.8 m) or in Nile crocodiles (*Crocodylus niloticus*) have never been described. We trialed a dose recommendation in nine Nile crocodiles using pancuronium bromide that was originally established for small- and medium-sized saltwater crocodiles, the effect of which can be reversed using neostigmine. We found that the recommended dose caused a prolonged recovery time in adult male Nile crocodiles, and we propose a weight-independent dose for Nile crocodiles with body weight ≥ 300 kg or total length ≥ 4.0 m, which we successfully trialed in 32 animals.

Abstract: (1) Background: Pancuronium bromide is a neuromuscular blocker used for immobilizing crocodiles that can be reversed with neostigmine. A recommended drug dose has only been established for saltwater crocodiles (*Crocodylus porosus*), mostly based on trials in juveniles and subadults. After trialing a dose recommendation in a small cohort of nine Nile crocodiles (*Crocodylus niloticus*), we developed and applied a new dose recommendation for large adult Nile crocodiles. (2) Methods: we trialed and adapted a pancuronium bromide (Pavulon 4 mg/2 mL) dose in Nile crocodiles originally established for saltwater crocodiles and applied the new dose for the immobilization of 32 Nile crocodiles destined for transport. Reversal was achieved with neostigmine (Stigmine 0.5 mg/mL). (3) Results: Nine crocodiles were included in the trial phase; the induction time was highly variable (average: 70 min; range: 20–143 min), and the recovery time was prolonged (average: 22 h; range: 50 min–5 days), especially in large animals after reversal with neostigmine. Based on these results, we established a dose-independent recommendation (3 mg pancuronium bromide and 2.5 mg neostigmine) for animals weighing ≥ 270 kg (TL ≥ ~3.8 m). When applied to 32 adult male crocodiles (BW range: 270–460 kg; TL range: 3.76–4.48 m), the shortest induction time was ~20 min and the longest ~45 min. (4) Conclusions: Pancuronium bromide and its antidote, neostigmine, are effective for the immobilization and reversal of adult male Nile crocodiles (TL ≥ 3.8 m or BW ≥ 270 kg) when given in a weight-independent fashion.

Keywords: crocodilians; chemical immobilization; Nile crocodile; pancuronium bromide; neostigmine; non-depolarizing neuromuscular blocking agents

Citation: Schilliger, L.; Najjar, C.; Paillusseau, C.; François, C.; Gandar, F.; Boughdiri, H.; Gansuana, M. Pancuronium Bromide for Chemical Immobilization of Adult Nile Crocodiles (*Crocodylus niloticus*): A Field Study. *Animals* **2023**, *13*, 1578. https://doi.org/10.3390/ani13101578

Academic Editor: Tom Hellebuyck

Received: 8 March 2023
Revised: 12 April 2023
Accepted: 6 May 2023
Published: 9 May 2023

1. Introduction

Capturing adult crocodiles is challenging and dangerous due to their significant size, aggressiveness, and potential to cause severe injury to themselves and to operators [1–4].

Gallamine triethiodide (Flaxedil 20 mg/mL), a nondepolarizing neuromuscular blocker, has been commonly used since the 1970s for immobilizing captive and wild saltwater crocodiles (*Crocodylus porosus*) and Nile crocodiles (*Crocodylus niloticus*) [5–9]. Following the discontinuation of gallamine triethiodide in Australia, pancuronium bromide has empirically proven to be an effective and affordable alternative for immobilizing wild and farmed saltwater crocodiles [4]. Research in mostly juvenile crocodiles (total length [TL] < 2.9 m and body weight [BW] < 246 kg) established a minimum effective dose of 0.019 mg/kg and a safety margin of 0.2 mg/kg for pancuronium bromide, with effective reversal obtained with neostigmine methylsulfate administered at 0.02 mg/kg [10,11]. Limited dose recommendation has, however, been published for adult saltwater crocodiles with a BW ≥ 246 kg despite it being inadvisable to extrapolate a dose from smaller individuals [4,10–12]. In addition, dose recommendations for pancuronium bromide have not been published for Nile crocodiles. Here, we report a dosing recommendation using pancuronium bromide and neostigmine methylsulfate for the immobilization of large adult captive Nile crocodiles.

2. Materials and Methods

In 2022, a total of 32 adult captive Nile crocodiles were scheduled for transportation from Djerba Explore Park, Djerba, Tunisia to the Dubai Crocodile Park, Dubai, United Arab Emirates. For this purpose, pancuronium bromide (Pavulon, 4 mg/2 mL, Neon Laboratories Ltd., Mumbai, India) was chosen as the drug of choice for its high safety profile, availability, low cost, and ability to be reversed with neostigmine methylsulfate (Stigmine, 0.5 mg/mL, Société Arabe des Industries Pharmaceutiques, Tunis, Tunisia).

Prior to transport, we trialed and fine-tuned a pancuronium bromide dose established for saltwater crocodiles by Bates et al. in subadult and adult Nile crocodiles [10,11]. All of the animals were reversed with neostigmine methylsulfate after one to two hours. On the basis of the results obtained from the trial, we established a dose for animals weighing ≥ 270 kg (TL ≥ ~3.8 m), consisting of 3 mg pancuronium bromide (1.5 mL Pavulon 4 mg/2 mL) and 2.5 mg neostigmine methylsulfate (5 mL Stigmine 0.5 mg/mL). Subsequent to the trial, 32 adult male crocodiles (distinct from those included in the trial group) were immobilized using the new recommendation and shipped to the Dubai Crocodile Park. For crocodiles being transported, the reversal agent was administered by intramuscular injection into the hind limb once immobilization was achieved and the crocodile was secured inside the transport crate (see below).

The depth of neuromuscular blockage was ascertained by prodding the base of the tail with a pole (absence of avoidance reflex), verifying that the mouth remained open in a relaxed position (the presence of the "Flaxedil (i.e., gallamine) reaction" [13]), and inserting a pole inside the mouth (absence of bite reflex). Full recovery was defined as crocodiles moving with a normal gait.

Before attempting injection during the trial and translocation phases, the enclosures were drained of water to eliminate any risk of drowning. Crocodiles were administered pancuronium bromide intramuscularly in the left lateral aspect of the base of the tail using an expandable (90–180 cm) pole syringe (Jabstick, Daninject, Olgas Allé 4, 6000 Kolding, Denmark) equipped with plastic syringes (Jabstick) and 2.0 × 50 mm (14G × 2") metallic needles (Kruuse A/S, Havretoften 4, DK-5550 Langeskov, Denmark). Only resting crocodiles were targeted to avoid injuring both the crocodile and the operator [14]. To calculate the pancuronium bromide dose, BW was first estimated based on the park's experience of extrapolating BW from TL. An accurate weight was obtained by weighing crocodiles during immobilization, except for four large crocodiles during the trial, which were too heavy to be weighed on site. All animals were paint-marked immediately after injection to avoid accidental double-dosing. For safety reasons, the jaws were secured with rope. Because nondepolarizing neuromuscular blockers do not abolish sensory input, the head was covered with a towel to reduce sensory stimuli, taking care not to obstruct the nostrils. Body temperature was measured immediately after induction, either using a cloacal thermometer inserted in the proctodeum (Checktemp, Cifec, 12bis rue du commandant Pilot, 92200 Neuilly-sur-Seine, France) during the trial, or a temperature gun (Ketotek, Xiamen Sizhi E-commerce Co., Ltd., 8E Mingyuan Building, 361000 Xiamen, China) aimed at the nuchal region before and during the transport phase [15,16]. Because the duration to induction and the duration of immobilization are expected to be inversely proportional to body temperature [10,11], we only proceeded with the trial and transport phases once ambient temperatures were above 19–20 °C. Following immobilization, crocodiles were handled carefully, kept out of direct sunlight (ambient temperature: 20–24 °C), and regularly doused with water [17].

Crocodiles marked for translocation were transported in individual wooden crates with foam-covered siding. The ambient temperature range measured in the crates was 13–20 °C during road transport and 22–24 °C during air transport. Transport was carried out in two separate shipments (16 animals per shipment) at a one-week interval. All of the transport specifications were in accordance with CITES and IATA (International Air Transport Association) guidelines [18,19]. The two translocations took an average of 43 h, consisting of crating (10 h 00 min and 8 h 15 min), followed by ground (22 h 30 min and 18 h 50 min), air (7 h 00 min and 6 h 40 min), and ground (7 h 10 min and 5 h 20 min) transport.

3. Results

The trial involved one subadult (BW: 54 kg; TL: 2.20 m) and eight adults (BW range: 130–370 kg; TL range: 2.75–4.30 m). The average cloacal temperature was 23 °C (range: 19–26.2 °C). The average induction time was 70 min (range: 20–143 min), and the average recovery time was 22 h and 14 min (median: 105 min; range: 50 min–5 days) (Table 1). The recovery time for three adults having received the highest pancuronium dose (0.011–0.015 mg/kg) was >24 h (two of which required an additional neostigmine injection), whereas the highest dose (0.019 mg/kg) administered to one subadult (54 kg) did not lead to delayed recovery. Four adult crocodiles were subsequently immobilized using a lower dose of 0.006–0.008 mg/kg, leading to a longer average induction time of 109 min (range: 84–143 min) but a shorter average recovery time of 120 min (range: 55–105 min). Based on these initial findings, and considering that the transfers to Dubai would only involve large adult specimens with a BW $\geq$ 270 kg (TL $\geq$ 3.8 m), it was decided to substantially reduce the pancuronium bromide dose in adults, deciding on a weight-independent dose of 3 mg pancuronium bromide (1.5 mL Pavulon) and 2.5 mg neostigmine methylsulfate (5 mL Stigmine) independent of BW or TL.

The transport phase involved 32 adult male crocodiles (aged approximately 25 years). The average BW was 365 kg (median: 371 kg; range: 270–460 kg), with an average TL of 4.17 m (range: 3.76–4.48 m). The average nuchal temperature at induction was 25.6 °C (range: 20.8–37.2 °C). The final pancuronium bromide dose range was 0.007–0.011 mg/kg (mean: 0.008 mg/kg, ±SD 0.001). For logistical reasons, individual induction time was not measured; however, the shortest induction time was ~20 min and the longest ~45 min. Recovery time could not be recorded because crocodiles were already inside crates by the time they would have recovered from immobilization. All of the 41 crocodiles included in the trial and transport phase were alive and apparently healthy approximately ten months after the intervention, with no reported instances of anorexia. No injuries during shipment were noted. Approximately ten males displayed mating behavior within a day of arrival at the Dubai Crocodile Park.

Table 1. Induction and recovery time for the immobilization of subadult and adult Nile crocodiles (*Crocodylus niloticus*) using pancuronium bromide (Pavulon, 2 mg/mL) and neostigmine (Stigmine, 0.5 mg/mL) based on existing recommendations in juvenile saltwater crocodiles (*Crocodylus porosus*), $n = 9$.

Sex	Body Weight (kg)	Total Length (cm)	Cloacal Temperature (°C)	Pancuronium Bromide (mg/kg)	Pavulon (mL)	Induction Time (min)	Neostigmine (mg/kg)	Stigmine (mL)	Recovery Time
M	54	220	19.0	0.019	0.5	20	0.014	1.50	50 min
F	130	275	20.5	0.015	1.0	30	0.034	8.75 §	5 days
F	172	315	21.6	0.008	0.7	50	0.015	5.00	120 min
M	260	380	22.6	0.015	2.0	45	0.019	10.00 §	2 days
M *	270	390	25.4	0.008	1.1	85	0.009	5.00	105 min
M *	330	410	26.0	0.007	1.1	84	0.008	5.50	90 min
M	353	425	20.4	0.011	2.0	55	0.007	5.00	1 day
M *	370	425	25.8	0.006	1.1	125	0.007	5.00	70 min
M *	370	430	26.2	0.006	1.1	143	0.007	5.00	55 min

* Approximate body weight; § received two doses.

4. Discussion

Immobilizing agents used in crocodiles include opioids (e.g., etorphine), dissociative anesthetics (e.g., ketamine, tiletamine–zolazepam), alpha-2 agonists (e.g., medetomidine), barbiturates (e.g., pentobarbital sodium), neuroactive steroids (e.g., alfaxalone), neuromuscular blocking agents (e.g., succinylcholine chloride, d-tubocurarine, atracurium besylate, gallamine triethiodide, and pancuronium bromide), and other agents (e.g., propofol, tricaine mesylate) [10,11,13,17,20–30].

Neuromuscular blocking agents (NBAs) can be separated into two classes depending on their mechanism of action. Depolarizing neuromuscular blockers (e.g., succinylcholine chloride) act by binding to postsynaptic cholinergic receptors on motor endplates, causing depolarization and fasciculation, which leads to flaccid paralysis. Nondepolarizing neuromuscular blockers (e.g., d-tubocurarine, atracurium besylate, gallamine triethiodide, and pancuronium bromide) are competitive acetylcholine antagonists that bind to nicotinic receptors on the postsynaptic membrane, blocking acetylcholine binding and thereby preventing motor endplate depolarization. Nondepolarizing neuromuscular blockers can be reversed with cholinesterase inhibitors, such as neostigmine [1,3,5,8,13].

There is a paucity of data on the immobilization of crocodiles using pancuronium bromide. Bates et al. found that the average induction time in juvenile (BL < 2.90 kg) and adult (BL < 90 kg) saltwater crocodiles immobilized with pancuronium bromide was 22 min when given at a dose of 0.02 mg/kg, which was lower than in our cohort [10,11]. The authors also showed that a higher dose of pancuronium bromide (>0.025 mg/kg) did not significantly decrease induction time. In addition, the duration of immobilization (without reversal with neostigmine) was proportional to the pancuronium bromide dose and exercise intensity prior to immobilization (possibly linked to muscle fatigue). In contrast, the duration of immobilization was inversely proportional to body temperature (possibly because of faster drug metabolism) and to induction time. The recovery time was shortened (<5 min) with a higher dose of neostigmine, but the time for the full recovery of all reflexes was not influenced by the dose of pancuronium bromide, sex, body weight, or induction time.

The translocation of 32 Nile crocodiles provided the opportunity to assess our dose recommendation for pancuronium bromide and neostigmine methylsulfate, for which no previous dose recommendations have been published. The results from our trial indicate that it unadvisable to extrapolate dose recommendation by weight, as already hypothesized by Bates et al. and as predicted by the principle of allometric scaling (i.e., body weight is inversely proportional to basal metabolic rate) [10,11]. Although our sample size was small, our observations indicate that extrapolating a dose recommendation established for juveniles in adult crocodiles resulted in a prolonged recovery time despite nearly doubling the recommended neostigmine dose. Using a lower, weight-independent dose of 3 mg pancuronium bromide for crocodiles with TL $\geq$ 3.8 m or BW $\geq$ 270 kg (corresponding to 0.007–0.011 mg/kg among transported subjects) resulted in a satisfactory induction time, a result that can be explained by allometric scaling. Based on the limited number of animals included in the trial, we hypothesize that using a higher dose would likely lead to a delayed recovery time. We were, however, unable to assess a lower, weight-independent dose of 2.5 mg neostigmine sulfate (corresponding to 0.006–0.009 mg/kg among transported subjects) because crocodiles were physically restrained and transported in crates for an average of 43 h.

Unlike general anesthetics, NBAs cause flaccid muscle paralysis without sedation or analgesia, thereby reducing the physical and physiologic risks associated with anesthesia, especially during transportation. Monitoring the respiration rate is nevertheless advised because NBAs can affect respiratory function in crocodiles, with a 66% decrease in respiratory rate noted for dose >0.2 mg/kg [4,10]. Our failure to measure the respiratory rate during the field trial constitutes a limitation of our report, although the survival of all 32 crocodiles during long-distance transport by road and air immediately after reversal suggests that respiratory depression, if present, is unlikely to have been severe.

Animals immobilized with NBA are fully conscious and sensitive to visual, auditory, and tactile stimuli, which can result in stress-induced physiologic reactions such as tachycardia and tachypnea [3,10]. In our case, immobilization was conducted with a minimum of people and reduced noise, and eyes and ears were covered to reduce stress from sensory input [13]. For this reason, any pain-inducing procedures should never be performed when a crocodile is immobilized with NBAs alone.

Physical techniques such as nets, snares, and traps can be used for capturing crocodiles, but they are a frequent source of stress and injury (e.g., fractures, ocular injuries, and skin abrasions), particularly due to fight or flight behaviors that can last until exhaustion (e.g., biting, tail thrashing, and body rolling) [3,4]. Anorexia of 18 months in duration has also been reported in a Mugger crocodile (*C. palustris*) following manual capture [31]. In instances of prolonged struggle (e.g., during protracted capture), physiological stress responses can result in the release of adrenaline and corticosterone, and increased anaerobic glucose catabolism [16,18,19,32]. This can lead to lactic acidemia, capture myopathy, and cardiac dysfunction, sometimes with fatal consequences [13,32–34]. Crocodilians, and reptiles in general, lack the ability to rapidly metabolize and correct acidemia, which can be severe [35]; blood pH levels of 6.6–6.8 (reference range: 7.0–7.4) have been recorded in Nile crocodiles after prolonged struggling [3]. Lactate levels, which can be considered as an indicator of anaerobic metabolism and physiological stress in crocodilians, were positively correlated with capture duration (longer handling time or struggling causes an increase in lactate), age (adult crocodilians had significantly higher lactate levels after capture than non-adults), and weight (possibly confounded with handling time because larger animals are generally harder to capture) [34]. Blood biochemistry values in chemically immobilized crocodiles have, however, not been published and might differ from physically caught crocodiles in terms of stress markers (e.g., blood lactate and pH) and physiological parameters (e.g., heart rate and respiratory rate).

5. Conclusions

Pancuronium bromide is effective for the immobilization of adult male Nile crocodiles (TL $\geq$ 3.8 m or BW $\geq$ 270 kg) when given in a weight-independent fashion. Reversal using neostigmine methylsulfate was equally effective when given in a weight-independent fashion, although the small cohort (n = 5) in which reversal was measured precludes any firm conclusions. Systematic studies in a larger population of animals (juveniles, subadults, and adults) are needed to confirm our observations that dose is inversely proportional to body weight and to monitor the effects of pancuronium bromide on physiological parameters and stress markers.

Author Contributions: Conceptualization, L.S., C.N., C.P., F.G., C.F., and M.G.; methodology, L.S., H.B., C.N., and M.G.; formal analysis, L.S., H.B., and M.G.; data curation, L.S., M.G., and C.N.; writing—original draft preparation, L.S.; writing—review and editing, L.S., C.P., F.G., C.F., and M.G.; and supervision, L.S. and M.G. All authors have read and agreed to the published version of the manuscript.

Funding: This research received no external funding.

Institutional Review Board Statement: The animal study protocol was evaluated by the Ethics Committee of VetAgro Sup, 1 avenue Bourgelat, 69280 Marcy l'Étoile, France (protocol 253 code 2333); it was considered to conform with common practices in the field of crocodile farming practices. This committee expressed a recommendation for future improvement to include stress reduction components in these immobilization protocols.

Informed Consent Statement: Not applicable.

Data Availability Statement: Data supporting the reported results can be found at Djerba Explore Crocodile Pars, Djerba, Tunisia: marc.gansuana@gmail.com.

Acknowledgments: The authors would like to thank Jesse Bonwitt for the translation and critical review of the manuscript.

Conflicts of Interest: The authors declare no conflict of interest.

References

1. Jacobson, E.R. Immobilization, blood sampling, necropsy techniques and diseases of crocodilians: A review. *J. Zoo Anim. Med.* **1984**, *15*, 38–45. [CrossRef]
2. Huchzermeyer, F.; Wyk, W. Crocodiles—Biology, husbandry and diseases. *J. S. Afr. Vet. Assoc.* **2003**, *74*, 529. [CrossRef]
3. Fleming, G.J. Crocodilian anesthesia. *Vet. Clin. N. Am. Exot. Anim. Pract.* **2001**, *4*, 119–145. [CrossRef] [PubMed]
4. Manolis, S.C.; Webb, G.J.W. *Best Management Practices for Crocodilian Farming*; Manolis, S.C., Webb, G.J.W., Eds.; IUCN-SSC Crocodile Specialist Group: Karama, Australia, 2016.
5. Woodford, M.H. The use of gallamine triethiodide as a chemical immobilizing agent for the Nile Crocodile (*Crocodilus niloticus*). *Afr. J. Ecol.* **1972**, *10*, 67–70. [CrossRef]
6. Morgan-Davies, A.M. Immobilization of the Nile crocodile (*Crocodilus niloticus*) with gallamine triethiodide. *J. Zoo Anim. Med.* **1980**, *11*, 85–87. [CrossRef]
7. Loveridge, J.P.; Blake, D.K. Techniques in the immobilisation and handling of the Nile crocodile, *Crocodylus niloticus*. *Arnoldia* **1972**, *40*, 1–14.
8. Whitaker, R.; Andrews, H. Chemical immobilization of the Mugger crocodile (*Crocodylus palustris*) with gallamine triethiodide. *Indian For.* **1989**, *115*, 355–356.
9. Lloyd, M.L.; Reichard, T.; Odum, R.A. Gallamine reversal in Cuban crocodiles (*Crocodilus rhombifer*) using neostigmine alone vs neostigmine with hyaluronidase. In Proceedings of the Annual Conference of the Association of Reptile and Amphibian Veterinarians, Mt. Juliet, TN, USA, 22–27 October 1994; pp. 117–120.
10. Bates, L.; Webb, G.J.W.; Richardson, K.C.; Bitton, A.R.C.; Bar-Lev, J.; Manolis, S.C. "Pancuronium bromide"—An immobilising agent for crocodiles. In Proceedings of the 17th Working Meeting of the IUCN-SSC Crocodile Specialist Group, Gland, Switzerland, 24–29 May 2004; pp. 447–451.
11. Bates, L. *Chemical Restraint of Saltwater Crocodiles: A Review, and the Testing of the Immobilising Agent—"Pancuronium Bromide"*; Murdoch University: Perth, Australia, 2001.
12. DCCEEW. *Code of Practice for the Humane Treatment of Wild and Farmed Australian Crocodiles*; The Natural Resource Management Ministerial Council (NRMMC): Canberra, Australia, 2009.
13. Flamand, J.R.; Rogers, P.S.; Blake, D.K. Immobilization of crocodiles. In Proceedings of the Wildlife Tranquilizer Symposium, Pretoria, South Africa, 17–18 March 1989; pp. 61–65.
14. Buys, A.C. Operator dangers associated with the use of immobilizing drugs—Accidents and emergency treatment. In *The Capture and Care of Wild Animals*; Young, E., Ed.; Human and Rousseau: Cape Town, South Africa, 1973; pp. 77–83.
15. Baines, F.; Cusack, L. Environmental Lighting. In *Mader's Reptile and Amphibian Medicine and Surgery*, 3rd ed.; Divers, S.J., Stahl, S.J., Eds.; W.B. Saunders: St. Louis, MO, USA, 2019; pp. 131–138.
16. Divers, S.J. Medical History and Physical Examination. In *Mader's Reptile and Amphibian Medicine and Surgery*, 3rd ed.; Divers, S.J., Stahl, S.J., Eds.; W.B. Saunders: St. Louis, MO, USA, 2019; pp. 385–404.

17. Loveridge, J.P.; Blake, D.K. Crocodile immobilization and anaesthesia. In *Wildlife Management: Crocodiles and Alligators*; Webb, G., Manolis, S.C., Whitehead, P.J., Eds.; Surrey Beatty & Sons: Chipping Norton, UK; the Conservation Commission of the Northern Territory: Darwin, Australia, 1987; pp. 259–267.
18. Transport of Live Specimens. Available online: https://cites.org/eng/prog/imp/Transport_of_live_specimens (accessed on 4 March 2023).
19. Live Animals Regulations (LAR). Available online: https://www.iata.org/en/publications/store/live-animals-regulations (accessed on 4 March 2023).
20. Loveridge, J.P. The immobilisation and anaesthesia of crocodilians. *Int. Zoo Yearb.* **1979**, *19*, 103–112. [CrossRef]
21. Bonath, K.H.; Bonath, I.; Haller, R.D.; Amelang, D. The influence of gallamine on immobilisation, cardiovascular and respiratory parameters of *Crocodylus niloticus*. In Proceedings of the 10th Working Meeting of the IUCN-SSC Crocodile Specialist Group, Gainesville, FL, USA, 23–27 April 1990; pp. 13–15.
22. Clyde, V.L.; Cardeilhac, P.T.; Jacobson, E.R. Chemical restraint of American alligators (*Alligator mississippiensis*) with atracurium or tiletamine-zolazepam. *J. Zoo Wildl. Med.* **1994**, *25*, 525–530.
23. Lloyd, M.L. Crocodilian anaesthesia. In *Zoo & Wild Animal Medicine: Current Therapy*, 4th ed.; Fowler, M.E., Miller, R.E., Eds.; W.B. Saunders: Phildelphia, PA, USA, 1999; pp. 205–216.
24. Smith, J.A.; Mitchell, M.A.; Backhues, K.A.; Tully, T.N.; Aguilar, R.F. Immobilization of American alligators with medetomidine and its reversal with atipamezole. *Vet. Surg.* **1999**, *28*, 133.
25. Heaton-Jones, T.G.; Ko, J.C.; Heaton-Jones, D.L. Evaluation of medetomidine-ketamine anesthesia with atipamezole reversal in American alligators (*Alligator mississippiensis*). *J. Zoo Wildl. Med.* **2002**, *33*, 36–44. [CrossRef] [PubMed]
26. Olsson, A.; Phalen, D. Preliminary studies of chemical immobilization of captive juvenile estuarine (*Crocodylus porosus*) and Australian freshwater (*C. johnstoni*) crocodiles with medetomidine and reversal with atipamezole. *Vet. Anaesth. Analg.* **2012**, *39*, 345–356. [CrossRef] [PubMed]
27. Olsson, A.; Phalen, D. Medetomidine immobilisation and atipamezole reversal in large estuarine crocodiles (*Crocodylus porosus*) using metabolically scaled dosages. *Aust. Vet. J.* **2012**, *90*, 240–244. [CrossRef] [PubMed]
28. Olsson, A.; Phalen, D. The effects of decreased body temperature on the onset, duration and action of medetomidine and its antagonist atipamezole in juvenile farmed estuarine crocodiles (*Crocodylus porosus*). *Vet. Anaesth. Analg.* **2013**, *40*, 272–279. [CrossRef] [PubMed]
29. Olsson, A.; Phalen, D.; Dart, C. Preliminary studies of alfaxalone for intravenous immobilization of juvenile captive estuarine crocodiles (*Crocodylus porosus*) and Australian freshwater crocodiles (*Crocodylus johnstoni*) at optimal and selected sub-optimal thermal zones. *Vet. Anaesth. Analg.* **2013**, *40*, 494–502. [CrossRef] [PubMed]
30. Mans, C.; Sladky, K.K.; Schumacher, J. General anesthesia. In *Mader's Reptile and Amphibian Medicine and Surgery*, 3rd ed.; Divers, S.J., Stahl, S.J., Eds.; W.B. Saunders: St. Louis, MO, USA, 2019; pp. 447–464.
31. Klide, A.M.; Klein, L.V. Chemical restraint of three reptilean species. *J. Zoo Anim. Med.* **1973**, *4*, 8–11. [CrossRef]
32. Seymour, R.S.; Webb, G.J.W.; Bennett, A.F.; Bradford, D. Effect of capture on the physiology of *Crocodylus porosus*. In *Wildlife Management: Crocodiles and Alligators*; Webb, G.J.W., Manolis, S.C., Whitehead, P.J., Eds.; S. Beatty & Sons: Chipping Norton, UK, 1987; pp. 253–256.
33. Lance, V.A.; Morici, L.A.; Elsy, R.M. Physiology and endocrinology of stress in crocodilians. In *Crocodilian Biology and Evolution*; Grigg, G.C., Seebacher, F., Franklin, C.E., Eds.; Surrey Beatty & Sons: Chipping Norton, UK, 2001; pp. 327–340.
34. Molinaro, H.G.; Anderson, G.S.; Gruny, L.; Sperou, E.S.; Heard, D.J. Use of Blood Lactate in Assessment of Manual Capture Techniques of Zoo-Housed Crocodilians. *Animals* **2022**, *12*, 397. [CrossRef] [PubMed]
35. Gleeson, T.T. Post-exercise lactate metabolism: A comparative review of sites, pathways, and regulation. *Annu. Rev. Physiol.* **1996**, *58*, 565–581. [CrossRef] [PubMed]

animals

Article

Histological Variants of Squamous and Basal Cell Carcinoma in Squamates and Chelonians: A Comprehensive Classification

Ferran Solanes Vilanova *,†, Tom Hellebuyck † and Koen Chiers

Department of Pathobiology, Pharmacology and Zoological Medicine, Faculty of Veterinary Medicine, Ghent University, Salisburylaan 133, B-9820 Merelbeke, Belgium
* Correspondence: ferransolanesvilanova@hotmail.com
† These authors contributed equally to this work.

Simple Summary: The present study investigated the histological characteristics of 35 tumors from 21 lizards, 1 snake, 10 tortoises and 3 turtles that were initially diagnosed as squamous or basal cell carcinoma. Based on in-depth re-evaluation of the tissue characteristics, eight tumors initially diagnosed as squamous cell carcinoma were re-classified as basal cell carcinomas and three squamous cell carcinomas proved to be non-neoplastic lesions. All squamous and basal cell carcinomas were classified into distinct histological variants. To date, basal cell carcinomas have only been described in two reptile species. In the present study, basal cell carcinomas were diagnosed in seven additional species. While immunohistochemical staining with cyclooxygenase-2 and E-cadherin showed significant differences between the examined squamous and basal cell carcinomas, no immunoreactivity was observed for epithelial antigen clone Ber-EP4 and epithelial membrane antigen. The results of this study provide a proposal classification that allows the differentiation of squamous and basal cell carcinoma and their histological variants in squamates and chelonians.

Abstract: In the present study, the histological characteristics of squamous cell carcinomas (SCCs) and basal cell carcinomas (BCCs) obtained from 22 squamate and 13 chelonian species were retrospectively evaluated. While the examined tissues were originally diagnosed as 28 SCCs and 7 BCCs based on histological evaluation by a specialty diagnostic service, eight SCCs could be re-classified as BCCs and three SCCs proved to be non-neoplastic lesions. In addition, all SCCs and BCCs were classified into distinct histological variants. The SCCs could be categorized as one SCC in situ, three moderately differentiated SCCs, seven well-differentiated SCCs, and six keratoacanthomas. BCCs were classified as five solid BCCs, four infiltrating BCCs, five keratotic BCCs, and one basosquamous cell carcinoma. In addition, the present study reports the occurrence of BCCs in seven reptile species for the first time. In contrast to what has been documented in humans, IHC staining with the commercially available epithelial membrane antigen and epithelial antigen clone Ber-EP4 does not allow differentiation of SCCs from BCCs in reptiles, while cyclooxygenase-2 and E-cadherin staining seem to have discriminating potential. Although the gross pathological features of the examined SCCs and BCCs were highly similar, each tumor could be unequivocally assigned to a distinct histological variant according to the observed histological characteristics. Based on the results of this study, a histopathological classification for SCCs and BCCs is proposed, allowing accurate identification and differentiation of SCCs and BCCs and their histological variants in the examined reptile species. Presumably, BCCs are severely underdiagnosed in squamates and chelonians.

Keywords: basal cell carcinoma; immunohistochemistry; neoplasm; reptiles; squamous cell carcinoma

Citation: Solanes Vilanova, F.; Hellebuyck, T.; Chiers, K. Histological Variants of Squamous and Basal Cell Carcinoma in Squamates and Chelonians: A Comprehensive Classification. *Animals* **2023**, *13*, 1327. https://doi.org/10.3390/ani13081327

Academic Editor: Francesca Millanta

Received: 21 March 2023
Revised: 10 April 2023
Accepted: 11 April 2023
Published: 12 April 2023

1. Introduction

Neoplasms are frequently encountered in the practice of reptile medicine, although they were once considered uncommon [1]. Most data about the occurrence of neoplasms in captive reptiles originate from specialty diagnostic services [1–5], and considerable variation in prevalence data, ranging from 9.8% to 26%, is reported [1–7]. In general, neoplasms are more frequently observed in snakes and lizards in comparison to chelonians and crocodilians [1,2,4–6], with the integumentary, hepatic, and musculoskeletal systems being the most commonly affected sites [1,2,4–6]. Reports of skin tumors in reptiles are largely derived from single cases, and mainly comprise squamous cell carcinomas (SCCs), papillomas, and chromatophoromas [1–3,5]. The increasing number of neoplastic disorders that are being diagnosed in captive reptiles can at least be partly attributed to the fact that reptile owners more readily seek veterinary advice as well as to the increasing availability and use of appropriate diagnostic tools. In addition, the increasing lifespan of reptile pets as well as predisposing environmental and genetic factors may also contribute to the seemingly increasing prevalence of neoplastic disorders [3,5,6,8–10]. Nevertheless, the diagnosis of neoplasms in reptile patients can easily be missed at initial presentation as associated clinical signs are often non-specific [4,5].

Basal cell carcinomas (BCCs) and SCCs account for 96% of skin neoplasms in humans. Human BCCs are the most commonly diagnosed non-melanoma skin tumors and are diagnosed three to four times more often than SCCs [11]. While SCCs are the most common skins neoplasm in cats and the second most common in dogs [12], most BCCs in cats and dogs have been re-classified as benign trichoblastomas and apocrine ductular adenomas, respectively [12,13]. Consequently, true BCCs are considered to be relatively rare neoplasms in these conventional domestic animals [12,13]. Furthermore, SCCs are among the most common integumentary neoplasms in reptilian species, while BCCs are rarely documented in these taxa. At present, reports of BCCs in reptile species are limited to a savannah monitor (*Varanus exanthematicus*), and two Hermann's tortoises (*Testudo hermanni*) [5,14]. Although the skin and oral cavity seem to be predilection sites for both SCCs and BCCs in reptiles, SCCs originating near the mucocutaneous junction (MCJ) are also frequently reported, particularly in bearded dragons (*Pogona vitticeps*) [3]. Recently, keratoacanthoma (KA) has been described as a new histological variant of dermal SCCs in lizards with a presumed species predisposition in panther chameleons (*Furcifer pardalis*) that often shows a multicentric distribution, including involvement of the MCJ of the eyelid [8].

The discrimination of dermal SCCs and BCCs and their histological variants is highly important towards prognosis estimation and establishing appropriate treatment protocols as the associated invasiveness, recurrence rate, and metastatic potential are strongly correlated with the involved histological variant [15,16]. Histological variants of SCCs and BCCs have been characterized in humans, dogs and cats and classifications have been established [12,15,17]. Several classifications for human SCCs and BCCs were adopted by the International Agency for Research on Cancer (IARC) and included in the World Health Organization (WHO) classification for skin tumors [18]. However, the differentiation and histological classification of certain variants remain challenging and, in some cases, even controversial [12,18]. Nevertheless, correct identification and classification of SCCs and BCC variants are essential towards the establishment of the most effective treatment protocols [12,19,20].

At present, commonly observed high-risk histological variants of SCCs include acantholytic SCCs, desmoplastic SCCs, and spindle cell SCCs, and low-risk histological variants include verrucous SCC and KA. For human BCCs, infiltrating BCCs and BSCCs represent high-risk histological variants, and solid BCCs, keratotic BCCs, and superficial BCCs represent low-risk histological variants [18]. While SCCs with varying degrees of differentiation are commonly observed in reptiles as well as birds [21–24], studies focused on in-depth comparative assessment of the histological characteristics of these neoplasms that allow the establishment of a classification system, are lacking.

In human medicine, immunohistochemistry (IHC) has proven to be an essential tool to differentiate certain histological variants of SCCs from BCCs [25,26]. In addition, IHC aids in tumor staging, selecting optimal treatment protocols, and identifying genetic variants in humans, dogs and cats [25–31]. Although two mammalian alpha-keratin markers were successfully used to characterize SCCs in loggerhead sea turtles (*Caretta caretta*), the limited availability of reptilian monoclonal antibodies and the lack of commercially available antibodies that cross-react with reptilian tissue continue to hamper the use of IHC in reptiles [7,32–34].

Taking into account the challenges that have been and are still encountered in humans and conventional pets towards the correct discrimination, identification, and classification of histological variants of SCCs and BCCs [35], the present study aims to provide a basis for the correct histological characterization and classification of SCCs and BCCs and their histological variants in squamates and chelonians.

2. Materials and Methods

2.1. Tissues

Formalin-fixed and paraffin-embedded tissues from lesions obtained in 35 unrelated, captive reptiles (22 squamates and 13 chelonians) that were presented at a veterinary teaching hospital between 2010 and 2022 were included. The paraffin-embedded tissues were selected based on their initial diagnosis as SCCs or BCCs by a specialty diagnostic service following routine histological examination (Table 1). The tumors either originated from the skin, the epidermis of the shell, the MCJ of the eyelid, or the oral mucosa. All tissues were collected antemortem following excisional biopsy or in toto during surgical resection of the lesion.

Table 1. Anatomic location and number of tumors from 35 reptile patients that were initially diagnosed as squamous cell carcinomas (SCCs) or basal cell carcinomas (BCCs) and re-classified into histological variants following retrospective histological and immunohistochemical characterization.

Species	Location	Initial Diagnosis	Re-Classification	
			Final Diagnosis	Histological Variant
Bearded dragon (*Pogona vitticeps*) (*n* = 9)	Skin	6 SCC	4 SCC 1 BCC 1 Non-neoplastic	3 WD SCC 1 KA 1 Infiltrating BCC Gingival fibrous hyperplasia
	MCJ	2 SCC	2 SCC	1 WD SCC 1 MD SCC
	Oral	1 BCC	1 BCC	1 Solid BCC
Panther chameleon (*Furcifer pardalis*) (*n* = 5)	Skin	4 SCC	4 SCC	4 KA
	MCJ	1 SCC	1 SCC	1 SCC in situ
Veiled chameleon (*Chamaeleo calyptratus*) (*n* = 2)	Skin	1 SCC	1 SCC	1 KA
	MCJ	1 SCC	1 BCC	1 Keratotic BCC
Brown anole (*Anolis sagrei*) (*n* = 2)	Skin	2 SCC	1 SCC 1 BCC	1 WD SCC 1 Keratotic BCC
Common blue-tongued skink (*Tiliqua scincoides*) (*n* = 1)	Skin	1 SCC	1 Non-neoplastic	Cystic mass lined by pseudostratified ciliated epithelium
Green iguana (*Iguana iguana*) (*n* = 1)	Skin	1 SCC	1 BCC	1 Keratotic BCC
Von Höhnel's chameleon (*Trioceros hoehnelii*) (*n* = 1)	Skin	1 SCC	1 BCC	1 Keratotic BCC
Boa constrictor (*Boa constrictor*) (*n* = 1)	Skin	1 SCC	1 BCC	1 Keratotic BCC
False map turtle (*Graptemys pseudogeographica*) (*n* = 1)	Skin	1 SCC	1 Non-neoplastic	Pyogranulomatous dermatitis

Table 1. *Cont.*

Species	Location	Initial Diagnosis	Re-Classification	
			Final Diagnosis	Histological Variant
Yellow-belied slider (*Trachemys scripta scripta*) (*n* = 1)	Skin	1 SCC	1 SCC	1 WD SCC
Hermann's tortoise (*Testudo hermanni*) (*n* = 9)	Skin	1 SCC	1 BCC	1 Infiltrating BCC
	Shell	2 SCC	1 SCC	1 WD SCC
		3 BCC	4 BCC	1 Solid BCC 2 Infiltrating BCC 1 BSCC
	MCJ	1 SCC	1 SCC	1 MD SCC
	Oral	2 BCC	2 BCC	2 Solid BCC
African spurred tortoise (*Centrochelys sulcata*) (*n* = 1)	Oral	1 SCC	1 SCC	1 MD SCC
Red-eared slider (*T. scripta elegans*) (*n* = 1)	Skin	1 BCC	1 BCC	1 Solid BCC

MCJ, mucocutaneous junction of the eyelid; WD, well-differentiated SCC; MD, moderately differentiated SCC; KA, keratoacanthoma; BSCC, basosquamous cell carcinoma.

2.2. Histopathology

Paraffin-embedded blocks were cut into 5-μm thick sections and stained with haematoxylin and eosin (HE). All sections were re-evaluated and neoplasms were further characterized. Mitotic figures were counted in 10 high-power fields (HPF) in randomly chosen areas and mean numbers were calculated. The mitotic index was graded as low (fewer or 2 mitoses per 10 HPFs), moderate (3 to 4 mitoses per 10 HPFs), or high (5 or more mitoses per 10 HPFs). The degree of nuclear atypia was graded as mild, moderate, or marked if less than 30%, between 30% and 60%, or more than 60% of the neoplastic cells had nuclear atypia, respectively.

2.3. Immunohistochemistry

IHC staining for epithelial antigen clone Ber-EP4, epithelial membrane antigen (EMA) clone E29, cyclooxygenase-2 (COX-2) clone 33, E-cadherin clone NCH-38, and cytokeratin (Pan-CK) clones AE1/AE3 were performed in all tumors. Tissues were cut into 5-μm thick sections and prepared on 3-aminopropyltriethoxysilane-coated (APES) slides. The slides were then deparaffinised and rehydrated in xylene and decreasing concentrations of alcohol in H2O (100, 96, 50, and 100% H_2O, respectively).

Antigen retrieval was performed by immersion in citrate-buffered (0.01 M, pH 6) distilled water and microwave treatment for 3.5 min at 850 W and 10 min at 450 W. Next, slides were allowed to cool down for 20 min and incubated with H2O2 (S202386-2, Agilent, Santa Clara, CA, USA) at room temperature for 5 min. Subsequently, slides were incubated with mouse primary monoclonal COX-2 (1/20, 610204, BD Biosciences, San José, CA, USA)/E-cadherin (1/100, M3612, Agilent, Santa Clara, CA, USA)/Pan-CK (1/50, M3515, Agilent, Santa Clara, CA, USA)/epithelial antigen clone Ber-EP4 (1/10, M0804, Agilent, Santa Clara, CA, USA)/EMA (1/10, M0613, Agilent) antibody at room temperature for 30 min with an antibody diluent solution with background-reducing components (S302283-2, Agilent). After incubation with a polymer-based secondary anti-mouse antibody (K400111, Agilent, Santa Clara, CA, USA) at room temperature for 30 min, visualization was performed in a 3.3-diaminobenzidine solution (K346811, Agilent) at room temperature for 5 min. Cell nuclei were counterstained with haematoxylin, rinsed in tap water, dehydrated and coverslips applied. Between all steps, the sections were washed extensively and repeatedly with phosphate-buffered saline.

Negative controls consisted of omitting the primary antibody in normal skin samples from a dog and a bearded dragon. Normal skin samples from a dog were used as positive controls for E-cadherin, COX-2 and pan-CK and mammary gland tissues were used as positive controls for epithelial antigen clone Ber-EP4 and EMA.

To evaluate the expression of epithelial antigen clone Ber-EP4, EMA, COX-2, E-cadherin, and Pan-CK, an immunoreactive score system (IRS), based on the percentage of positive cells and intensity of staining according to Fedchenko et al. [36], was used (Table 2).

Table 2. Immunoreactive score system (IRS) [36].

A (Percentage of Positive Cells)	B (Intensity of Staining)	IRS Score (A × B)
0 = no positive cells 1 ≤ 10% positive cells 2 = 10–50% positive cells 3 = 51–80% positive cells	0 = no color reaction 1 = mild reaction 2 = moderate reaction 3 = intense reaction	0–1 = negative expression 2–3 = poor expression 4–8 = moderate expression 9–12 = strong expression
4 ≥ 80% positive cells	Final IRS score (A × B): 0–12 [1]	

[1] The immunoreactive score (IRS) is calculated by multiplying the positive cells proportion score (0–4) and the staining intensity score (0–3).

2.4. Statistical Analysis

Statistical analysis was performed by using the SPSS statistical software (IBM SPSS Statistics version 27.0). A p-value < 0.05 with a 95% confidence interval was considered statistically significant.

The Mann–Whitney U test was used to compare the median ranks of continuous variables between two independent groups. The Kruskal–Wallis test was used to compare the median ranks of continuous variables between three and more independent groups, and the Friedman test was used to compare the median ranks of continuous variables between three and more related groups.

3. Results

3.1. Histological Re-Classification

Based on the retrospective histological characterization of tissues obtained from 35 individual reptile patients, 3 tissues that were initially diagnosed as SCCs proved to be non-neoplastic lesions and 8 SCCs were re-classified as BCCs. The three non-neoplastic lesions consisted of a cystic mass lined by pseudostratified ciliated epithelium, pyogranulomatous dermatitis with irregular epidermal hyperplasia, and gingival fibrous hyperplasia. In total, 17 out of the remaining 32 tissues were identified as SCCs (53.1%) and 15 were identified as BCCs (46.9%) (Table 1). A total of 13 out of 17 (76.5%) SCCs were obtained from squamates and 4 out of 17 (23.5%) from chelonians, while 7 out of 15 (46.7%) BCCs were obtained from squamates and 8 out of 15 (53.3%) from chelonians.

Sixteen of the SCCs (94.1%) had a dermal origin and one (5.9%) originated from the oral mucosa. Eleven dermal SCCs originated from the skin, one from the epidermis of the shell and four from the MCJ of the eyelid (Table 1). A total of 12 out of 15 BCCs (80%) had a dermal origin and 3 (20%) originated from the oral mucosa. Of the 12 dermal BCCs, 7 originated from the skin, 4 from the epidermis of the shell, and 1 from the MCJ of the eyelid.

3.2. Squamous Cell Carcinoma and Its Histological Variants

Generally, all SCCs could be defined as malignant neoplasms originating from the stratified squamous epithelium of the oral mucosa, MCJ of the eyelid or epidermis that presented as irregular proliferations of tumor cells with various degrees of differentiation. The neoplastic cells were often large with abundant eosinophilic cytoplasm and large nuclei, resulting in a low nuclear-to-cytoplasm ratio. In a few SCCs, areas of necrosis and cystic degeneration were noted. A variable degree of synchronous differentiation of peripheral basal-type cells to central squamous epithelial cells was observed. The most advanced stage of differentiation resulted in advanced keratinization, presenting as keratin pearls. Consequently, the presence and number of keratin pearls could be related to the degree of differentiation of the involved neoplasm. In several cases, ulceration of the neoplastic nests with infiltration of inflammatory cells from the dermis into the epidermis or subcutis was present.

The histological SCC variants included 1 SCC in situ (5.9%), 10 conventional SCCs (58.8%) and 6 KA (35.3%). Conventional SCCs could be further classified into two histological grades, representing either well-differentiated SCCs (7 cases; 41.2%) or moderately differentiated SCCs (3 cases; 17.6%). An overview of the histological variants with their mitotic index and degree of nuclear atypia are provided in Table 3.

The SCC in situ was characterized by epidermal dysplasia with enlarged and pleomorphic squamous cells that showed high mitotic activity and mild nuclear atypia, replacing the entire thickness of the epidermis without invading the basal membrane (Figure 1).

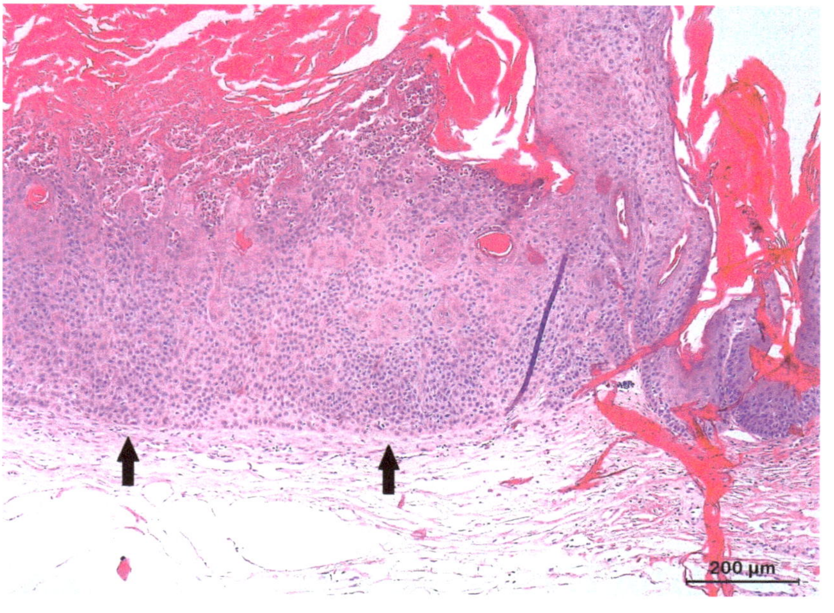

Figure 1. Histological section of a squamous cell carcinoma in situ from the skin of the eyelid of a panther chameleon (*Furcifer pardalis*). Prominent epidermal dysplasia with enlarged and pleomorphic squamous cells replacing the entire thickness of the epidermis can be noted. The basal membrane remains intact (arrows), without invasion of the dermis.

Conventional SCCs infiltrated the papillary dermis and, in some cases, the reticular dermis and the subcutis. Seven conventional SCCs were graded as well-differentiated SCCs and were characterized by tumor cells containing slightly enlarged and hyperchromatic nuclei with abundant eosinophilic cytoplasm (Figure 2). They exhibited low mitotic activity, mild to moderate nuclear atypia, and evidence of synchronous differentiation of peripheral basal-type cells to central squamous epithelial cells, which resulted in the formation of extracellular keratin pearls in most cases.

Table 3. Histological characteristics of 17 squamous cell carcinomas (SCCs) and 15 basal cell carcinomas (BCCs) histological variants obtained from squamates and chelonians. The mitotic index and degree of nuclear atypia are specified for each histological variant.

Neoplasm	Total	Mitotic Index			Degree of Nuclear Atypia		
		0–2	3–4	≥5	<30%	30–60%	>60%
SCC	17	15 (88.2%)	1 (5.9%)	1 (5.9%)	11 (64.7%)	4 (23.5%)	2 (11.8%)
• SCC in situ	1	0	0	1	1	0	0
• Conventional SCC							
∘ WD	7	6	1	0	3	4	0
∘ MD	3	3	0	0	1	0	2
• KA	6	6	0	0	6	0	0
BCC	15	7 (46.7%)	7 (46.7%)	1 (6.7%)	8 (53.3%)	3 (20%)	4 (26.7%)
• Solid	5	4	1	0	4	0	1
• Keratotic	5	2	2	1	4	1	0
• Infiltrating	4	1	3	0	0	2	2
• BSCC	1	0	1	0	0	0	1

WD, well-differentiated SCC; MD, moderately differentiated SCC; KA, keratoacanthoma; BSCC, basosquamous cell carcinoma.

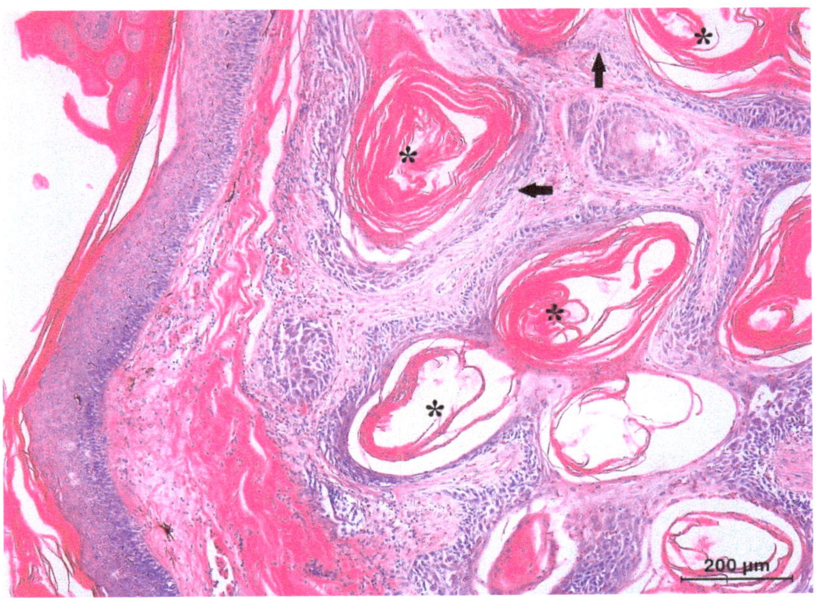

Figure 2. Histological section of a well-differentiated squamous cell carcinomas (SCCs) from the skin of the body wall in a bearded dragon (*Pogona vitticeps*) with tumor cells containing abundant eosinophilic cytoplasm. Synchronous differentiation of peripheral basal-type cells to central squamous epithelial cells (arrows) can be observed, eventually resulting in the formation of keratin pearls (asterisks).

Three conventional SCCs were graded as moderately differentiated SCCs based on a low mitotic index and marked nuclear atypia (Figure 3). Keratin pearl formation in moderately differentiated SCCs was a less prominent feature in comparison to well-differentiated SCCs.

Six SCCs could be unambiguously classified as KAs and were exclusively diagnosed in lizards. They presented as an exo–endophytic, cyst-like epidermal proliferation that creates a crateriform lesion with a central keratinous plug (Figure 4). Areas of pseudoepitheliomatous hyperplasia with minimally infiltrating well-differentiated squamous cells formed folds inside the crater and the adjacent dermis. Mild nuclear atypia and a low mitotic index were invariably present.

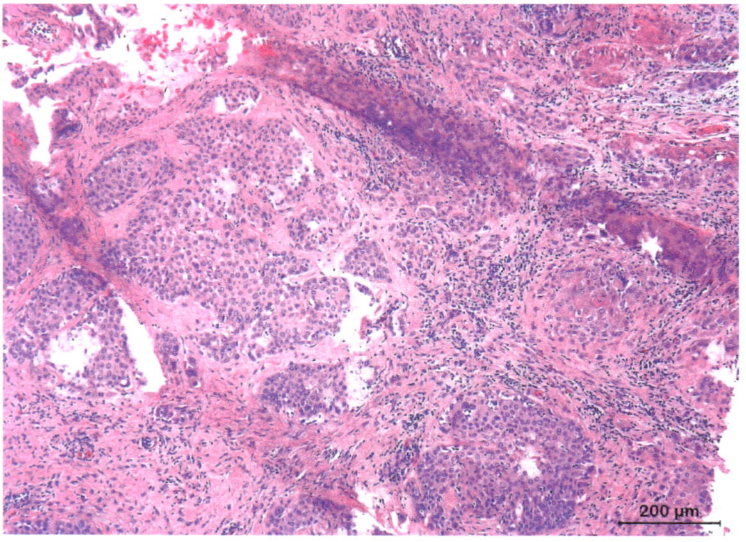

Figure 3. Histological section of a moderately differentiated squamous cell carcinoma from the oral mucosa of an African spurred tortoise (*Centrochelys sulcata*) showing tumor cells containing scant eosinophilic cytoplasm, haphazard squamous differentiation and marked nuclear atypia.

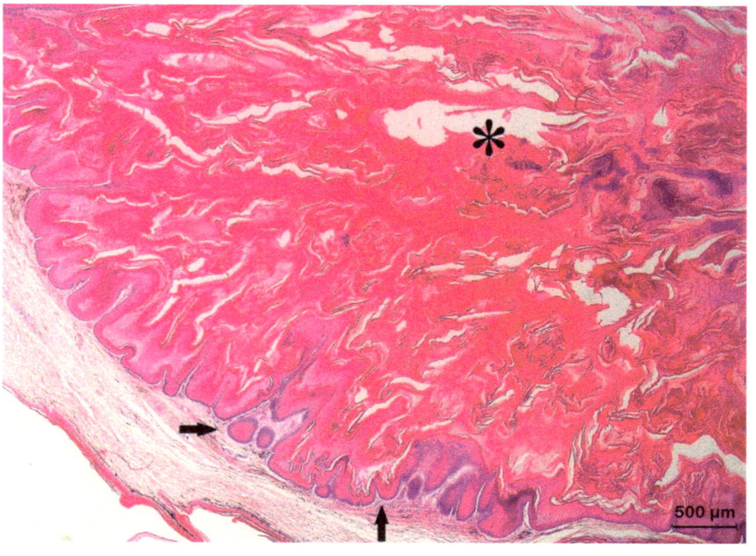

Figure 4. Histological section of a keratoacanthoma from the skin of the body wall of a bearded dragon (*Pogona vitticeps*) showing a characteristic architectural pattern consisting of an exo–endophytic, cyst-like invagination of the epidermis that creates a crateriform lesion with a central keratinous plug (asterisk) and minimally infiltrating borders (arrows).

3.3. Basal Cell Carcinoma and Its Histological Variants

BCCs were identified as malignant proliferations of the skin, the MCJ of the eyelid or the oral mucosa originating from the epidermal basal cells. All BCC histological variants typically contained islands or nests of cuboidal basaloid cells with a central, haphazard, hyperchromatic nuclear arrangement and scant amount of slightly basophilic cytoplasm, resulting in a high nuclear-to-cytoplasm ratio. Keratinization varied according to the histological variant. Peripheral palisade formation and stromal cleft formation was considered a rare and inconsistent histological feature. Sporadic areas of necrosis and cystic degeneration were noted. In several cases, ulceration with infiltration of inflammatory cells from the dermis into the epidermis or subcutis was present. Four distinct histological variants could be identified: solid BCCs (5 cases; 33.3%), keratotic BCCs (5 cases; 33.3%), infiltrating BCCs (4 cases; 26.7%), and basosquamous cell carcinoma (1 case; 6.7%). An overview of the histological variants and their mitotic index and degree of nuclear atypia are provided in Table 3.

Solid BCCs (also referred to as nodular BCCs) were characterized by cords of small polyhedral basaloid cells extending into the dermis that contained small foci of squamous metaplasia (Figure 5). Advanced keratinization with frank extracellular keratin production resulting in the formation of keratin pearls in the squamoid foci was occasionally seen. The mitotic activity, the number of atypical mitotic figures and the degree of nuclear atypia were low, except for one solid BCC that showed high nuclear atypia and moderate mitotic activity.

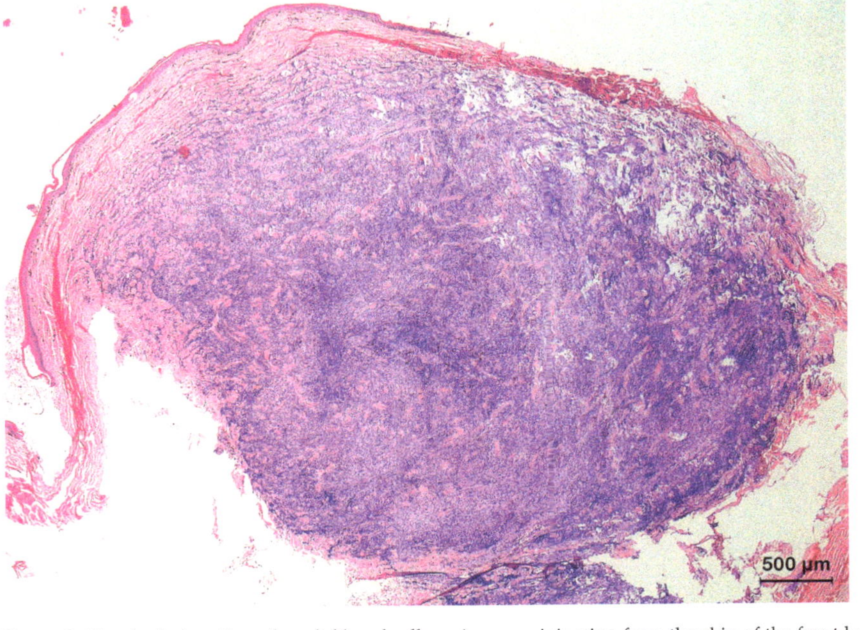

Figure 5. Histological section of a solid basal cell carcinoma originating from the skin of the front leg of a red-eared slider (*Trachemys scripta elegans*) characterized by a nodular pattern composed of cords of small polyhedral basaloid cells extending into the dermis.

Keratotic BCCs presented a similar architecture as solid BCCs, but most epithelial islands contained central or peripheral foci of abrupt squamous differentiation with central mature keratinization (Figure 6). In addition, the squamous cells possessed large, vesicular nuclei with small nucleoli and atypical mitotic figures. The neoplastic cells exhibited variable mitotic activity, ranging from low to high, with minimal nuclear atypia.

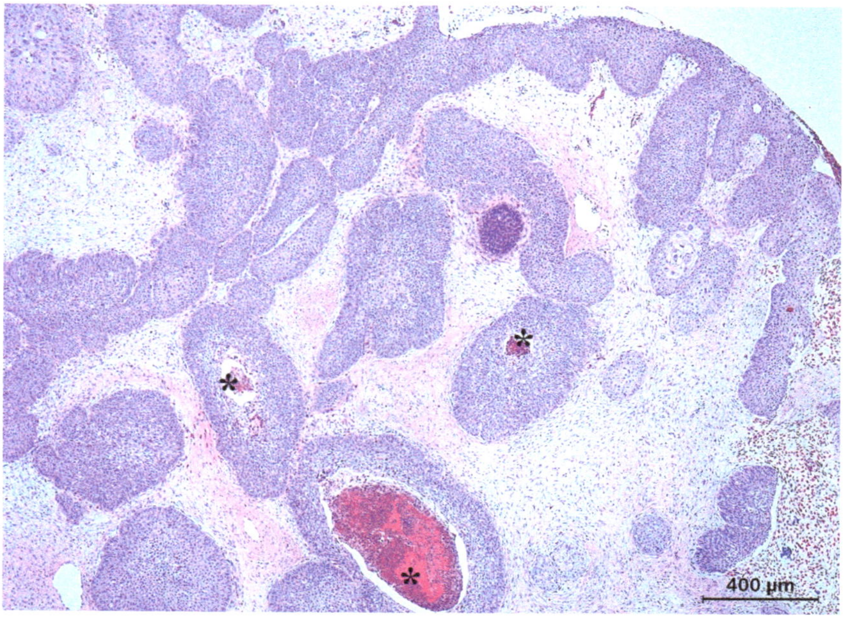

Figure 6. Histological section of a keratotic basal cell carcinoma from the skin of the body wall of a boa constrictor (*Boa constrictor*) characterized by cords of small polyhedral basaloid cells extending into the dermis and foci of abrupt squamous differentiation with central mature keratinization (asterisks).

Infiltrating BCCs were mainly characterized by irregular, narrow (<8 cells thick), and elongated cords of small, atypical basophilic basal tumor cells, without differentiation into squamous epithelium that were deeply infiltrating into the dermis and subcutis (Figure 7). The neoplastic cells showed moderate to marked nuclear atypia, and mitotic activity was consistently observed. Extensive fibroblast proliferation of the dermis was often observed in response to the infiltrating neoplastic cords.

Basosquamous cell carcinomas (BSCCs) (also referred to as metatypical BCCs) presented areas with BCC as well SCC features (Figure 8). Atypical squamous cells formed scattered islands, trabeculae, and nests that often had angular, irregular profiles. Similar to what is seen in conventional SCCs and solid BCCs, a scarce stroma with moderate cellularity surrounded the epithelial structures. Architecturally, the histological characteristics of the BSCCs resembled those of keratotic BCCs but within the squamous component, malignant histological features as observed for SCCs were prominent. Frequently, keratin pearls were present.

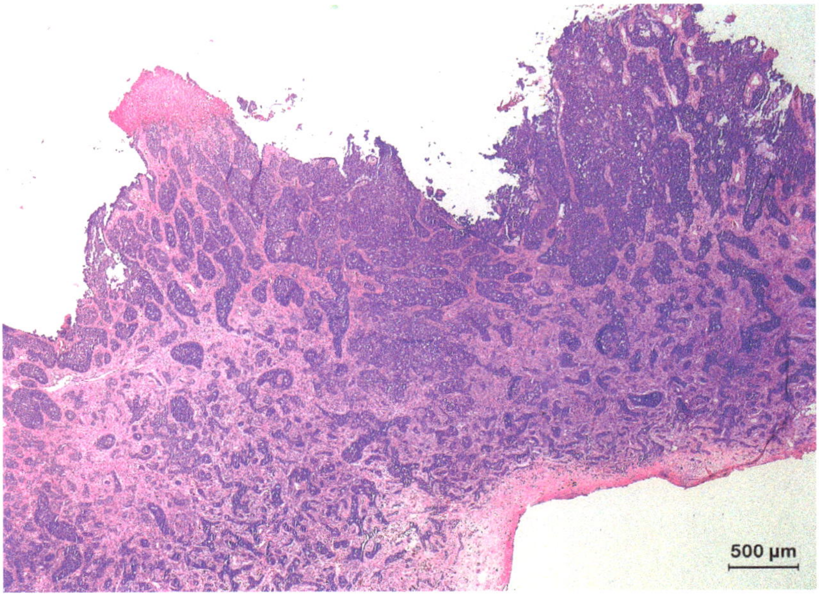

Figure 7. Histological section of an infiltrating basal cell carcinoma originating from the epidermis of the shell of a Hermann's tortoise (*Testudo hermanni*) characterized by irregular, narrow, and elongated cords of small, atypical basophilic basal tumor cells.

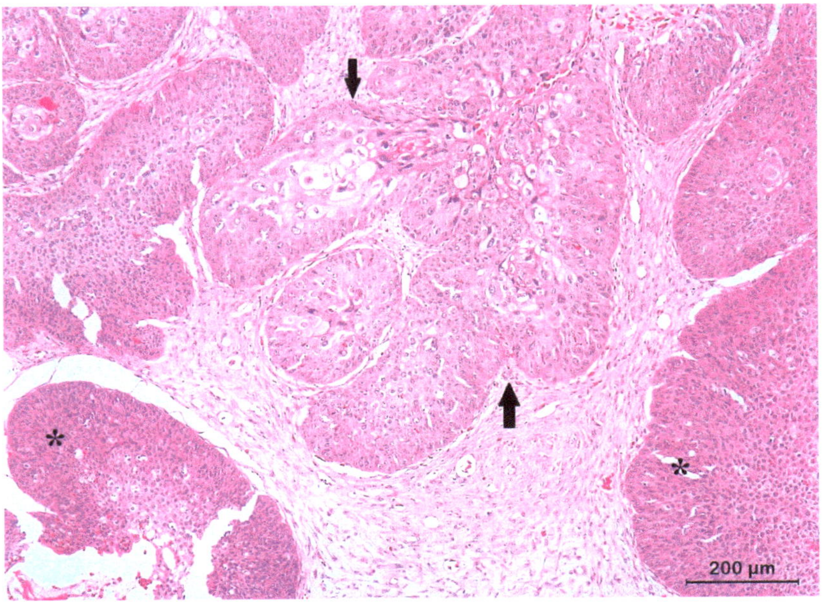

Figure 8. Histological section of a basosquamous cell carcinoma originating from the epidermis of the shell of a Hermann's tortoise (*Testudo hermanni*) presenting as an invasive front showing BCCs (asterisks) as well as SCCs (arrows) histological features.

3.4. Immunohistochemistry

While 70.6% of the SCCs exhibited strong expression levels for E-cadherin, the remaining 29.4% showed moderate expression levels (Table 4). In contrast, 80% of the BCCs showed moderate E-cadherin expression and the remaining 20% showed poor expression levels, without overrepresentation of a particular histological variant. Statistical analysis demonstrated that the observed difference in expression of E-cadherin between SCCs and BCCs was significant ($p < 0.05$), with SCCs generally displaying higher expression levels compared to BCCs (Figure 9). While E-cadherin expression levels were significantly higher in the KAs compared to the well-differentiated SCCs ($p < 0.05$), no significant differences in E-cadherin expression could be observed between the BCC histological variants.

Table 4. Expression of E-Cadherin and COX-2 according to the IRS score system (0–12) in 17 squamous cell carcinomas (SCCs) and 15 basal cell carcinomas (BCCs) from squamate and chelonian species.

Neoplasm	Total	E-Cadherin (IRS Score) (%)				COX-2 (IRS Score) (%)			
		Neg	Poor	Mod	Strong	Neg	Poor	Mod	Strong
SCC	17	0 (0%)	0 (0%)	5 (29.4%)	12 (70.6%)	0 (0%)	0 (0%)	10 (58.8%)	7 (41.2%)
• SCC in situ	1	0	0	0	1	0	0	0	1
• Conventional SCC									
○ WD	7	0	0	5	2	0	0	3	4
○ MD	3	0	0	0	3	0	0	1	2
• KA	6	0	0	0	6	0	0	6	0
BCC	15	0 (0%)	3 (20%)	12 (80%)	0 (0%)	7 (46.7%)	4 (26.7%)	4 (26.7%)	0 (0%)
• Solid	5	0	1	4	0	4	1	0	0
• Keratotic	5	0	1	4	0	0	1	4	0
• Infiltrating	4	0	1	3	0	3	1	0	0
• BSCC	1	0	0	1	0	0	1	0	0

WD, well-differentiated SCC; MD, moderately differentiated SCC; KA, keratoacanthoma; BSCC, basosquamous cell carcinoma; Neg, negative; Mod, moderate.

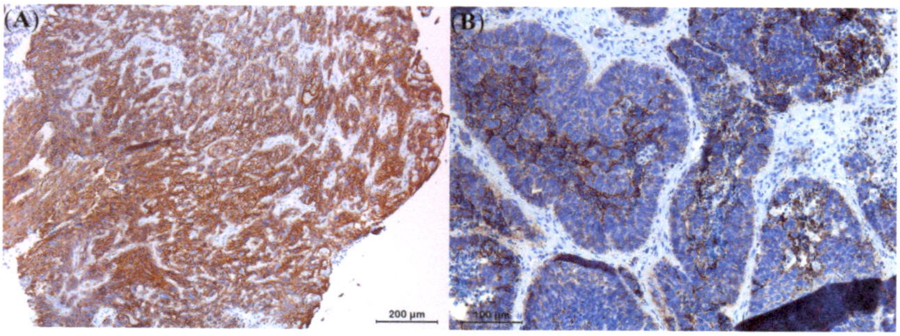

Figure 9. Strong E-cadherin expression with immunoreactivity at the level of the plasma membrane could be noted in all moderately differentiated squamous cell carcinoma cells (**A**), in contrast to moderate to poor expression levels in cells of a solid basal cell carcinoma (**B**).

The proportion of SCCs that showed moderate or strong COX-2 expression was higher (58.8% and 41.2%, respectively) than what was observed for the BCCs that almost in all cases showed poor to negative expression (26.7% and 46.7%, respectively). Only in keratotic BCCs, moderate COX-2 expression could be noted (Table 4). Based on the performed statistical analysis, a significantly higher COX-2 expression was observed in the SCCs than in the BCCs ($p < 0.05$) (Figure 10). While KA histological variants exhibited lower COX-2 expression compared to the other SCC variants ($p < 0.05$), the keratotic BCC histological variant exhibited higher expression levels than the other BCC variants ($p < 0.05$).

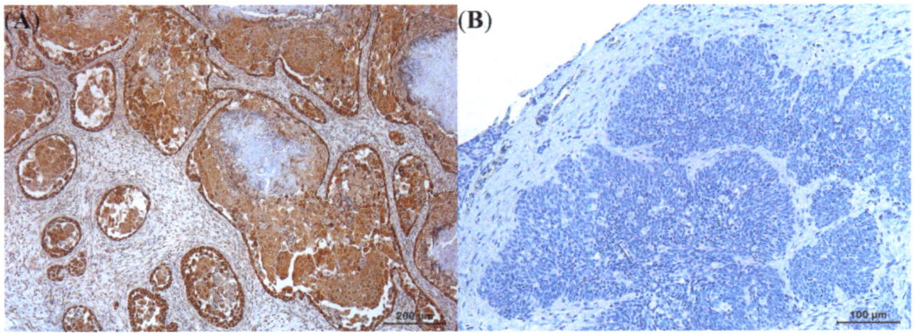

Figure 10. Diffuse and strongly positive intracytoplasmic cyclooxygenase-2 expression could be noted in a well-differentiated squamous cell carcinoma (**A**) in contrast to negative cyclooxygenase-2 expression in a solid basal cell carcinoma (**B**).

All SCCs and BCCs described in this study showed strong pan-CK reactivity, which allowed sharp delineation of the neoplastic processes from the surrounding tissue and assessment of the integrity of the basement membrane.

Immunohistochemical analysis utilizing antisera against epithelial antigen clone Ber-EP4 and EMA did not demonstrate positive staining in either the tumor cells or the control skin tissues from reptiles. Nonetheless, positive controls consisting of mammary gland tissue from dogs showed positive immunoperoxidase labelling for epithelial antigen clone Ber-EP4 and EMA.

4. Discussion

Taking into account the existing human, canine, and feline classifications, we propose a histological classification for SCCs and BCCs in squamates and chelonians based on the results of the present study. As for humans and conventional pets, such a classification may facilitate adapting the clinical and therapeutic approach according to the involved SCC or BCC histological variant. Considering the existing controversy regarding the definition and classification of certain SCC and BCC histological variants in humans as well as dogs and cats [12,20,21,37–39], the proposed classification for squamates and chelonians should be considered as a dynamic concept that needs to be subjected to regular evaluation and revision based on scientific progress and new insights.

Even in humans, accurate identification of BCC and SCC histological variants [8,40], might be challenging because of their similar basic histological in addition to their often highly comparable clinical appearance as well as their mutual predilection sites. Moreover, certain species predispositions are reported in reptiles, notably for certain SCC histological variants in bearded dragons and panther chameleons [3,8,10]. The latter may strongly bias the differential diagnostic approach and the differentiation of SCCs from BCCs as illustrated by the initial misdiagnosis of eight SCCs that were re-classified as five keratotic BCCs, two infiltrative BCCs, and one solid BCC in the present study. As normal reptile skin typically contains higher amounts of keratin due to the presence of alpha and beta keratin epidermal layers in comparison to humans and other mammals [41], especially keratotic BCCs can be easily misdiagnosed as conventional SCCs or BSCCs because of the abundant presence of keratin pearls [12]. For this reason, it was fundamental to classify keratotic BCCs as a distinct histological variant from solid BCCs in the present study. Misdiagnosis of infiltrative BCCs is presumably related to the limited amount of neoplastic basal cells in the typically narrow neoplastic cords of this histological variant [1–3]. The solid BCC that was misdiagnosed as SCC originated from the epidermis of the shell. The high concentration of rigid beta-keratin layers in the shell of most chelonian species makes its epidermis particularly hard to process for histological sectioning [42]. This can potentially lead to loss of keratin layers or the architectural pattern of the sample, which are essential for histological characterization of BCCs and SCCs. For the solid BCC in the present case, repeated sectioning of the paraffin-embedded tissue was necessary to obtain a correct histological diagnosis. In conclusion, BCCs are presumably highly underdiagnosed in reptiles because of their misdiagnosis as SCCs and the results of the present study may raise awareness to correctly identify BCCs in reptiles.

Based on the histological examination of SCCs and BCCs from squamates and chelonians, the characteristics observed for all histological variants that were identified in the present study, fully comply with the definitions applied for these histological variants in humans [20]. In addition, KA characteristics fully complied with those described by Solanes et al. [8]. Other commonly observed SCC histological variants in humans, such as acantholytic SCCs, desmoplastic SCCs, spindle cell SCCs, and verrucous SCCs, could not be demonstrated in the present study. Nevertheless, it cannot be excluded that these histological variants may occur in reptiles.

Although poorly differentiated SCCs were not demonstrated in the present study, this third histological grade of conventional SCCs may be encountered in reptiles. Histological characteristics of poorly differentiated SCCs in humans include enlarged, pleomorphic nuclei with asynchronous differentiation from basal to central squamous cells, marked nuclear atypia, and frequent mitoses [20,22]. Keratin pearls are a highly unusual finding in poorly differentiated SCCs.

At present, infiltrating BCCs and sclerosing BCCs are classified as distinct histological variants in the WHO classification of skin tumors [20]. In the present study, we did not classify sclerosing BCCs as a distinct histological variant as the clinical and microscopic features of both variants in human medicine are highly similar. Additionally, the limited number of infiltrating BCCs did not allow for the identification of distinguishing features. Identification and characterization of additional cases from affected reptiles would be necessary to allow their future classification as distinct histological variants.

Superficial BCCs are a highly prevalent histological variant in humans that presents as a well-defined, red, scaly patch or plaque with a central clearing and thin rolled edges [21]. This histological variant of BCCs was not identified in this study and has not been previously reported in other animal species. It should be considered, however, that the diagnosis of superficial BCCs may be easily missed in reptiles and other animals due to its discrete gross pathological features and low tendency to invade or ulcerate, in comparison to other BCC histological variants that mostly have a prominent invasive nodular appearance. Consequently, these lesions are presumably often not routinely submitted for histological evaluation or misdiagnosed as SCCs in situ or non-neoplastic lesions [43].

In the present study, a certain discriminative value of IHC staining using E-cadherin and COX-2 towards differentiating SCCs from BCCs, including KA from well-differentiated SCCs and keratotic BCCs from other BCC histological variants, was demonstrated. The use of epithelial antigen clone Ber-EP4 and EMA IHC staining does not seem to have a value towards distinguishing reptilian SCCs and BCCs as immunoreactivity was absent, in contrast to their highly discriminative value in human SCCs and BCCs. The latter can presumably be attributed to lack of cross-reactivity with mammal antibodies or different protein expression patterns between reptilian and mammal tissues [24]. With regard to EMA, it has been documented that the EMA gene emerged during evolution from poikilothermic reptiles to homoeothermic mammals and for this reason, the epitopes for the EMA antibodies are presumably lacking in reptiles [44]. It might have been interesting to explore the value of CD10 and Bcl-2 markers in reptiles, as they are highly reliable markers to distinguish SCCs from BCCs in humans [27,45]. However, a similar lack of immunoreactivity, as observed for epithelial antigen clone Ber-EP4 and EMA markers, is to be expected [46–48].

In humans and dogs, it has been demonstrated that upregulated COX-2 in SCCs enhances prostaglandin synthesis, which increases cell proliferation, promotes angiogenesis, inhibits immunosurveillance, and enhances invasiveness [38,49,50]. Based on the results of this study, it can be presumed that COX-2 overexpression in reptile BCCs and SCCs, as previously demonstrated in KAs in lizards, can also be used as a marker for invasiveness and it might be interesting to explore the development of therapeutic strategies relying on the effect of specific COX-2 enzyme inhibitors for the adjuvant treatment of SCCs and keratotic BCCs in reptiles [8,15]. Intercellular adhesion is mediated by E-cadherin, which is typically downregulated in SCCs with high invasive and metastatic potential [51,52]. Despite evidence in humans and dogs indicating that BCCs are generally less invasive than SCCs [12,51,52], BCC histological variants with low E-cadherin expression in squamates and chelonians, such as solid, keratotic, and infiltrative BCCs, may exhibit unusually rapid and infiltrative growth, as reported in a terminal BCC in a Hermann's tortoise [16]. Further studies, correlating the IHC staining patterns with the clinical course and therapeutic results are needed to further elucidate the value of these IHC markers.

5. Conclusions

Based on the established human, canine, and feline classifications for skin neoplasms, a classification for SCCs and BCCs, and their histological variants, in squamates and chelonians is proposed in the present study. This proposed classification should be subjected to continuous evaluation and revision as new scientific insights emerge. Accurate histological classification of SCC and BCC histological variants is considered crucial to predict their biological behavior and guide treatment decisions, particularly for histological variants with high invasiveness and metastatic potential. The results of the present study allow the correct identification of SCCs and BCCs in squamates and chelonians and might be highly valuable in diagnosing and differentiating challenging histological variants, such as keratotic BCCs and BCSCCs. IHC staining with E-cadherin and COX-2 markers aids in the differentiation of SCCs from BCCs and their histological variants in squamates and chelonians. COX-2 overexpression in reptile SCCs implies that specific COX-2 enzyme inhibitors could be included as a part of the adjuvant therapy of these neoplastic disorders. Further research should warrant the association of the distinct histological variants of SCCs and BCCs of squamates and chelonians to their respective biological behavior.

Author Contributions: Conceptualization, F.S.V. and T.H.; data curation, F.S.V. and T.H.; project administration, F.S.V., K.C. and T.H.; supervision, K.C.; validation, F.S.V. and T.H.; writing—original draft, F.S.V. and T.H.; writing—review and editing, F.S.V., K.C. and T.H. All authors have read and agreed to the published version of the manuscript.

Funding: This research received no external funding.

Institutional Review Board Statement: Not applicable.

Data Availability Statement: The data presented in this study are available on request from the corresponding author.

Acknowledgments: We thank Glenn Strypsteen for his aid in the statistical analyses and the Department of Pathobiology, Pharmacology and Zoological Medicine, especially Sarah Loomans, Delphine Ameye and Joachim Christiaens, for the aid and technical assistance required to process histopathological samples.

Conflicts of Interest: The authors declare no conflict of interest.

References

1. Garner, M.M.; Hernandez-Divers, S.M.; Raymond, J.T. Reptile neoplasia: A retrospective study of case submissions to a specialty diagnostic service. *Vet. Clin. N. Am. Exot. Anim. Pract.* **2004**, *7*, 653–671. [CrossRef] [PubMed]
2. Hernandez-Divers, S.M.; Garner, M.M. Neoplasia of reptiles with an emphasis on lizards. *Vet. Clin. N. Am. Exot. Anim. Pract.* **2003**, *6*, 251–273. [CrossRef] [PubMed]
3. Hannon, D.E.; Garner, M.M.; Reavill, D.R. Squamous cell carcinomas in inland bearded dragons (*Pogona vitticeps*). *J. Herpetol. Med. Surg.* **2012**, *21*, 101–108. [CrossRef]
4. Sykes, J.M.; Trupkiewicz, J.G. Reptile neoplasia at the Philadelphia zoological garden, 1901–2002. *J. Zoo Wildl. Med.* **2006**, *37*, 11–19. [CrossRef]
5. Kubiak, M.; Denk, D.; Stidworthy, M.F. Retrospective review of neoplasms of captive lizards in the United Kingdom. *Vet. Rec.* **2020**, *186*, 28–37. [CrossRef]

6. Christman, J.; Devau, M.; Wilson-Robles, H.; Hoppes, S.; Rech, R.; Russell, K.E.; Heatley, J.J. Oncology of reptiles: Diseases, diagnosis, and treatment. *Vet. Clin. N. Am. Exot. Anim. Pract.* **2017**, *20*, 87–110. [CrossRef]
7. Heckers, K.O.; Aupperle, H.; Schmidt, V.; Pees, M. Melanophoromas and iridophoromas in reptiles. *J. Comp. Pathol.* **2012**, *146*, 258–268. [CrossRef]
8. Solanes, F.; Chiers, K.; Kik, M.J.L.; Hellebuyck, T. Gross, histologic and immunohistochemical characteristics of keratoacanthomas in lizards. *Animals* **2023**, *13*, 398–413. [CrossRef]
9. Monahan, C.F.; Garner, M.M.; Kiupel, M. Chromatophoromas in Reptiles. *Vet. Sci.* **2022**, *9*, 115–137. [CrossRef]
10. Meyer, J.; Kolodziejek, J.; Häbich, A.C.; Dinhopl, N.; Richter, B. Multicentric Squamous Cell Tumors in Panther Chameleons (*Furcifer pardalis*). *J. Exot. Pet. Med.* **2019**, *29*, 166–172. [CrossRef]
11. Gordon, R. Skin cancer: An overview of epidemiology and risk factors. *Semin. Oncol. Nurs.* **2013**, *29*, 160–169. [CrossRef] [PubMed]
12. Gross, T.L.; Ihrke, P.J.; Walder, E.J.; Affolter, V.K. Neoplasms and other tumors. In *Skin Diseases of the Dog and Cat: Clinical and Histopathologic Diagnosis*, 2nd ed.; Gross, T.L., Ihrke, P.J., Walder, E.J., Affolter, V.K., Eds.; Blackwell Science: Oxford, UK, 2005; Volume 1, pp. 581–598.
13. Ho, N.T.; Smith, K.C.; Dobromylskyj, M.J. Retrospective study of more than 9000 feline cutaneous tumours in the UK: 2006–2013. *J. Feline Med. Surg.* **2018**, *20*, 128–134. [CrossRef] [PubMed]
14. Hellebuyck, T.; Ducatelle, R.; Bosseler, L.; Van Caelenberg, A.; Versnaeyen, H.; Chiers, K.; Martel, A. Basal cell carcinoma in two Hermann's tortoises (*Testudo hermanni*). *J. Vet. Diagn. Investig.* **2016**, *28*, 750–754. [CrossRef] [PubMed]
15. Paolino, G.; Donati, M.; Didona, D.; Mercuri, S.; Cantisani, C. Histology of non-melanoma skin cancers: An update. *Biomedicines* **2017**, *5*, 71–83. [CrossRef] [PubMed]
16. Burton, K.A.; Ashack, K.A.; Khachemoune, A. Cutaneous squamous cell carcinoma: A review of high-risk and metastatic disease. *Am. J. Clin. Dermatol.* **2016**, *17*, 491–508. [CrossRef]
17. Zargaran, M.; Baghaei, F. A clinical, histopathological and immunohistochemical approach to the bewildering diagnosis of keratoacanthoma. *J. Dent.* **2014**, *15*, 91–97.
18. Murphy, G.; Beer, T.; Cerio, R.; Kao, G.; Nagore, E.; Pulitzer, M. Keratinocytic/epidermal tumours. In *World Health Organization (WHO) Classification of Skin Tumours*, 4th ed.; Elder, D., Massi, D., Scolyer, R., Willemze, R., Eds.; International Agency for Research on Cancer (IARC): Argonay, France, 2018; Volume 11, pp. 36–38.
19. Vantuchová, Y.; Čuřík, R. Histological types of basal cell carcinoma. *Scr. Med. Fac. Med. Univ. Brun Masaryk* **2006**, *79*, 261–270.
20. Stratigos, A.; Garbe, C.; Lebbe, C.; Malvehy, J.; Del Marmol, V.; Pehamberger, H.; Peris, K.; Becker, J.C.; Zalaudek, I.; Saiag, P.; et al. Diagnosis and treatment of invasive squamous cell carcinoma of the skin: European consensus-based interdisciplinary guideline. *Eur. J. Cancer* **2015**, *51*, 1989–2007. [CrossRef]
21. Zehnder, A.M.; Swift, L.A.; Sundaram, A.; Speer, B.L.; Olsen, G.P.; Hawkins, M.G.; Paul-Murphy, J. Clinical features, treatment, and outcomes of cutaneous and oral squamous cell carcinoma in avian species. *J. Am. Vet. Med. Assoc.* **2018**, *252*, 309–315. [CrossRef]
22. Orós, J.; López-Yánez, M.; Rodríguez, F.; Calabuig, P.; Castro, P.L. Immunohistochemical staining patterns of alpha-keratins in normal tissues from two reptile species: Implications for characterization of squamous cell carcinomas. *BMC Vet. Res.* **2018**, *14*, 219. [CrossRef]
23. Erdogan Bamac, O.; Seckin Arun, S. Histological and immunohistochemical evaluation of epithelial and mesenchymal tumors of psittacines. *Med. Weter.* **2020**, *76*, 165–169.
24. Jones, A.L.; Suárez-Bonnet, A.; Mitchell, J.A.; Ramirez, G.A.; Stidworthy, M.F.; Priestnall, S.L. Avian papilloma and squamous cell carcinoma: A histopathological, immunohistochemical and virological study. *J. Comp. Pathol.* **2020**, *175*, 13–23. [CrossRef]
25. Beer, T.W.; Shepherd, P.; Theaker, J.M. Ber EP4 and epithelial membrane antigen aid distinction of basal cell, squamous cell and basosquamous carcinomas of the skin. *Histopathology* **2000**, *37*, 218–223. [CrossRef] [PubMed]
26. Ramezani, M.; Zavattaro, E.; Sadeghi, M. Immunohistochemistry expression of EMA, CD10, CEA, and Bcl-2 in distinguishing cutaneous basal cell from squamous cell carcinoma: A systematic review. *Gulhane Med. J.* **2020**, *62*, 63–71. [CrossRef]
27. Alhumaidi, A. Practical immunohistochemistry of epithelial skin tumor. *Indian J. Dermatol. Venereol. Leprol.* **2012**, *78*, 698–708. [CrossRef] [PubMed]
28. Compton, L.A.; Murphy, G.F.; Lian, C.G. Diagnostic immunohistochemistry in cutaneous neoplasia: An update. *Dermatopathology* **2015**, *2*, 15–42. [CrossRef]
29. Nagamine, E.; Hirayama, K.; Matsuda, K.; Okamoto, M.; Ohmachi, T.; Uchida, K.; Kadosawa, T.; Taniyama, H. Invasive front grading and epithelial-mesenchymal transition in canine oral and cutaneous squamous cell carcinomas. *Vet. Pathol.* **2017**, *54*, 783–791. [CrossRef]
30. Thaiwong, T.; Sledge, D.G.; Collins-Webb, A.; Kiupel, M. Immunohistochemical characterization of canine oral papillary squamous cell carcinoma. *Vet. Pathol.* **2018**, *55*, 224–232. [CrossRef]
31. Sanz Ressel, B.L.; Massone, A.R.; Barbeito, C.G. Immunohistochemical expression of selected phosphoproteins of the mTOR signalling pathway in canine cutaneous squamous cell carcinoma. *Vet. J.* **2019**, *245*, 41–48. [CrossRef]
32. Gál, J.; Mándoki, M. Adenoma of the cloacal scent gland in a California Kingsnake (*Lampropeltis getulus californiae*). *Acta Vet. Hung.* **2012**, *60*, 459–463. [CrossRef]

33. Petterino, C.; Bedin, M.; Podestà, G.; Ratto, A. Undifferentiated tumor in the ovary of a corn snake (*Elaphe guttata guttata*). *Vet. Clin. Pathol.* **2006**, *35*, 95–100. [CrossRef] [PubMed]
34. Ritter, J.M.; Garner, M.M.; Chilton, J.A.; Jacobson, E.R.; Kiupel, M. Gastric neuroendocrine carcinomas in bearded dragons (*Pogona vitticeps*). *Vet. Pathol.* **2009**, *46*, 1109–1116. [CrossRef] [PubMed]
35. Vargo, N. Basal cell and squamous cell carcinoma. *Semin. Oncol. Nurs.* **2003**, *19*, 12–21. [CrossRef] [PubMed]
36. Fedchenko, N.; Reifenrath, J. Different approaches for interpretation and reporting of immunohistochemistry analysis results in the bone tissue—A review. *Diagn. Pathol.* **2014**, *9*, 221–233. [CrossRef] [PubMed]
37. Goldschmidt, M.H.; Goldschmidt, K.H. Epithelial and melanocytic tumors of the skin. In *Tumors in Domestic Animals*, 5th ed.; Meuten, D.J., Ed.; John Wiley & Sons: Oxford, UK, 2017; Volume 1, pp. 88–148.
38. Misago, N.; Inoue, T.; Koba, S.; Narisawa, Y. Keratoacanthoma and other types of squamous cell carcinoma with crateriform architecture: Classification and identification. *J. Dermatol.* **2013**, *40*, 443–452. [CrossRef]
39. Yanofsky, V.R.; Mercer, S.E.; Phelps, R.G. Histopathological variants of cutaneous squamous cell carcinoma: A review. *J. Skin Cancer* **2011**, *2011*, 210813. [CrossRef]
40. Putti, T.C.; Teh, M.; Lee, Y.S. Biological behavior of keratoacanthoma and squamous cell carcinoma: Telomerase activity and COX-2 as potential markers. *Mod. Pathol.* **2004**, *17*, 468–475. [CrossRef]
41. Alibardi, L.; Dalla Valle, L.; Toffolo, V.; Toni, M. Scale keratin in lizard epidermis reveals amino acid regions homologous with avian and mammalian epidermal proteins. *Anat. Rec.* **2006**, *288*, 734–752. [CrossRef]
42. Alibardi, L.; Toni, M. Immunolocalization and characterization of beta-keratins in growing epidermis of chelonians. *Tissue Cell* **2006**, *38*, 53–63. [CrossRef]
43. Marzuka, A.G.; Book, S.E. Basal cell carcinoma: Pathogenesis, epidemiology, clinical features, diagnosis, histopathology, and management. *Yale J. Biol. Med.* **2015**, *88*, 167–179.
44. Rajabi, H.; Kufe, D. MUC1-C oncoprotein integrates a program of EMT, epigenetic reprogramming and immune evasion in human carcinomas. *Biochim. Biophys. Acta* **2017**, *1868*, 117–122. [CrossRef] [PubMed]
45. Ramezani, M.; Mohamadzaheri, E.; Khazaei, S.; Najafi, F.; Vaisi-Raygani, A.; Rahbar, M.; Sadeghi, M. Comparison of EMA, CEA, CD10 and Bcl-2 biomarkers by immunohistochemistry in squamous cell carcinoma and basal cell carcinoma of the skin. *Asian Pac. J. Cancer Prev.* **2016**, *17*, 1379–1383. [CrossRef] [PubMed]
46. Ordi, J.; Romagosa, C.; Tavassoli, F.A.; Nogales, F.; Palacin, A.; Condom, E.; Torné, A.; Cardesa, A. CD10 expression in epithelial tissues and tumors of the gynecologic tract: A useful marker in the diagnosis of mesonephric, trophoblastic, and clear cell tumors. *Am. J. Surg. Pathol.* **2003**, *27*, 178–186. [CrossRef] [PubMed]
47. Peat, T.J.; Edmondson, E.F.; Miller, M.A.; DuSold, D.M.; Ramos-Vara, J.A. Pax8, Napsin A, and CD10 as immunohistochemical markers of canine renal cell carcinoma. *Vet. Pathol.* **2017**, *54*, 588–594. [CrossRef]
48. Aouacheria, A.; Navratil, V.; Combet, C. Database and bioinformatic analysis of Bcl-2 family proteins and BH3-Only proteins. *Methods Mol. Biol.* **2019**, *1877*, 23–43.
49. Tsujii, M.; Kawano, S.; Dubois, R.N. Cyclooxygenase-2 expression in human colon cancer cells increases metastatic potential. *Proc. Natl. Acad. Sci. USA* **1997**, *94*, 3336–3340. [CrossRef]
50. Karagece Yalçin, Ü.; Seçkin, S. The expression of p53 and Cox-2 in basal cell carcinoma, squamous cell carcinoma and actinic keratosis cases. *Turkish J. Pathol.* **2012**, *28*, 119–127. [CrossRef]
51. Lyakhovitsky, A.; Barzilai, A.; Fogel, M.; Trau, H.; Huszar, M. Expression of E-cadherin and beta-catenin in cutaneous squamous cell carcinoma and its precursors. *Am. J. Dermatopathol.* **2004**, *26*, 372–378. [CrossRef]
52. Papadavid, E.; Pignatelli, M.; Zakynthinos, S.; Krausz, T.; Chu, A.C. The potential role of abnormal E-cadherin and a-, b- and g -catenin immunoreactivity in the determination of the biological behaviour of keratoacanthoma. *Br. J. Dermatol.* **2001**, *145*, 582–589. [CrossRef]

animals

MDPI

Article

Gross, Histologic and Immunohistochemical Characteristics of Keratoacanthomas in Lizards

Ferran Solanes [1,*], Koen Chiers [1], Marja J. L. Kik [2] and Tom Hellebuyck [1]

[1] Department of Pathobiology, Pharmacology and Zoological Medicine, Faculty of Veterinary Medicine, Ghent University, Salisburylaan 133, B-9820 Merelbeke, Belgium
[2] Department of Biomedical Health Sciences, Pathology Division, Pathology Exotic Animals and Wildlife, Faculty of Veterinary Medicine, Utrecht University, Yalelaan 1, 3584 CL Utrecht, The Netherlands
* Correspondence: ferransolanesvilanova@hotmail.com; Tel.: +32-9-264-74 -42; Fax: +32-9-264-7490

Simple Summary: Tumors of the skin are one of the most commonly observed neoplasms in captive lizards. The current study characterizes keratoacanthoma, a previously undescribed skin tumor, in five male lizards (one bearded dragon, one veiled chameleon, and three panther chameleons) with an average to high age. In all lizards, keratoacanthomas presented as cystic nodules with a central keratin pearl that was predominantly located at the body wall. In all chameleons, a multicentric distribution was observed. Following surgical removal of the keratoacanthomas in all lizards, a follow-up period of one to two years was established. While the skin neoplasia reappeared in the bearded dragon and the veiled chameleon, no recurrence was seen in the panther chameleons. Keratoacanthoma constitutes a rather benign histologic variant of squamous cell carcinoma, representing a non-invasive but rapidly growing skin neoplasia that may be associated with the inappropriate use of ultraviolet lighting in the captive environment. In addition, panther chameleons may show a species predisposition as well as a tendency to develop multicentric keratoacanthomas. The present study delivers pertinent results for the diagnosis, prevention, and treatment of keratoacanthomas in lizards.

Abstract: The present study describes the clinical behavior as well as the histopathologic and immunohistochemical characteristics of keratoacanthomas (Kas) in three different saurian species. While Kas presented as two dermal lesions in a bearded dragon (*Pogona vitticeps*), multicentric Kas were observed in three panther chameleons (*Furcifer pardalis*) and a veiled chameleon (*Chamaeleo calyptratus*). Macroscopically, Kas presented as dome-shaped skin tumors with a centralized keratinous pearl and a diameter ranging from 0.1–1.5 cm. In all lizards, Kas were predominantly located at the dorsolateral body wall, and KA of the eyelid was additionally observed in three out of four chameleons. Histologically, KAs presented as relatively well-defined, circumscribed epidermal proliferations that consisted of a crateriform lesion containing a central keratinous pearl with minimally infiltrating borders. In all KAs, a consistent immunohistochemical pattern was observed, with the expression of cyclooxygenase-2, E-cadherin, and pan-cytokeratin. A follow-up period of one to two years was established in all lizards. While no recurrence was observed in the panther chameleons, recurrence of a single keratoacanthoma was observed in the bearded dragon after one year, and in the veiled chameleon, multicentric keratoacanthomas reappeared during a follow-up period of two years. We describe KA as a previously unrecognized neoplastic entity in lizards that constitutes a low-grade, non-invasive but rapidly growing skin tumor that may show a multicentric appearance, especially in chameleons. As previously postulated for dermal squamous cell carcinomas (SCC), artificial ultraviolet lighting may play an important role in the oncogenesis of KAs in lizards. Although dermal SCCs in lizards show similar predilection sites and gross pathologic features, our results suggest that KA should be considered as a histologic variant of SCC that represents a rather benign squamous proliferation in comparison to conventional SCCs. Early diagnosis of KA and reliable discrimination from SCCs are essential for the prognosis of this neoplastic entity in lizards.

Citation: Solanes, F.; Chiers, K.; Kik, M.J.L.; Hellebuyck, T. Gross, Histologic and Immunohistochemical Characteristics of Keratoacanthomas in Lizards. *Animals* **2023**, *13*, 398. https://doi.org/10.3390/ani13030398

Academic Editor: Volker Schmidt

Received: 4 January 2023
Revised: 18 January 2023
Accepted: 23 January 2023
Published: 24 January 2023

Keywords: cutaneous; immunohistochemistry; keratoacanthoma; lizards; neoplasia; squamous cell carcinoma

1. Introduction

Although once considered to be uncommon, neoplasia in reptiles is routinely encountered in veterinary practice, with the hematopoietic, hepatobiliary, and integumentary systems being most frequently affected [1,2]. Chromatophoromas and squamous cell carcinomas (SCCs) represent the predominant skin neoplasia in captive squamates [3,4], with SCC showing a presumptive species predisposition in bearded dragons (*Pogona vitticeps*) and panther chameleons (*Furcifer pardalis*) [2,5–7]. Keratoacanthoma (KA) has been described in humans, dogs, and birds, especially in broiler chickens, as a well-differentiated histologic variant of SCC [8–11]. In dogs, KA is either referred to as infundibular keratinizing acanthoma (IKA) or subungual KA depending on its localization [10–12]. Although correct histologic characterization allows differentiation of KA from SCC, some controversy remains concerning the correct classification of KAs, as some consider it to be a precancerous stage of dermal SCC or a pseudo-cancerous lesion [13–16]. The present study describes the clinical, histologic, and immunohistochemical (IHC) characteristics of KAs in lizards.

2. Materials and Methods

2.1. Animals

Five unrelated lizards that were part of captive collections were presented at a veterinary teaching hospital between 2020 and 2022 because of showing nodular skin lesions. The lizards included one bearded dragon, one veiled chameleon (*Chamaeleo calyptratus*), and three panther chameleons (Table 1). In all cases, tissue samples were collected from the skin nodules following in toto excision after intravenous (IV, jugular vein) induction of anesthesia with 10 mg/kg alfaxalone (Alfaxan® Multidose, 10 mg/mL, Jurox Limited, Crawley, UK). Anesthesia was maintained with 1.5–2.0% isoflurane (Isoflo®, Abbott Logistics B.V., Breda, The Netherlands) in 1 L medical oxygen with intermittent positive-pressure ventilation. All dermal nodules were surgically removed with resection margins of, on average, 1 mm in small lesions and 3 mm in the larger lesions. Routine closure of the skin was performed using a simple everting pattern with a 5-0 absorbable suture (Monocryl®, Ethicon, Raritan, NJ, USA). All samples were fixed in 10% neutral buffered formalin for 24–36 hours for histopathological evaluation and IHC staining.

2.2. Histopathology

Following dehydration and embedding of tissues into paraffin blocks, 5-µm-thick sections were cut and stained with hematoxylin and eosin (HE). All histological sections were confirmed as neoplastic and further characterized histologically. Mitotic figures were counted in 10 high-power fields (HPF) in randomly selected areas, and the mean numbers were calculated. The degree of nuclear atypia was categorized as mild, moderate, or marked if less than 30%, between 30–60%, or more than 60% of the neoplastic cells showed nuclear atypia, respectively.

2.3. Immunohistochemistry

IHC staining for cyclooxygenase-2 (COX-2), E-cadherin, and pan-cytokeratin (Pan-CK) in all eyelid lesions and one body wall lesion per lizard was performed. Paraffin-embedded dermal tissue blocks were cut into 5 µm sections and mounted on 3-aminopropyltriethoxysilane-coated slides. Next, slides were deparaffinized and rehydrated in xylene and decreasing concentrations of alcohol in H_2O (100, 96, 50, and 100% H_2O, respectively).

Antigen retrieval was performed by immersion in citrate-buffered (0.01 M, pH 6) distilled water and microwaving for 3.5 min at 850 W and 10 min at 450 W. Next, slides were allowed to cool down for 20min and incubated with H_2O_2 (S202386-2; Agilent,

Santa Clara, CA, USA) at room temperature for 5 min. Subsequently, incubation with the primary monoclonal mouse COX-2 (1/20, 610204, BD Biosciences, Franklin Lakes, NJ, USA)/E-cadherin (1/100, M3612, Agilent)/pan-CK (1/50, M3515, Agilent) antibodies (1:200; ab7778; Abcam, Cambridge, UK) was performed at room temperature for 30 min with background-reducing components (S302283-2; Agilent). Followed by incubation with a polymer-based anti-mouse secondary antibody (K400111, Agilent) at room temperature for 30 min, visualization was performed in a 3.3-diaminobenzidine solution (K346811, Agilent) at room temperature for 5 min. The cell nuclei were counterstained with hematoxylin, rinsed in tap water and dehydrated and coverslips were applied. In between all steps, the sections were washed extensively and repeatedly with phosphate-buffered saline.

Table 1. Signalment of five lizards that were presented with nodular skin lesions. The number and distribution of skin lesions, the interval between detection of the skin lesions and initial presentation, the follow-up period, and recurrence are specified for each case, as well as the type of ultraviolet B (UV-B) source that was provided and the average basking distance to the UV-B lighting source.

Species	**Panther Chameleon 1 (*Furcifer pardalis*)**	**Panther Chameleon 2 (*Furcifer pardalis*)**	**Panther Chameleon 3 (*Furcifer pardalis*)**	**Veiled Chameleon (*Chamaeleo calyptratus*)**	**Bearded Dragon (*Pogona vitticeps*)**
Age	5 years	4 years	4 years	3 years	6 years
N° of nodular skin lesions at initial presentation	18	15	3	4	2
Location of the nodular skin lesions	Eyelid, dorsolateral body wall	Dorsolateral body wall	Eyelid, dorsolateral body wall	Eyelid, head, dorsolateral body wall	Dorsolateral body wall
Time between first detection and initial presentation	2 years	6 months	3 months	2 months	4 months
Follow-up period	2 years	2 years	1 year	2 years	1 year
Recurrence	No	No	No	16	1
UV-B source	ExoTerra Solar Glo 80 W	ZooMed Powersun 100 W	ZooMed Powersun 100 W	Arcadia D3 Forest	JBL UV-B Spot Plus 80 W
Average basking distance	16 cm	15 cm	18 cm	20 cm	26 cm

Negative controls consisted of omitting the primary antibody in normal skin samples from a dog and a bearded dragon.

To evaluate the expression of COX-2, E-cadherin, and cytokeratin, an immunoreactive score system (IRS), based on the percentage of positive cells and intensity of staining according to Fedchenko et al., was used (Table 2) [17].

Table 2. Immunoreactive score system (IRS) [17].

A (Percentage of Positive Cells)	**B (Intensity of Staining)**	**IRS Score (Multiplication of A and B)**
0 = no positive cells	0 = no color reaction	0–1 = negative
1 = <10% positive cells	1 = mild reaction	2–3 = mild
2 = 10–50% positive cells	2 = moderate reaction	4–8 = moderate
3 = 51–80% positive cells	3 = intense reaction	9–12 = strongly positive
4 = >80% positive cells	Final IRS score (A × B): 0–12 *	

* The immunoreactive score (IRS) is calculated by multiplying the positive cells' proportion score (0–4) and the staining intensity score (0–3).

3. Results

3.1. Clinical History and Gross Pathology

All examined lizards were males with an age ranging from 3–6 years. In general, the captive management of all cases was deemed adequate. In all lizards, however, the minimum safe distance to the ultraviolet (UV-B) source was considered inappropriate, taking into account manufacturer recommendations based on the natural UV-B exposure values that have been described for the involved species [18]. Based on the clinical history, first detection of dermal lesions by the owner ranged from two months to two years prior to initial presentation (Table 1). In all cases, the lesions initially presented as skin nodules with a smooth surface and greyish discoloration that rapidly developed into whitish-greyish crateriform nodules with a keratinous core. In most cases, the owners did not seek veterinary advice until multiple dermal nodules were noticed, or until one or more dermal nodules reached a considerable size.

While 18, 15, and 3 lesions were noted in panther chameleon 1, 2, and 3, respectively, the veiled chameleon was presented with four dermal nodules, and two dermal nodules were detected in the bearded dragon (Table 1). The number of dermal lesions seemed to be positively correlated with the interval between first detection of the lesions and initial presentation. Besides panther chameleon 1, which showed a poor body condition and sunken eyes, indicating more than 8–10% of dehydration [19], none of the remaining lizards displayed other clinical signs in addition to the skin lesions.

While the dermal lesions were located at the dorsolateral body wall in all lizards (Figure 1A–D), a single, unilateral nodular lesion of the eyelid was additionally present in the veiled chameleon, panther chameleon 1, and panther chameleon 3 (Figure 2). The diameter of the lesions ranged from 0.1–1.5 cm, with an average diameter of 0.33 cm. In the bearded dragon and panther chameleon 3, the average diameter of the lesions was 0.56 cm. In the chameleons with multicentric distribution, on average two large lesions with an average diameter of 1 cm (Figures 1A and 2) were present and the remaining lesions consisted of small nodular lesions with a diameter of 0.1–0.2 cm (Figure 1B,C).

During surgical removal, it was noted that all dermal nodules were well-demarcated and did not infiltrate the subcutis. Especially in the smaller nodules, the central keratin plug detached spontaneously when pressure was performed during surgical removal (Figure 1B).

Recovery from anesthesia was uneventful in all cases and a follow-up period of one to two years was established in all lizards (Table 1). When the lizards returned to their owners, management advice was provided with an emphasis on the optimization of the provision of UV-B lighting according to the recommended values. While the owners of panther chameleons followed these recommendations, the owners of the veiled chameleon and the bearded dragon did not make adjustments. In the bearded dragon, a new dermal nodule developed in approximately the same location as the first tumor after 1 year, and in the veiled chameleon, multicentric skin nodules located at the lateral body wall and the head, including both eyelids, reappeared during a 2-year follow-up period (Figure 3). As the nodules interfered with the normal feeding behavior, the chameleon deteriorated and was euthanized by a local veterinarian.

3.2. Histopathology and IHC

Histologic examination of eleven skin lesions revealed a characteristic architectural pattern in all five cases. Nodular lesions presented as a relatively well-defined, circumscribed epidermal proliferation including a multilobular, exo-endophytic cyst-like invagination of the epidermis that creates a crateriform lesion with a central keratinous plug with minimally infiltrating borders (Figure 4). Peripheral to the keratin-filled crater, lip-like borders of well-differentiated squamous cells with a low degree of pleomorphism were observed (Figure 5). Areas of pseudoepitheliomatous hyperplasia that formed folds inside the crater and the adjacent dermis were noticed. Depending on the section, those folds presented as buds, cords, or isolated islands, centered by orthokeratotic pearls. Based on these findings, a final histological diagnosis of KA was made for all examined skin nodules that were obtained from the lizards.

Figure 1. Nodular, crateriform dermal lesions located at the dorsolateral body wall in two panther chameleons (*Furcifer pardalis*) (**A**–**C**) and a bearded dragon (*Pogona vitticeps*) (**D**) that were histologically diagnosed as keratoacanthoma (KA). KAs presented as whitish-greyish crateriform nodules with a keratinous core. Especially in smaller lesions, the central keratin plug detached spontaneously when digital pressure was applied (arrows) (**B**).

Figure 2. Macroscopic multicentric nodular, crateriform dermal lesion located at the dorsal part of the eyelid in panther chameleon 1 (*Furcifer pardalis*) that was histologically diagnosed as a keratoacanthoma with malignant transformation.

Figure 3. Macroscopic multicentric nodular, crateriform dermal lesions located at the head, including both eyelids, in a veiled chameleon (*Chamaeleo calyptratus*).

Neoplastic cells were typically enlarged and showed a pale eosinophilic ground glass-like cytoplasm (Figure 5). All examined KAs showed a mitotic index ranging from 0–2 and absent to mild nuclear atypia. At the base of 80% of the KAs, a mild inflammatory infiltrate with heterophiles, lymphocytes, and macrophages, as well as small keratin pearls, was present.

Histological sections of the eyelid nodule in panther chameleon 1 showed two components. While the central area of the process clearly showed KA features with a mitotic index of 1, low cellular atypia, and pleomorphism (Figure 6A), the characteristics of the neoplastic cells at the periphery were fully compatible with well-differentiated SCC, including a mitotic index of 4, marked nuclear atypia, and pleomorphism (Figure 6B). A final histopathological diagnosis of KA with malignant transformation was made, similar to what has been previously described in humans by Sánchez Yus et al. [20] and Weedon et al. [21]

The results of the IHC study are presented in Table 3. Strong staining with E-cadherin (IRS score 9–12) was noted along the plasma membrane of the six examined Kas, indicating high integrity of cell-cell adhesion (Figure 7A). Moreover, strong positivity for Pan-CK (IRS 9–12) (Figure 7B) and moderate cytoplasmic immunoexpression for COX-2 (IRS 4–8) were noted (Figure 8A). In the KA with malignant transformation, strongly positive immunoexpression for COX-2 (IRS 9–12) was present in the SCC area of the process (Figure 8B) in contrast to the central part that showed KA features.

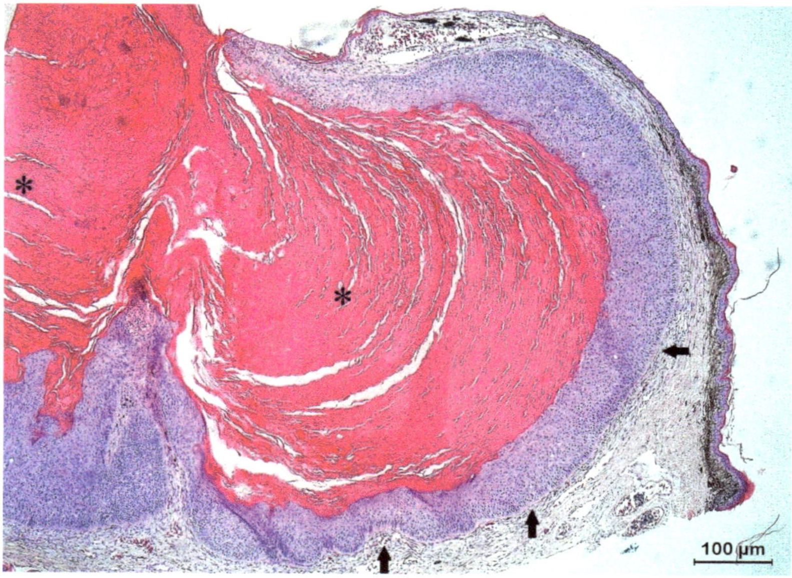

Figure 4. Histologic section of a keratoacanthoma (KA) from the dorsolateral body wall in a veiled chameleon (*Chamaeleo calyptratus*) showing a characteristic architectural pattern consisting of an exo-endophytic, cyst-like invagination of the epidermis that creates a crateriform lesion with a central keratinous plug (asterisks) and minimally infiltrating borders (arrows).

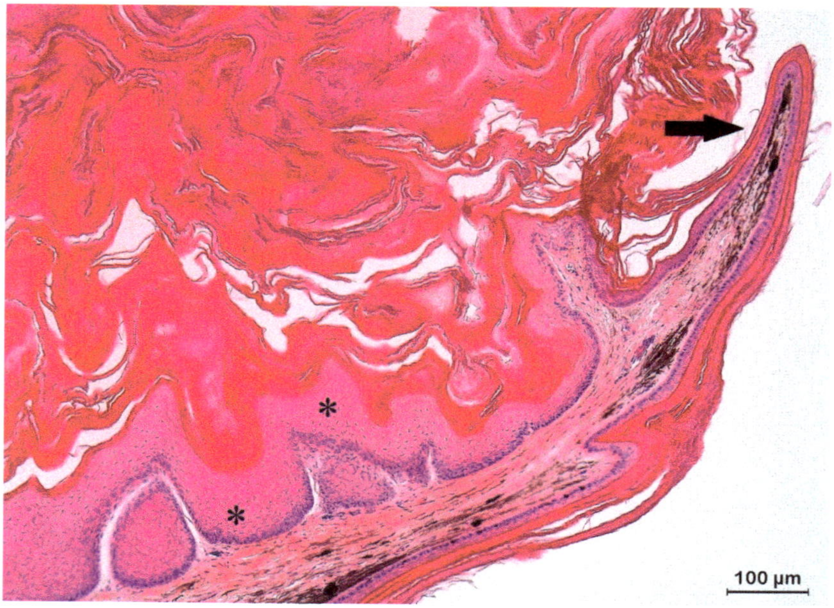

Figure 5. Histologic section of a keratoacanthoma in a panther chameleon (*Furcifer pardalis*) from the dorsolateral body wall. A characteristic epithelial lip (arrow) is present at the periphery, which partially extends over the central keratin plug. Note the typical large, pale pink cells with a glassy appearance (asterisks) surrounded by a thin layer of basophilic cells of the epithelial lip.

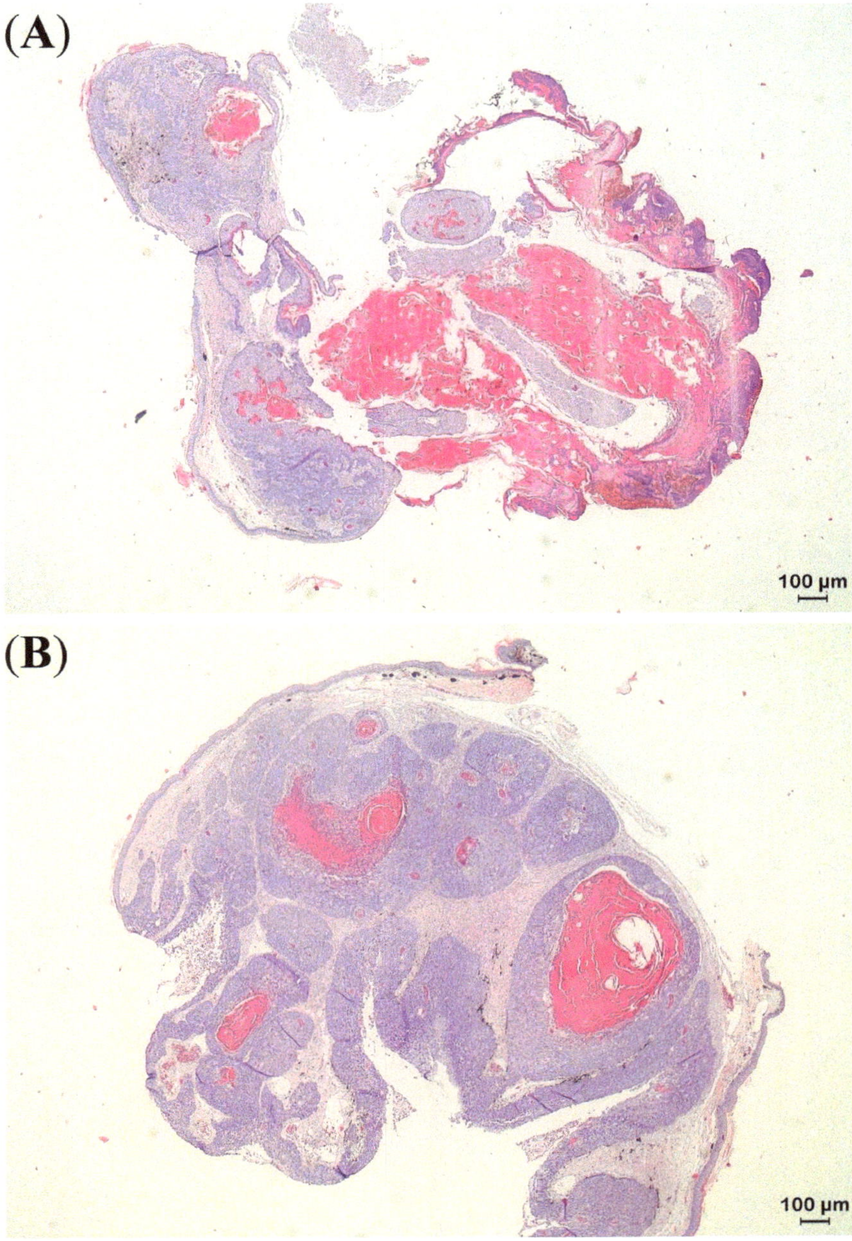

Figure 6. Histologic sections of a keratoacanthoma (KA) with malignant transformation obtained from the eyelid in panther chameleon 1 (*Furcifer pardalis*) with KA features in the central area (**A**) and a well-differentiated SCC at the periphery of the process (**B**).

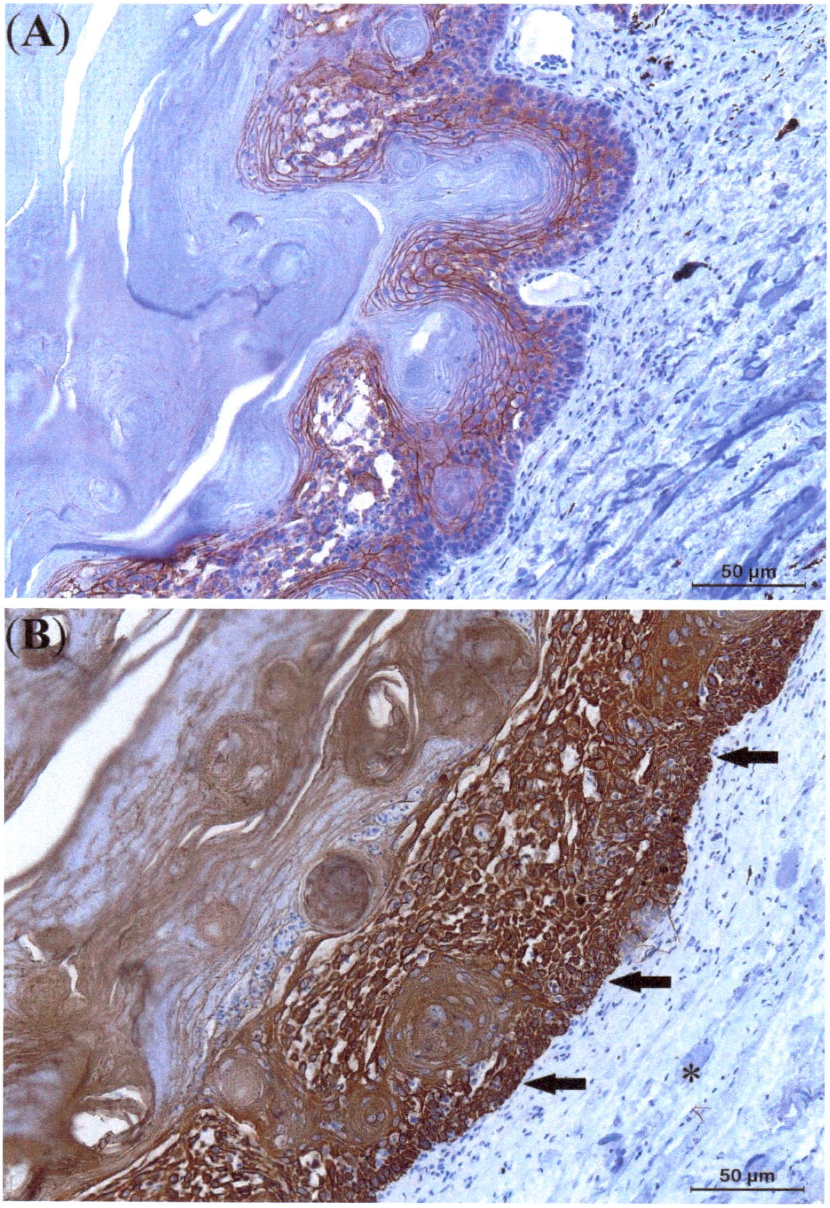

Figure 7. Immunohistochemical results of a keratoacanthoma (KA) from the lateral body wall of a panther chameleon (*Furcifer pardalis*) (**A**,**B**). Intense E-cadherin staining with immunoreactivity at the level of the plasma membrane of all neoplastic cells (**A**). Strongly positive pan-cytokeratin staining showing specific intracytoplasmic immunoreactivity abruptly delineating the neoplastic tissue (arrows) from the healthy skin (asterisk) (**B**).

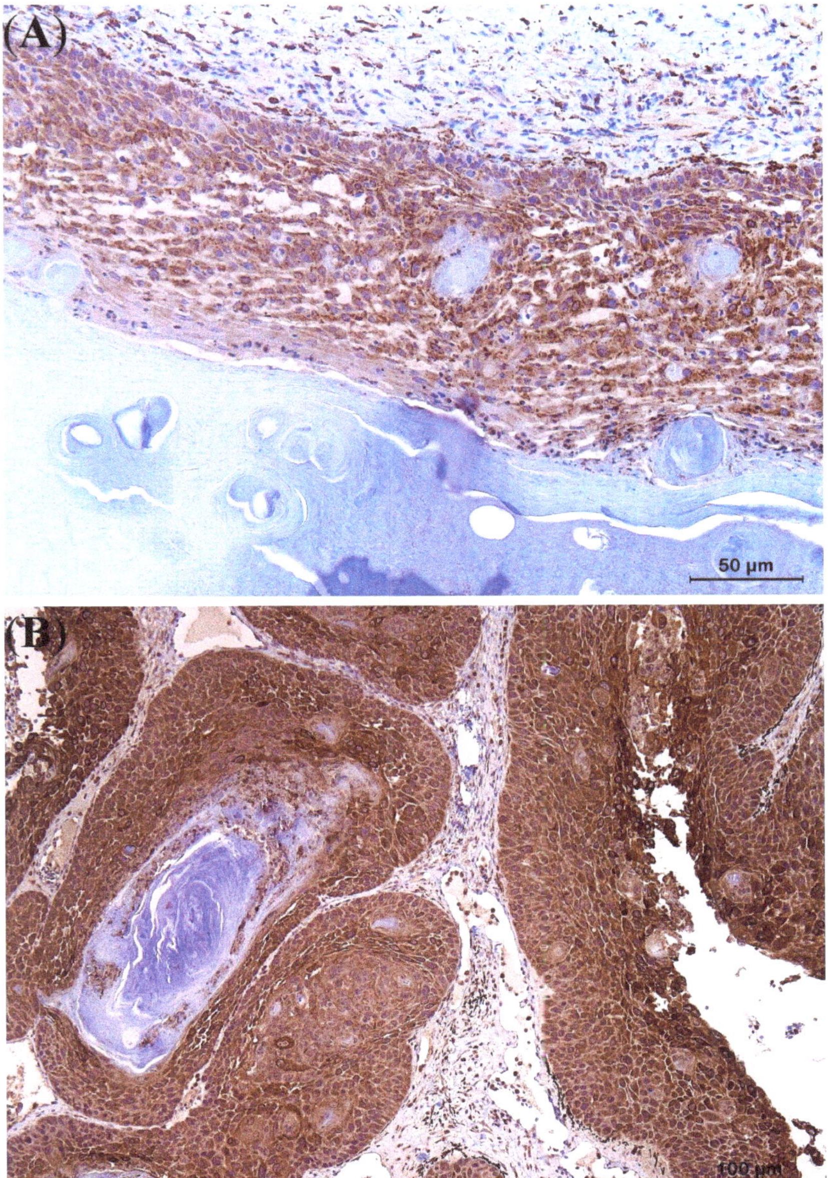

Figure 8. Moderate intracytoplasmic COX-2 staining of a keratoacanthoma (KA) from the lateral body wall of a panther chameleon (*Furcifer pardalis*) (**A**). Strong intracytoplasmic COX-2 staining of the squamous cell carcinoma arising at the periphery of a KA with malignant transformation, obtained from the eyelid of a panther chameleon (**B**).

Table 3. Six keratoacanthomas from lizards (n = 6) were categorized according to the percentage of positive cells (0–4), the intensity of immunostaining (0–3), and the IRS score (0–12) (Table 2) for e-cadherin, cyclooxygenase-2 (COX-2), and pan-cytokeratin (Pan-CK).

Immunomarker	Percentage of Positive Cells					Intensity of Immunostaining				IRS Score			
	0	1	2	3	4	0	1	2	3	0–1	2–3	4–8	9–12
E-cadherin (n = 6)	0	0	0	0	6	0	0	0	6	0	0	0	6
COX-2 (n = 6)	0	0	1	4	1	0	1	4	1	0	0	6	0
Pan-CK (n = 6)	0	0	0	0	6	0	0	0	6	0	0	0	6

4. Discussion

The present study describes the clinical behavior and macroscopic appearance as well as the histopathologic and IHC characteristics of KAs in three different saurian species. Besides KA of the spectacle in a boa (*Boa constrictor*) [22], KA has not previously been described in reptiles. Based on the findings of the present study, however, the description of the disorder of the spectacle in the boa does not seem to meet the clinical, macroscopic, and histologic criteria that are fundamental to obtaining a definitive diagnosis of KA, and the lesion rather resembled retention of multiple layers and ulceration of the spectacle associated with impaired shedding.

It is noteworthy that all affected lizards in the present study were males with an adult to relatively old age according to physiological lifespan reference intervals [23–25]; this is similar to what has been reported for KA in humans and dogs [26–29]. In avian species, however, KA is almost exclusively observed in juvenile broiler chickens without an obvious gender predisposition [9]. Despite the relatively limited number of cases in the present study, male panther chameleons with an average to old age may be predisposed to the development of KA. A similar age and gender predisposition seems to exist for multicentric SCCs in panther chameleons [6].

Dermal SCC is a malignant neoplasia of epidermal keratinocytes that may exhibit various degrees of differentiation and invasiveness, as well as metastatic potential [30]. Based on this variation, the World Health Organization (WHO) Classification of skin tumors proposed by the International Agency for Research on Cancer (IARC) categorizes different histologic variants of dermal SCC with acantholytic SCC, desmoplastic SCC and spindle cell SCC representing high-risk histologic variants, and verrucous SCC and KA representing low-risk histologic variants [8,30,31]. KA is defined as a well-differentiated, minimally invasive neoplasia with low metastatic potential [9,32,33]. In contrast to dogs and birds, KA in humans often shows a spontaneous regression phase that may occur within 4–6 months of the initial diagnosis [34]. In the present study, signs of spontaneous regression were observed in none of the lizards, even in the cases that had a 6-month to 2-year interval between the first detection of skin lesions and initial presentation; in two cases, recurrence was observed. Although spontaneous regression of human KA is often observed, cases wherein KA is observed as a pre-stage of malignant SCCs [14] have also been documented [35]. If malignant transformation of KA into SCC takes place in humans, it seems to occur within a couple of months of KA development. This transformation is considered to occur spontaneously and is mainly seen in senile skin with actinic degeneration or immunocompromised persons [33]. It is unknown, however, what proportion of human KAs eventually transform into SCC and what the deciding factors for transformation are [34,36,37]. In the present study, a skin nodule of the eyelid was diagnosed as a KA with malignant transformation in panther chameleon 1, which was presented 2 years after the first detection of skin lesions (Table 1 and Figure 2). In the other cases, the interval between detection and initial presentation ranged from 2 months to 6 months, and no signs of transformation into SCC could be demonstrated. The latter findings suggest that transformation of KA into SCC does not seem to rapidly occur in the examined lizard species, and if it occurs, it can rather be expected in chronic cases. Multicentric SCC has been reported in panther chameleons, a veiled chameleon, and a bearded dragon [2,5,6]. The multicentric SCCs that were described in panther chameleons by Meyer et al. [6] may have initially comprised Kas, and the case referred to as a multicentric in situ carcinoma was most likely a KA, based on the description that is provided. As the time between the development of the primary skin lesions and the diagnosis of the multicentric SCCs was not reported, it is impossible to state whether if malignant transformation had occurred it would have initially comprised KAs in the described cases [6].

Based on the distribution of KAs in the lizards in the present study, mainly areas of skin that are most frequently exposed to UV-B irradiation were affected. From a behavioral point of view, chameleons typically can be classified as occasional sun baskers living in leafy shrubs [18,38]. They create a larger body surface by basking in a lateral position and flattening the coelom to a maximum in order to increase their exposure to UV-B irradiation and/or their body temperature [38]. Although artificial UV-B lighting is considered an essential part of captive management, especially for insect- and herbivorous reptiles, exposure to high-intensity focused bundles of artificial UV-B irradiation often surpasses the exposure values in the wild [6,18]. As previously postulated for dermal SCCs, BCCs, chromatophoromas, and hemangiomas, the inappropriate use of UV-B lighting sources is considered one of the most important predisposing factors towards the development of these dermal neoplasia in reptiles [3,4,6,7]. In all cases in the present study, UV-B exposure was considered inappropriate based on the recommended UV-B values for the involved species [18,38]. As the latter may play an important role in the oncogenesis of KA in lizards, we consider the provision of well-balanced UV-B irradiation essential for the prevention of this neoplastic disorder in captive lizards.

In humans, the IHC study of keratin 16 (K16) expression demonstrated that KA originates from the outer root sheath cells below the infundibulum of the hair follicle [14,39]. While reptiles do not possess hair follicles, lacunar cells are typically seen as an intermediate, undifferentiated keratinocyte cell type that appears during the epidermal renewal phase (the second phase of the physiological shedding cycle) and regeneration of the skin following traumatic insults [40]. As these lacunar cells also express K16 [41], they may also play an important role in the development of KA in lizards, especially when lacunar cells are exposed to excessive levels of UV-B irradiation and/or other traumatic insults to the skin [41,42].

All KAs in this study showed strongly positive cell membrane reactivity (IRS score 9–12) for E-cadherin, indicating high integrity cell-cell adhesion and thus low migratory potential, in contrast to what is seen in SCCs and other malignant neoplasia [31,43,44]. The strong Pan-CK immunostaining (IRS score 9–12) contributes to the delineation of KA margins and assesses the integrity of the basement membrane [45]. In contrast to the moderate COX-2 expression of KAs in the lizards in this study (Figure 7A), strong COX-2 staining was observed in the KA with malignant transformation in panther chameleon 1, similar to what has been described in humans and bird SCCs, indicating increased cell proliferation, angiogenesis, and invasiveness [33,37,43,44,46–48]. Based on these IHC results, COX-2 and E-cadherin staining especially can serve as an important tool in the diagnosis of KA and its differentiation from KA with malignant transformation and SCC in lizards.

If left untreated, large KAs may severely impair normal vision if located at the eyelid, and may eventually interfere with normal foraging and feeding behavior. This study demonstrates that complete excision of KA in lizards is a highly feasible and safe treatment, even in cases that show large and multicentric KAs, KAs at delicate sites such as the skin of the eyelid in chameleons, or KA with malignant transformation. Long-term follow-up was achieved in all cases of the present study (Table 1). While no recurrence was seen in the three panther chameleons, KAs reappeared in the veiled chameleon and the bearded dragon. Presumably, recurrence was associated with continued exposure to inappropriate levels of UV-B irradiation. Although longer follow-up periods are necessary, we consider the provision of optimal UV-B irradiation in captive lizards essential for the prevention of the development and recurrence of KA.

5. Conclusions

This study describes the biological behavior and histopathological and IHC characteristics of KA, a previously undescribed neoplastic entity in lizards. KA seems to mainly affect adult male lizards and the dorsolateral body wall seems to be a predilection site. Chameleon species that show lateral basking behavior might make them predisposed to the development of multicentric KAs if they are exposed to inappropriate UV-B levels in the captive environment. As demonstrated in one of the panther chameleons, malignant transformation of KA can occur in a chronic stage. Even in chronic stages and cases with multicentric KAs, surgical treatment carries a good prognosis. Correct histologic characterization and IHC staining allow the differentiation of KA from conventional SCCs in lizards, although in some cases the exact classification of KA with a variable degree of malignant transformation may remain challenging; this is a similar situation to that encountered in humans.

Author Contributions: Conceptualization, F.S. and T.H.; Data curation, F.S.; Project administration, F.S., K.C. and T.H.; Supervision, T.H.; Validation, T.H.; Writing—original draft, F.S. and T.H.; Writing—review & editing, F.S., K.C., M.J.L.K. and T.H. All authors have read and agreed to the published version of the manuscript.

Funding: This research received no external funding.

Institutional Review Board Statement: Not applicable.

Informed Consent Statement: Not applicable.

Data Availability Statement: The data presented in this study are available on request from the corresponding author.

Acknowledgments: We thank Sergi Solanes Vilanova for his aid in macroscopic photography and the Department of Pathobiology, Pharmacology and Zoological Medicine, especially Delphine Ameye, Joachim Christiaens, and Sarah Loomans, for aid and technical assistance required to process histopathological and immunohistochemical samples.

Conflicts of Interest: The authors declare no conflict of interest.

References

1. Christman, J.; Devau, M.; Wilson-Robles, H.; Hoppes, S.; Rech, R.; Russell, K.E.; Heatley, J.J. Oncology of Reptiles: Diseases, Diagnosis, and Treatment. *Vet. Clin. N. Am. Exot. Anim. Pract.* **2017**, *20*, 87–110. [CrossRef] [PubMed]
2. Garner, M.M.; Hernandez-Divers, S.M.; Raymond, J.T. Reptile neoplasia: A retrospective study of case submissions to a specialty diagnostic service. *Vet. Clin. N. Am. Exot. Anim. Pract.* **2004**, *7*, 653–671. [CrossRef] [PubMed]
3. Monahan, C.F.; Garner, M.M.; Kiupel, M. Chromatophoromas in Reptiles. *Vet. Sci.* **2022**, *9*, 115. [CrossRef] [PubMed]
4. Hannon, D.E.; Garner, M.M.; Reavill, D.R. Squamous Cell Carcinomas in Inland Bearded Dragons (*Pogona vitticeps*). *J. Herpetol. Med. Surg.* **2011**, *21*, 101–106. [CrossRef]
5. Hellebuyck, T.; Ducatelle, R.; Bosseler, L.; Van Caelenberg, A.; Versnaeyen, H.; Chiers, K.; Martel, A. Basal cell carcinoma in two Hermann's tortoises (*Testudo hermanni*). *J. Vet. Diagn. Investig.* **2016**, *28*, 750–754. [CrossRef]
6. Kubiak, M.; Denk, D.; Stidworthy, M.F. Retrospective review of neoplasms of captive lizards in the United Kingdom. *Vet. Rec.* **2020**, *186*, 28. [CrossRef]
7. Meyer, J.; Kolodziejek, J.; Häbich, A.C.; Dinhopl, N.; Richter, B. Multicentric Squamous Cell Tumors in Panther Chameleons (*Furcifer pardalis*). *J. Exot. Pet Med.* **2019**, *29*, 166–172. [CrossRef]
8. Murphy, G.; Beer, T.; Cerio, R.; Kao, G.; Nagore, E.; Pulitzer, M. Keratinocytic/epidermal tumours. In *World Health Organization (WHO) Classification of Skin Tumours*, 4th ed.; Elder, D., Massi, D., Scolyer, R., Willemze, R., Eds.; International Agency for Research on Cancer (IARC): Argonay, France, 2018; Volume 11, pp. 36–38.
9. Williams, S.M.; Reece, R.L.; Hafner, S. Neoplastic diseases. In *Diseases of Poultry*, 14th ed.; Swayne, D.E., Glisson, J.R., McDougald, L.R., Nolan, L.K., Suarez, D.L., Nair, V., Eds.; John Wiley & Sons: Hoboken, NJ, USA, 2019; Volume 2, pp. 652–653.
10. Abramo, F.; Pratesi, F.; Cantile, C.; Sozzi, S.; Poli, A. Survey of canine and feline follicular tumours and tumour-like lesions in central Italy. *J. Small Anim. Pract.* **1999**, *40*, 479–481. [CrossRef]

11. Scott, D.W.; Miller, W.H.; Griffin, C.E. Neoplastic and Non-Neoplastic Tumors. In *Muller & Kirk's Small Animal Dermatology*, 6th ed.; Scott, D.W., Miller, W.H., Griffin, C.E., Eds.; Elsevier Health Sciences: St. Louis, MO, USA, 2001; Volume 1, pp. 1236–1414.
12. Mauldin, E.A.; Peters-Kennedy, J. Integumentary System. In *Jubb, Kennedy & Palmer's Pathology of Domestic Animals*, 6th ed.; Maxie, M.G., Ed.; Elsevier Health Sciences: St. Louis, MO, USA, 2015; Volume 1, pp. 509–736.
13. Gleich, T.; Chiticariu, E.; Huber, M.; Hohl, D. Keratoacanthoma: A distinct entity? *Exp. Dermatol.* **2016**, *25*, 85–91. [CrossRef]
14. Cerroni, L.; Kerl, H. Keratoacanthoma. In *Fitzpatrick's Dermatology in General Medicine*, 7th ed.; Goldsmith, L.A., Katz, S.I., Gilchrest, B.A., Paller, A.S., Leffell, D.J., Wolff, K., Eds.; McGraw-Hill: New York, NY, USA, 2012; Volume 2, pp. 1049–1053.
15. Lonsdorf, A.; Hadaschick, E. Squamous cell carcinoma and keratoacanthoma. In *Fitzpatrick's Dermatology*, 9th ed.; Kang, S., Amagai, M., Bruckner, A.L., Enk, A.H., Margolis, D.J., McMichael, A.J., Orringer, J.S., Eds.; McGraw-Hill: New York, NY, USA, 2019; Volume 2, pp. 1901–1916.
16. Wobeser, B.K.; Kidney, B.A.; Powers, B.E.; Withrow, S.J.; Mayer, M.N.; Spinato, M.T.; Allen, A.L. Agreement among surgical pathologists evaluating routine histologic sections of digits amputated from cats and dogs. *J. Vet. Diagn. Investig.* **2007**, *19*, 439–443. [CrossRef]
17. Fedchenko, N.; Reifenrath, J. Different approaches for interpretation and reporting of immunohistochemistry analysis results in the bone tissue—A review. *Diagn. Pathol.* **2014**, *9*, 221. [CrossRef]
18. Baines, F.; Chattell, J.; Dale, J.; Garrick, D.; Gill, I.; Goetz, M.; Skelton, T.; Swatman, M. How much UV-B does my reptile need? The UV-Tool, a guide to the selection of UV lighting for reptiles and amphibians in captivity. *J. Zoo Aquar. Res.* **2016**, *4*, 42–63.
19. Boyer, T.H.; Scott, P.W. Nutritional therapy. In *Mader's Reptile and Amphibian Medicine and Surgery*, 3rd ed.; Divers, S.J., Stahl, S.J., Eds.; Elsevier Health Sciences: St. Louis, MO, USA, 2019; Volume 1, pp. 1173–1176.
20. Sánchez Yus, E.; Simón, P.; Requena, L.; Ambrojo, P.; de Eusebio, E. Solitary keratoacanthoma: A self-healing proliferation that frequently becomes malignant. *Am. J. Dermatopathol.* **2000**, *22*, 305–310.
21. Weedon, D.D.; Malo, J.; Brooks, D.; Williamson, R. Squamous cell carcinoma arising in keratoacanthoma: A neglected phenomenon in the elderly. *Am. J. Dermatopathol.* **2010**, *32*, 423–426. [CrossRef] [PubMed]
22. Hardon, T.; Fledelius, B.; Heegaard, S. Keratoacanthoma of the spectacle in a *Boa constrictor*. *Vet. Ophthalmol.* **2007**, *10*, 320–322. [CrossRef] [PubMed]
23. McGeough, R. *Furcifer pardalis* (Panther Chameleon)—A brief species description and details on captive husbandry. *Biol. Eng. Med. Sci. Reports.* **2016**, *2*, 27–38. [CrossRef]
24. Schmid-Brunclik, N.; Stefka, S.C.; Madeleine, K.B.; Max, G.; Jean-Michel, H. Liposarcoma in a Veiled Chameleon, *Chamaeleo calyptratus*. *J. Herpetol. Med. Surg.* **2008**, *17*, 132–135. [CrossRef]
25. Paré, J.A.; Lentini, A.M. Reptile Geriatrics. *Vet. Clin. N. Am. Exot. Anim. Pract.* **2010**, *13*, 15–25. [CrossRef]
26. Zargaran, M.; Baghaei, F. A clinical, histopathological and immunohistochemical approach to the bewildering diagnosis of keratoacanthoma. *J. Dent.* **2014**, *15*, 91–97.
27. Ogita, A.; Ansai, S.I. What is a solitary keratoacanthoma? A benign follicular neoplasm, frequently associated with squamous cell carcinoma. *Diagnostics* **2021**, *11*, 1848. [CrossRef]
28. Mitrache, C.; Benea, V.; Tovaru, M.; Georgescu, S.R.; Tudose., I. Clinical, epidemiological, physiopathological and histopathological aspects of keratoacanthoma. *DermatoVenerol* **2011**, *56*, 209–215.
29. Wiener, D.J. Histologic features of hair follicle neoplasms and cysts in dogs and cats: A diagnostic guide. *J. Vet. Diagn. Investig.* **2021**, *33*, 479–497. [CrossRef] [PubMed]
30. Waldman, A.; Schmults, C. Cutaneous Squamous Cell Carcinoma. *Hematol. Oncol. Clin. N. Am.* **2019**, *33*, 1–12. [CrossRef]
31. Burton, K.A.; Ashack, K.A.; Khachemoune, A. Cutaneous Squamous Cell Carcinoma: A Review of High-Risk and Metastatic Disease. *Am. J. Clin. Dermatol.* **2016**, *17*, 491–508. [CrossRef] [PubMed]
32. Hafner, S.; Harmon, B.G.; Stewart, R.G.; Rowland, G.N. Avian Keratoacanthoma (Dermal Squamous Cell Carcinoma) in Broiler Chicken Carcasses. *Vet. Pathol.* **1993**, *30*, 265–270. [CrossRef]
33. Vîlcea, A.M.; Stoica, L.E.; Georgescu, C.V.; Popescu, F.C.; Ciurea, R.N.; Vîlcea, I.D.; Mirea, C.S. Clinical, histopathological and immunohistochemical study of keratoacanthoma. *Rom. J. Morphol. Embryol.* **2021**, *62*, 445–456. [CrossRef]
34. Schwartz, R.A. Keratoacanthoma. *J. Am. Acad. Dermatol.* **1994**, *30*, 1–19. [CrossRef]
35. Vargo, N. Basal cell and squamous cell carcinoma. *Semin. Oncol. Nurs.* **2003**, *19*, 12–21. [CrossRef]
36. Rook, A.; Whimster, I. Keratoacanthoma—A thirty year retrospect. *Br. J. Dermatol.* **1979**, *100*, 41–47. [CrossRef]
37. Kwiek, B.; Schwartz, R.A. Keratoacanthoma (KA): An update and review. *J. Am. Acad. Dermatol.* **2016**, *74*, 1220–1233. [CrossRef]
38. Kubiak, M. Chameleons. In *Handbook of Exotic Pet Medicine*, 1st ed.; Kubiak, M., Ed.; John Wiley & Sons: Hoboken, NJ, USA, 2020; Volume 1, pp. 263–281.
39. Ito, Y.; Kurokawa, I.; Nishimura, K.; Hakamada, A.; Isoda, K.I.; Yamanaka, K.I.; Tsubura, A.; Mizutani, H. Keratin and filaggrin expression in keratoacanthoma. *J. Eur. Acad. Dermatol. Venereol.* **2008**, *22*, 353–355. [CrossRef]
40. Chang, C.; Wu, P.; Baker, R.E.; Maini, P.K.; Alibardi, L.; Chuong, C.M. Reptile scale paradigm: Evo-Devo, pattern formation and regeneration. *Int. J. Dev. Biol.* **2009**, *53*, 813–826. [CrossRef] [PubMed]
41. Alibardi, L.; Toni, M. Wound keratins in the regenerating epidermis of lizard suggest that the wound reaction is similar in the tail and limb. *J. Exp. Zool. A Comp. Exp. Biol.* **2005**, *303*, 845–860. [CrossRef]
42. Alibardi, L.; Maurizii, M.; Taddei, C. Immunocytochemical and electrophoretic distribution of cytokeratins in the resting stage epidermis of the lizard *Podarcis sicula*. *J. Exp. Zool.* **2001**, *289*, 409–418. [CrossRef]

43. Jones, A.L.; Suárez-Bonnet, A.; Mitchell, J.A.; Ramirez, G.A.; Stidworthy, M.F.; Priestnall, S.L. Avian Papilloma and Squamous Cell Carcinoma: A Histopathological, Immunohistochemical and Virological study. *J. Comp. Pathol.* **2020**, *175*, 13–23. [CrossRef]
44. Papadavid, E.; Pignatelli, M.; Zakynthinos, S.; Krausz, T.; Chu, A.C. The potential role of abnormal E-cadherin and a-, b- and g-catenin immunoreactivity in the determination of the biological behaviour of keratoacanthoma. *Br. J. Dermatol.* **2001**, *145*, 582–589. [CrossRef]
45. Orós, J.; López-Yánez, M.; Rodríguez, F.; Calabuig, P.; Castro, P.L. Immunohistochemical staining patterns of alpha-keratins in normal tissues from two reptile species: Implications for characterization of squamous cell carcinomas. *BMC Vet. Res.* **2018**, *14*, 219. [CrossRef]
46. Slater, M.; Barden, J.A. Differentiating keratoacanthoma from squamous cell carcinoma by the use of apoptotic and cell adhesion markers. *Histopathology* **2005**, *47*, 170–178. [CrossRef]
47. Putti, T.C.; Teh, M.; Lee, Y.S. Biological behavior of keratoacanthoma and squamous cell carcinoma: Telomerase activity and COX-2 as potential markers. *Mod. Pathol.* **2004**, *17*, 468–475. [CrossRef]
48. Hua, H.K.; Jin, C.; Yang, L.J.; Tao, S.Q.; Zhu, X.H. Expression of Cyclooxygenase-2 in Squamous Cell Carcinoma and Keratoacanthoma and its Clinical Significance. *Cell Biochem. Biophys.* **2015**, *72*, 475–480. [CrossRef]

animals MDPI

Article

Evaluation of Neoplasia, Treatments, and Survival in Lizard Species

Frank Willig [1,2,†], Fred J. Torpy [2,3,†], Scott H. Harrison [4], Elizabeth G. Duke [2,5], Brigid Troan [2,6], Amy M. Boddy [2,7], Lisa M. Abegglen [2,8] and Tara M. Harrison [2,5,*]

1 Virginia Maryland College of Veterinary Medicine, Blacksburg, VA 24061, USA
2 Exotic Species Cancer Research Alliance, North Carolina State University College of Veterinary Medicine, Raleigh, NC 27607, USA; ecgraebe@ncsu.edu (E.G.D.); bvtroan@ncsu.edu (B.T.); amyboddy@ucsb.edu (A.M.B.); lisa.abegglen@hci.utah.edu (L.M.A.)
3 Department of Small Animal Medicine and Surgery, University of Georgia College of Veterinary Medicine, Athens, GA 30602, USA
4 Department of Biology, North Carolina Agriculture and Technical State University, Greensboro, NC 27411, USA; scotth@ncat.edu
5 Department of Clinical Sciences, North Carolina State University College of Veterinary Medicine, Raleigh, NC 27607, USA
6 Department of Molecular and Biomedical Sciences, North Carolina State University College of Veterinary Medicine, Raleigh, NC 27607, USA
7 Department of Anthropology, University of California Santa Barbara, Santa Barbara, CA 93106, USA
8 Department of Pediatrics, University of Utah, Salt Lake City, UT 84112, USA
* Correspondence: tara_harrison@ncsu.edu
† These authors have contributed equally to this work.

Simple Summary: Neoplasia is a complex disease that affects many species across the animal kingdom, including lizards. Currently, cancer in lizard species is an understudied part of veterinary medicine. In this study, we focused on identifying factors that could aid in improving patient care and quality of life for lizards with neoplasia. We identified multiple factors including species, type of neoplasia, and type of treatment significantly associated with both positive and negative outcomes for lizards affected by different types of neoplasia. Specifically, we tested for statistical associations between eight clinical factors and patient outcomes. We used reported cases of neoplasia in lizards from published papers, as well as a clinical oncology database for exotic animal species. We also identified a subset of neoplasia types that were not associated with death due to their neoplasia. Our results highlight the importance of determining variables that aid veterinarians in deciding the most appropriate care for their patients. We expect that future research in this area will improve our understanding of neoplasia in lizards and better improve the identification of predictor variables for improving patient outcomes.

Citation: Willig, F.; Torpy, F.J.; Harrison, S.H.; Duke, E.G.; Troan, B.; Boddy, A.M.; Abegglen, L.M.; Harrison, T.M. Evaluation of Neoplasia, Treatments, and Survival in Lizard Species. *Animals* **2024**, *14*, 1395. https://doi.org/10.3390/ani14101395

Academic Editor: Josep Pastor

Received: 10 March 2024
Revised: 18 April 2024
Accepted: 30 April 2024
Published: 7 May 2024

Abstract: Neoplasia has been reported in lizards, but more research is needed to accurately document the prevalence and prognosis of the various known neoplasms that affect lizards. This study reviewed medical records from an online database, the Exotic Species Cancer Research Alliance (ESCRA), and reviewed published literature to determine the prevalence of neoplasia, malignancy, metastasis, treatment strategies, and outcomes by species and sex. Records from 55 individual lizards, 20 different species, and 37 different tumors were identified. In the literature, 219 lizards, 59 species, and 86 unique tumors were identified from 72 published case reports. Potential signalment factors such as age, sex, and species were evaluated to see if they affected case outcome. Additional factors including neoplasia type, presence of metastasis, and types of pursued treatments were also evaluated. Statistical analysis was performed to determine whether a factor was significantly associated with animal death due to the identified neoplasia or with animal survival or death due to other causes (non-neoplastic outcomes). Komodo dragons and savannah monitors were more likely to die from neoplasia compared to other lizard species. Cases where the status of metastasis was unknown were significantly associated with death due to neoplasia. Having an unknown status of male versus female was significantly associated with non-neoplastic outcomes of death. Leukemia and islet cell carcinoma were significantly associated with death due to neoplastic causes. Chondrosarcoma, myxosarcoma, osteosarcoma, and squamous cell carcinoma were significantly associated with non-neoplastic outcomes of death. Surgery alone and radiation therapy alone each were significantly associated with non-neoplastic outcomes of death, while lizards not receiving treatment were significantly associated with death due to neoplasia. Benign neoplasia was significantly associated with non-neoplastic outcomes of death. These results will aid in the improved diagnosis and management of neoplasia in lizard species, as well as expanding our understanding of prognostic indicators of neoplasia in lizards.

Keywords: neoplasia; lizard; surgery; chemotherapy; bearded dragon; iguana; monitor; lymphoma; carcinoma

1. Introduction

Neoplasia in lizards has been extensively documented going back well into the previous century consisting of mostly single case reports and a handful of reviews [1,2]. Neoplasia is common in lizards and a wide variety of neoplasms have been diagnosed. Similar to other taxa, various etiologies have been proposed, such as environmental, genetic, and infectious causes [3–7]. Several reports hypothesize species predispositions for certain neoplasms such as gastric neuroendocrine carcinoma in bearded dragons and pancreatic masses in Komodo dragons [4,8]. As advanced care is increasingly utilized in lizards, treatment is also more commonly being pursued for neoplasia. The therapeutic approaches for cancer reported in lizards include chemotherapeutics, radiation therapy, and surgical excision [9–11]. Though treatment modalities are described, comprehensive data on treatment efficacy and outcomes are lacking. Efforts to gather further information on neoplasia prevalence, treatment, and outcomes are ongoing. The Exotic Species Cancer Research Alliance (ESCRA) is a multi-institution database that collects clinical data on neoplasia in zoological species. The existence of such databases in other fields, such as the National Cancer Institute Data Catalogue, increases access to relevant information for researchers and promotes broader analyses. This database also promotes the standardization of information recording and reporting within the field.

This study combined case data from the ESCRA database with available published cases to determine the described prevalence of specific neoplasms, as well as to document the efficacy of treatment approaches by measuring outcomes. In addition, this summary of lizard neoplasia data from the literature can benefit researchers and clinicians alike.

2. Materials and Methods

2.1. Case Selection

2.1.1. Literature Review

Databases including PubMed, CAB Abstracts, and Web of Science were searched for lizard neoplasia cases from the dates of database inception (1983 for PubMed, 1973 for CAB, and 1865 for Web of Science). Final searches for appropriate publications were performed between 27 July and 6 August 2021. Lizards that had experimentally induced tumors were excluded from this study. Search criteria were related to neoplasia in lizards, using terms for species within the order Squamata and including scientific names, common names, synonyms, and alternate spellings. The list of terms for neoplasia based on PubMed's cancer subset search strategy was also utilized in this search. The described search criteria yielded manuscript results as follows: 28 manuscripts from PubMed met the inclusion criteria for this study, of which 20 were unique from CAB Abstracts and Web of Science; 51 manuscripts were identified from Web of Science, of which 13 were unique from the CAB Abstracts and PubMed manuscripts; and 73 manuscripts were identified from CAB Abstracts, of which 39 were unique from the PubMed and Web of Science manuscripts. This yielded a combined total of 72 unique manuscripts ranging from 1978 to 2021.

Data were imported from these lizard neoplasia case searches into Microsoft Excel (Microsoft 365 MSO (Version 2307)). Collected publication information included PubMed identification number/CAB ID code/Web of Science ID, title, authors, year of publication, and journal of publication. Collected animal information included the common and scientific name of the animal, age (in months), maturity (adult vs. juvenile (<3 months old)), sex (if known), body area/system of the primary mass, the histological diagnosis of the primary mass, primary mass malignancy, presence and location of any metastases, presence and identity of concurrent but non-metastatic masses, if treatment was pursued, treatment method pursued, case outcome, and survival from time of diagnosis (in months).

Diagnoses were grouped by primary histologic diagnosis, as reported by the primary investigator. The designation of benign versus malignant was primarily based on the reported diagnosis. Lymphoma and leukemia cases were categorically defined as malignant. In cases in which the reported diagnosis was ambiguous, neoplasms were classified as "malignant" in cases with observed distal (metastatic disease) or local invasion with tissue destruction [12]. If neither the diagnosis nor the description were sufficient, the neoplasm was classified as "undetermined." Treatments were categorized as follows: surgery only (including complete and marginal excision), chemotherapy only, radiation therapy only (including external beam and strontium techniques), surgery and chemotherapy, surgery and radiation, unknown treatment, and no treatment provided. The diagnostic sampling of masses such as fine needle aspirates and incision biopsies was not categorized as a treatment.

2.1.2. ESCRA Database Review

The ESCRA database was searched for submitted lizard neoplasia cases. Cases were confirmed to be neoplasia by board-certified veterinary pathologists working at either independent laboratories or university settings before entry per ESCRA submission guidelines.

Data were summarized from these lizard neoplasia case searches using Microsoft Excel (Microsoft 365 MSO (Version 2307)). The same animal data were collected from each submitted case as from the literature review. Diagnoses were grouped also by primary histologic diagnosis and behavior as reported by submitting clinicians and the benign/malignancy parameters applied to the literature review.

Treatments were categorized into groups per the ESCRA submission form as follows: surgery only (which included surgical excision, cryosurgery, or cryotherapy), chemotherapy only (which included electrochemotherapy, steroid therapies, and NSAIDs), radiation therapy only (including photoradiation and phototherapy), surgery and either radiation therapy or chemotherapy, surgery with radiation therapy and chemotherapy, supportive care (which included systemic or topical antibiotics and NSAIDs), unknown treatment, and no treatment provided. The diagnostic sampling of masses such as fine needle aspirates and incision biopsies were not categorized as treatments.

2.2. Data Analysis

Identified case information was analyzed using IBM Corp. IBM SPSS Statistics for Windows (released 2019; Version 26.0, Armonk, NY, USA) and R statistical computing software (version 4.3.2; R Core Team 2023. R: A language and environment for statistical computing. R foundation for Statistical Computing, Vienna, Austria, https://www.Rproject.org/, accessed on 7 December 2023) [13]. Boosting analysis was used to evaluate the data. For the statistical method of boosting, the mboost package (Model-based boosting, R package version 2.9-9, https://CRAN.R-project.org/package=mboost, accessed on 7 December 2023) was used. Neoplasms were grouped into more general classification categories by behavior (benign vs. malignant) as well as by tissue and tumor type for the mboost evaluation of the survival of animals affected by these types of neoplasms. For boosting analysis, predictor variables were modeled with the use of brandom as a base-learner. For each data set, a set of predictor variables was utilized. The literature data and ESCRA data sets were evaluated together and also separately using the eight variables of species, sex, life stage (juvenile or adult), primary neoplasia diagnosis, neoplasia malignancy, presence of metastasis, tumor location, and treatment type. Outcomes were assigned to lizards, with "1" designated for those that died or were euthanized due to the neoplasm and "0" designated for those that died due to another cause. Lizards without a known cause of death were included in the final analysis. The modeled effects of the predictor variables with outcomes for each animal evaluated were compared to a set of 2000 null model distributions of effects generated from modeling performed with permutated outcomes, with $p < 0.05$ being the threshold of significance regarding the tails of the null model distribution (two-sided hypothesis), similar in method to Mayr et al. [13]. Significance was calculated only for variables with two or more individuals in the represented population.

3. Results

3.1. Lizard Population

There were 274 individual lizards included in this study. There were 219 individual lizard cases identified in the literature review. The described search criteria applied to the ESCRA database yielded a total of 55 unique individual lizard cases. The individual cases represented 65 different species, with 59 species identified in the literature review, 20 species identified in the ESCRA database, and 14 species being common between the two.

3.1.1. Literature

Green iguanas (*Iguana iguana*, $n = 55$) were the most represented species, followed by central bearded dragons (*Pogona vitticeps*, $n = 32$) and North African spiny-tailed lizard (*Varanus exanthematicus*, $n = 10$) (Table 1). Males ($n = 40$) were slightly more represented than females ($n = 41$), with the majority of the animals' sex being unknown ($n = 138$). Life stage was generally unknown ($n = 141$), with adults ($n = 52$) being more represented than juveniles ($n = 6$). Individual age ranged from 5 months to 300 months (12 years) where age was available ($n = 50$), with a mean of 77.7 months (6.5 years) and a median of 60 months (5 years). Age information was not available for the remaining 169 individuals.

Table 1. Frequency of the most prevalent lizard species with neoplasia from the literature and ESCRA sources.

Literature Species (Scientific Name)	Frequency	ESCRA Species (Scientific Name)	Frequency
Beaded lizard (*Heloderma exasperatum/horridum*)	2/219 (0.91%)	Central bearded dragon (*Pogona vitticeps*)	21/55 (45.5%)
Broad headed skink (*Eumeces laticeps*)	2/219 (0.91%)	Blotched blue-tongued lizard (*Tiliqua nigrolutea*)	2/55 (3.6%)
Central bearded dragon (*Pogona vitticeps*)	32/219 (14.6%)	Green iguana (*Iguana iguana*)	3/55 (5.5%)
Crocodile lizard (*Shinisaurus crocodilurus*)	2/219 (0.91%)	Leopard gecko (*Eublepharis macularius*)	4/55 (7.3%)
Desert grassland whiptail lizard (*Cnemidophorus uniparens*)	2/219 (0.91%)	Panther chameleon (*Furcifer pardalis*)	4/55 (7.3%)
East Indian water lizard (*Hydrosaurus amboinensis*)	2/219 (0.91%)	Veiled chameleon (*Chamaeleo calyptratus*)	3/55 (5.5%)
European green lizard (*Lacerta viridis*)	4/219 (1.8%)		
Gila monster (*Heloderma suspectum*)	5/219 (2.3%)		
Green anole (*Anolis carolinensis*)	2/219 (0.91%)		
Green iguana (*Iguana iguana*)	55/219 (28%)		

Table 1. *Cont.*

Literature Species (Scientific Name)	Frequency	ESCRA Species (Scientific Name)	Frequency
Komodo dragon (*Varanus komodoensis*)	3/219 (1.4%)		
Leopard gecko (*Eublepharis macularius*)	5/219 (2.3%)		
Mexican beaded lizard (*Heloderma horridum*)	2/219 (0.91%)		
Panther chameleon (*Furcifer pardalis*)	2/219 (0.91%)		
Round island skink (*Leiolopisma telfairii*)	2/219 (0.91%)		
Savannah monitor (*Varanus exanthematicus*)	10/219 (4.6%)		
Spiny-tailed lizard (*Uromastyx acanthinura*)	10/219 (4.6%)		
Spiny-tailed monitor (*Varanus acanthurus*)	3/219 (1.4%)		
Veiled chameleon (*Chamaeleo calyptratus*)	8/219 (3.7%)		

3.1.2. ESCRA

Central bearded dragons (*Pogona vitticeps*, n = 25) were the most represented species, followed by Panther chameleons (*Furcifer pardalis*, n = 4) and Leopard geckos (*Eublepharis macularius*, n = 4) (Table 1). Males (n = 26) were more represented than females (n = 20), and a minority of individuals' sex were unknown (n = 9). All individuals were either adults (n = 42) or their life stage was unknown (n = 13). Individual age ranged from 24 (2 years) to 173 months (14.4 years) for individuals where age was available (n = 34), with a median age of 72 months (6 years) and mean age of 81.5 months (6.8 years). Age information was not available for the remaining 21 individuals.

3.2. Neoplasia Information

There were 108 neoplasms included in this study. A total of 85 neoplasms were identified in the literature review sources, and 38 neoplasms were identified in the ESCRA database cases. Sixteen neoplasms were common between the two sources.

3.2.1. Literature

Lymphoma was the most represented neoplasia (n = 18) followed by squamous cell carcinoma (n = 13), chromatophoroma (n = 10), and teratoma (n = 10) (Table 2). Malignant neoplasia (n = 133) was more represented than benign neoplasia (n = 58). Some cases (n = 28) did not have an identified location of neoplasia. The liver was the most commonly identified site (n = 15). An unknown presence of metastasis (n = 156) was more common than confirmed presence of metastasis (n = 13) or lack of metastasis (n = 50).

Table 2. Frequency of the most prevalent neoplasias from reviewed literature sourced with identified affected lizard species, malignancy behavior, and represented individuals reported.

Neoplasia (Malignancy Behavior)	Frequency	Affected Species (Scientific Name)	Represented Individuals
Biliary adenocarcinoma (malignant)	6/219 (2.7%)	Broad headed skink (*Eumeces laticeps*)	1
		Green iguana (*Iguana iguana*)	3
		Plumed basilisk (*Basiliscus plumifrons*)	1
		Spiny tailed iguana (*Ctenosaura pectinate*)	1
Carcinoma (malignant)	3/219 (1.4%)	Central bearded dragon (*Pogona vitticeps*)	2
		Warren's girdled lizard (*Cordylus warren*)	1
Cholangiocarcinoma (malignant)	4/219 (1.8%)	Central bearded dragon (*Pogona vitticeps*)	1
		Green iguana (*Iguana iguana*)	2
		Texas horned lizard (*Phrynosoma cornutum*)	1
Chondrosarcoma (malignant)	2/219 (0.91%)	Spiny-tailed monitor (*Varanus acanthurus*)	1
		Storr's monitor (*Varanus storri*)	1
Chromatophoroma (benign)	3/219 (1.6%)	Green iguana (*Iguana iguana*)	1
		Savannah monitor (*Varanus exanthematicus*)	1
		Veiled chameleon (*Chamaeleo calyptratus*)	1
Chromatophoroma (malignant)	6/219 (2.7%)	Central bearded dragon (*Pogona vitticeps*)	1
		Day Gecko (*Gekko phelsuma*)	1
		Dwarf bearded dragon (*Pogona henrylawsoni*)	1
		Leopard gecko (*Eublepharis macularius*)	1
		Veiled chameleon (*Chamaeleo calyptratus*)	2
Colon adenocarcinoma (malignant)	2/219 (0.91%)	Leopard gecko (*Eublepharis macularius*)	1
		Mexican beaded lizard (*Heloderma horridum*)	1
Fibroma (benign)	6/219 (2.7%)	Green iguana (*Iguana iguana*)	4
		Veiled chameleon (*Chamaeleo calyptratus*)	2

Table 2. *Cont.*

Neoplasia (Malignancy Behavior)	Frequency	Affected Species (Scientific Name)	Represented Individuals
Fibropapilloma (benign)	2/219 (0.91%)	Green iguana (*Iguana iguana*)	1
		Sailfin lizard (*Hydrosaurus postulatus*)	1
Fibrosarcoma (malignant)	3/219 (1.4%)	Central bearded dragon (*Pogona vitticeps*)	1
		Spiny-tailed lizard (*Ctenosaura pectinata*)	1
		Unknown water dragon (*Physignathus* sp.)	1
Gastric carcinoid (malignant)	3/219 (1.4%)	Central bearded dragon (*Pogona vitticeps*)	3
Granulosa cell tumor (malignant)	3/219 (1.4%)	Green iguana (*Iguana iguana*)	2
		Savannah monitor (*Varanus exanthematicus*)	1
Hemangiosarcoma (malignant)	2/219 (0.91%)	Green iguana (*Iguana iguana*)	1
		Hispaniolan curly-tailed lizard (*Leiocephalus schreibersii*)	1
Hepatocellular carcinoma (malignant)	3/219 (1.4%)	Green iguana (*Iguana iguana*)	4
Hepatoma (benign)	2/219 (0.91%)	Broad headed lizard (*Eumeces laticeps*)	1
		Gila monster (*Heloderma suspectum*)	1
Interstitial cell adenoma (benign)	2/219 (0.91%)	Cuban iguana (*Cyclura nubila*)	1
		Green iguana (*Iguana iguana*)	1
Islet cell carcinoma (malignant)	3/219 (1.4%)	Komodo dragon (*Varanus komodoensis*)	3
Leiomyosarcoma (malignant)	5/219 (1.4%)	Central bearded dragon (*Pogona vitticeps*)	3
Leukemia (malignant)	5/219 (2.3%)	Central bearded dragon (*Pogona vitticeps*)	2
		Green iguana (*Iguana iguana*)	2
		Green tree monitor (*Varanus prasinus*)	1

Table 2. *Cont.*

Neoplasia (Malignancy Behavior)	Frequency	Affected Species (Scientific Name)	Represented Individuals
Lymphoma (malignant)	18/219 (8.2%)	East Indian water lizard (*Hydrosaurus amboinensis*)	1
		Green iguana (*Iguana iguana*)	5
		Italian wall lizard (*Lacerta sicula*)	1
		Savannah monitor (*Varanus exanthematicus*)	3
		Spiny-tailed lizard (*Uromastyx acanthinura*)	8
Melanophoroma (benign)	2/219 (0.91%)	Green iguana (*Iguana iguana*)	2
Melanophoroma (malignant)	3/219 (1.4%)	Beaded lizard	1
		Leopard gecko (*Eublepharis macularius*)	1
		Veiled chameleon (*Chamaeleo calyptratus*)	1
Myxoma (benign)	3/219 (1.4%)	Central bearded dragon (*Pogona vitticeps*)	1
		Green iguana (*Iguana iguana*)	2
Osteosarcoma (malignant)	5/219 (2.3%)	Central bearded dragon (*Pogona vitticeps*)	1
		Green iguana (*Iguana iguana*)	2
		Spiny-tailed monitor (*Varanus acanthurus*)	2
Pancreatic adenocarcinoma (malignant)	2/219 (0.91%)	Solomon island skink (*Corucia zebrata*)	1
		Veiled chameleon (*Chamaeleo calyptratus*)	1
Papilloma (benign)	3/219 (1.4%)	Ocellated lizard (*Lacerta lepida*)	1
		European green lizard (*Lacerta viridis*)	2
Parathyroid adenoma (benign)	4/219 (1.8%)	Green iguana (*Iguana iguana*)	5
Renal adenoma (benign)	2/219 (0.91%)	Green iguana (*Iguana iguana*)	2

Table 2. *Cont.*

Neoplasia (Malignancy Behavior)	Frequency	Affected Species (Scientific Name)	Represented Individuals
Squamous cell carcinoma (malignant)	13/219 (5.9%)	Blue-tongued skink (*Tiliqua scincoides*)	1
		Central bearded dragon (*Pogona vitticeps*)	4
		European green lizard (*Lacerta viridis*)	1
		Leopard gecko (*Eublepharis macularius*)	2
		Panther chameleon (*Furcifer pardalis*)	2
		Round island skink (*Leiolopisma telfairii*)	2
		Spiny-tailed monitor (*Varanus acanthurus*)	1
Teratoma (benign)	7/219 (3.2%)	Central bearded dragon (*Pogona vitticeps*)	1
		Desert grassland whiptail lizard (*Cnemidophorus uniparens*)	2
		Green iguana (*Iguana iguana*)	3
		Lau banded iguana (*Brachylophus fasciatus*)	1
Teratoma (malignant)	3/219 (1.4%)	Central bearded dragon (*Pogona vitticeps*)	1
		Green iguana (*Iguana iguana*)	2
Thyroid carcinoma (malignant)	3/219 (1.4%)	Marbled velvet gecko (*Oedura marmorata*)	1
		Centralian rough-knob tail gecko (*Nephrusus amyae*)	1
		Smooth knob-tail gecko (*Nephrusus levis*)	1
Undifferentiated sarcoma (malignant)	2/219 (0.091%)	Chuckwalla (*Sauromalus obesus*)	1
		Five-lined skink (*Eumeces fasciatus*)	1

[1–11,14–112].

3.2.2. ESCRA

Squamous cell carcinoma was the most represented neoplasia (n = 8) followed by spindle cell sarcoma (n = 4) (Table 3). Skin (n = 6) and skin/scales of the head region (n = 6) were the most commonly identified sites of neoplasia. Malignant neoplasia (n = 40) was more represented than benign neoplasia (n = 7), with many other cases not having identified malignancy information (n = 9). A lack of metastasis (n = 7) was slightly more common than the presence of metastasis (n = 5), with an unknown presence of metastasis being most common (n = 43).

Table 3. Frequency of the most prevalent neoplasms from the ESCRA database with identified affected lizard species, malignancy behavior, and represented individuals reported.

Neoplasia (Malignancy Behavior)	Frequency	Affected Species (Scientific Name)	Represented Individuals
Chromatophoroma (benign)	1/55 (1.8%)	Veiled chameleon (*Chamaeleo calyptratus*)	1
Chromatophoroma (malignant)	1/55 (1.8%)	Central bearded dragon (*Pogona vitticeps*)	1
Leukemia (malignant)	2/55 (3.6%)	Central bearded dragon (*Pogona vitticeps*)	2
Lymphosarcoma (malignant)	2/55 (3.6%)	Central bearded dragon (*Pogona vitticeps*)	1
		Western banded gecko (*Coleonyx variegatus*)	1
Myxosarcoma (malignant)	2/55 (3.6%)	Central bearded dragon (*Pogona vitticeps*)	2
Papilloma (benign)	1/55 (1.8%)	Green iguana (*Iguana iguana*)	1
Papilloma (unknown)	1/55 (1.8%)	Veiled chameleon (*Chamaeleo calyptratus*)	1
Sarcoma (malignant)	1/55 (1.8%)	Central bearded dragon (*Pogona vitticeps*)	1
Sarcoma (unknown)	1/55 (1.8%)	Central bearded dragon (*Pogona vitticeps*)	1
Spindle cell sarcoma (malignant)	4/55 (7.3%)	Blotched blue-tongued lizard (*Tiliqua nigrolutea*)	1
		Central bearded dragon (*Pogona vitticeps*)	2
		Leopard gecko (*Eublepharis macularius*)	1
Squamous cell carcinoma (malignant)	7/55 (12.7%)	Central bearded dragon (*Pogona vitticeps*)	4
		Leopard gecko (*Eublepharis macularius*)	1
		Panther chameleon (*Furcifer pardalis*)	2

3.3. *Treatment Information*

3.3.1. Literature

Treatment was more commonly not pursued (n = 161) than pursued (n = 31), with surgery being the most common treatment performed (n = 25) followed by chemotherapy alone (n = 3), radiation therapy alone (n = 2), and surgery with chemotherapy (n = 1) (Table 4).

Table 4. Frequency of the most prevalent treatments sought for lizards with neoplasia from literature sources with average survival time, species, and identified neoplasia reported.

Treatment	Frequency	Average Survival Time (Months)	Species (Scientific Name)	Neoplasia
Chemotherapy only	3/219 (1.4%)	3.5	Central bearded dragon (*Pogona vitticeps*)	Squamous cell carcinoma
			Emerald monitor (*Varanus prasinus*)	Leukemia
Radiation only	2/219 (0.91%)	19.5	Central bearded dragon (*Pogona vitticeps*)	Myxosarcoma
			Green iguana (*Iguana iguana*)	Lymphoma
Surgery only	25/219 (11.4%)	10.4	Beaded lizard (*Heloderma horridum*)	Adenocarcinoma
			Blue-tongued skink (*Tiliqua scincoides*)	Squamous cell carcinoma
			Central bearded dragon (*Pogona vitticeps*)	Carcinoma, anaplastic sarcoma, myxoma, peripheral nerve sheath tumor, teratoma, squamous cell carcinoma
			Day Gecko (*Gekko phelsuma*)	Chromatophoroma
			European Green lizard (*Lacerta viridis*)	Papilloma
			Green iguana (*Iguana iguana*)	Adenoma, benign melanophoroma, chromatophoroma, granulosa cell tumor, adrenal cortical adenocarcinoma, adrenal cortical adenocarcinomas, multifocalc cholangiocarcinoma, teratoma
			Lau banded iguana (*Brachylophus fasciatus*)	Teratoma
			Leopard gecko (*Eublepharis macularius*)	Chromatophoroma
			Marbled velvet gecko (*Oedura marmorata*)	Thyroid carcinoma
			Panther chameleon (*Furcifer pardalis*)	Squamous cell carcinoma
			Savannah monitor (*Varanus exanthematicus*)	Mesothelioma
			Veiled chameleon (*Chamaeleo calyptratus*)	Chromatophoroma

3.3.2. ESCRA

The pursuit of treatment was either unknown or not attempted for the majority of cases (n = 36), compared to instances of treatment (n = 19). Surgery was the most common treatment performed (n = 17), followed by chemotherapy alone (n = 1) and surgery combined with radiation therapy (n = 1) (Table 5).

Table 5. Frequency of the most prevalent treatments sought from the ESCRA database of lizards with neoplasia with average survival time, species, and identified neoplasia reported.

<table>
<tr><th>Treatment</th><th>Frequency</th><th>Average Survival Time (Months)</th><th>Species (Scientific Name)</th><th>Neoplasia</th></tr>
<tr><td>Chemotherapy only</td><td>1/55 (1.8%)</td><td>Not reported</td><td>Central bearded dragon (Pogona vitticeps)</td><td>Leukemia</td></tr>
<tr><td>Surgery and radiation</td><td>1/55 (1.8%)</td><td>Not reported</td><td>Central bearded dragon (Pogona vitticeps)</td><td>Anaplastic sarcoma</td></tr>
<tr><td rowspan="7">Surgery only</td><td rowspan="7">17/55 (30.9%)</td><td rowspan="7">9.75</td><td>Central bearded dragon (Pogona vitticeps)</td><td>Fibrosarcoma, myxosarcoma, sarcoma, spindle cell sarcoma, squamous cell carcinoma, chondrosarcoma</td></tr>
<tr><td>Fat-tailed gecko (Hemitheconyx caudicinctus)</td><td>Neurofibrosarcoma</td></tr>
<tr><td>Green anole (Anolis carolinensis)</td><td>Carcinoma</td></tr>
<tr><td>Green iguana (Iguana iguana)</td><td>Sertoli cell tumor</td></tr>
<tr><td>Leopard gecko (Eublepharis macularius)</td><td>Squamous cell carcinoma</td></tr>
<tr><td>Panther chameleon (Furcifer pardalis)</td><td>Squamous cell carcinoma, papilloma</td></tr>
<tr><td>Veiled chameleon (Chamaeleo calyptratus)</td><td>Chromatophoroma</td></tr>
</table>

3.4. Case Outcome and Survival Times

3.4.1. Literature

Case outcome information was available for 89 of the identified cases. Death or euthanasia due to neoplasia (n = 25) and the individual still being alive (n = 25) were the most commonly known outcomes. Death due to unknown causes (n = 11) and death not due to neoplasia (n = 1) were also identified. Case outcomes were unknown for the majority of cases (n = 130). Survival time information was available for 108 of the identified cases, ranging from 0 months to 56 months (4.6 years) after the time of diagnosis with neoplasia until death. Mean survival time was 20.0 months (1.7 years), with a median survival time of 0 months.

3.4.2. ESCRA

Case outcome information was available for 41 of the identified cases. Death or euthanasia due to neoplasia was the most common outcome identified (n = 17), with death or euthanasia due to non-neoplastic causes (n = 14) and unknown status of survival (n = 14) as the next largest groups, followed by the individuals still being alive (n = 9). Only 1 individual was identified to have died from unknown causes. Survival time information was available for 13 of the identified cases, ranging from 0 months to 30 months (2.5 years) after the time of diagnosis with neoplasia until death. Mean survival time was 10.3 months, with a median survival time of 2 months.

3.5. Prognosis

The significance of predictor variables on individual outcomes was estimated using boosting and permutation as described above and applied to each of the literature and ESCRA sample populations. This provided estimations for the impact of a given variable on whether an individual case would result in death from diagnosed neoplasia or survival or death from other causes (referred to as "non-neoplastic outcomes" in this paper).

3.5.1. Combined Data from Literature and ESCRA

Komodo dragons and savannah monitors were the species identified as being significantly more likely to die due to neoplasia ($p < 0.05$). Panther chameleons and round island skinks were significantly associated with non-neoplastic outcomes of death ($p < 0.05$). Leukemia and islet cell carcinoma were significantly associated with death due to neoplasia ($p < 0.05$). Chondrosarcoma, myxosarcoma, osteosarcoma, and squamous cell carcinoma were significantly associated with non-neoplastic outcomes ($p < 0.05$). Surgery alone and radiation therapy alone each were significantly associated with non-neoplastic outcomes, while lizards not receiving treatment were significantly associated with death due to neoplasia ($p < 0.05$). Unknown sex was significantly associated with non-neoplastic outcomes of death ($p < 0.05$), whereas unknown life stage was significantly associated with death due to neoplasia ($p < 0.05$). Benign neoplasia was significantly associated with non-neoplastic outcomes of death ($p < 0.05$), and an unknown status of being benign or malignant was associated with death due to neoplasia ($p < 0.05$). The tumor location of the liver was significantly associated with death due to neoplasia, while some of the other tumor locations were significantly associated with non-neoplastic outcomes of death ($p < 0.05$) (Table 6). A contrast in survival times relating to death being due to neoplasia (with or without treatment) compared to non-neoplastic outcomes is shown in Figure 1. For both benign and malignant neoplasia, neoplasia without treatment exhibits a faster rate of mortality in comparison to neoplasia with treatment. Malignant neoplasia also exhibits a faster rate of mortality in comparison to benign neoplasia. For malignant neoplasia, non-neoplastic outcomes of death were intermediate in the rate of mortality compared to neoplasia without treatment and neoplasia with treatment.

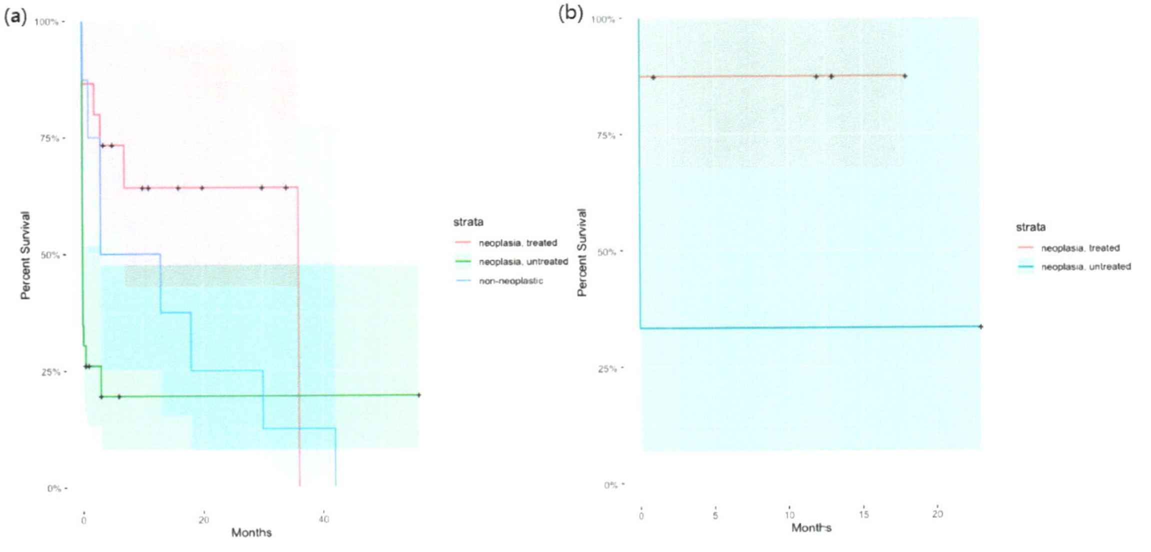

Figure 1. Kaplan–Meier survival plot of survival percentages according to status of neoplasia and treatment. + symbols are for right-censored data (i.e., still alive). Shaded areas show 95% confidence intervals. (**a**) Cases of malignant neoplasia in lizards, including those dying from non-neoplastic cause. (**b**) Cases of benign neoplasia in lizards.

Table 6. Predictor variable values of lizards with neoplasia that had significant association with treatment outcomes based on boosted modeling from the combined data of reviewed literature sources and the ESCRA database.

Predictor Variable	Variable Value	Sample Size	Outcome [a]
Benign vs. malignant	Benign	20	0.10 [b]
	Unknown	6	−0.073 [b]
Histological diagnosis	Chondrosarcoma	2	0.028 [b]
	Islet cell carcinoma	3	−0.038 [b]
	Leukemia	4	−0.060 [b]
	Myxosarcoma	2	0.027 [b]
	Osteosarcoma	3	0.040 [b]
	Squamous cell carcinoma	3	0.12 [b]
Life stage	Unknown	7	−0.077 [b]
Sex	Unknown	17	0.074 [b]
Species	Komodo dragon (*Varanus komodoensis*)	3	−0.038 [b]
	Panther chameleon (*Furcifer pardalis*)	4	0.046 [b]
	Round island skink (*Leiolopisma telfairii*)	2	0.025 [b]
	Savannah monitor (*Varanus exanthematicus*)	5	−0.070 [b]
	Spiny-tailed monitor (*Varanus acanthurus*)	2	0.027 [b]
Treatment	Surgery alone	35	0.12 [b]
	Radiation alone	2	0.031 [b]
	No treatment	52	−0.13 [b]
Tumor location	Abdominal organ (gastrointestinal and alimentary)	3	0.046
	Arm/leg (joint/bone/cartilage)	2	0.028 [b]
	Jaw	2	0.027 [b]
	Kidney	2	0.031 [b]
	Liver	4	−0.053 [b]
	Pancreas	3	0.042 [b]
	Skin/scales (head region)	4	0.049 [b]

[a] Outcome effect is relative to positive or negative outcomes with each clinical report scored as +1 and 0, respectively. Negative outcome is death due to neoplasia, and positive outcome is non-neoplastic (survival or death to non-neoplastic cause). [b] Significance of outcome effect based on $p < 0.05$.

3.5.2. Literature

Malignant neoplasia was significantly associated with death due to neoplasia ($p < 0.05$). Radiation alone and surgery alone were significantly associated ($p < 0.05$) with non-neoplastic outcomes of death, while no treatment was significantly associated with death due to neoplasia ($p < 0.05$). Chondrosarcoma, osteosarcoma, and squamous cell carcinoma were significantly associated with non-neoplastic outcomes ($p < 0.05$), while papilloma was significantly associated with death due to neoplasia ($p < 0.05$). Tumor location in the hematopoietic/lymphatic system was significantly associated with death ($p < 0.05$). No metastasis was significantly associated with death due to neoplasia ($p < 0.05$), while unknown sex was significantly associated with non-neoplastic outcomes of death (Table 7). Round island skinks were significantly associated with non-neoplastic outcomes of death ($p < 0.05$), while savannah monitors were significantly associated with death due to neoplasia ($p < 0.05$). Jaw and skin tumor locations were significantly associated with non-neoplastic outcomes of death ($p < 0.05$), while bone marrow and liver tumor locations were significantly associated with death due to neoplasia ($p < 0.05$).

Table 7. Predictor variable values of lizards with neoplasia that had significant association with treatment outcomes based on boosted modeling from the reviewed literature sources.

Predictor Variable	Variable Value	Sample Size	Outcome [a]
Benign vs. malignant	Malignant	41	−0.076 [b]
Histological diagnosis	Chondrosarcoma	2	0.046 [b]
	Osteosarcoma	3	0.062 [b]
	Papilloma	2	−0.068 [b]
	Squamous cell carcinoma	7	0.14 [b]
Metastasis	No	39	−0.077 [b]
Sex	Unknown	16	0.095 [b]
Species	Round island skink (*Leiolopisma telfairii*)	2	0.039 [b]
	Savannah monitor (*Varanus exanthematicus*)	4	−0.083 [b]
Treatment	Radiation only	2	0.054 [b]
	Surgery only	20	0.095 [b]
	No treatment	29	−0.14 [b]
Tumor location	Bone marrow	3	−0.069 [b]
	Jaw	2	0.042 [b]
	Liver	4	−0.079 [b]
	Skin	5	0.093 [b]

[a] Outcome effect is relative to positive or negative outcomes with each clinical report scored as +1 and 0, respectively. Negative outcome is death due to neoplasia, and positive outcome is non-neoplastic (survival or death from non-neoplastic cause). [b] Significance of outcome effect based on $p < 0.05$.

3.5.3. ESCRA

Cases where the status of metastasis was unknown or the life stage (juvenile or adult) was unknown were significantly more likely to die due to non-neoplastic causes of death ($p < 0.05$). Benign neoplasia and the adult life stage were significantly associated ($p < 0.05$) with death due to non-neoplastic causes of death. Papilloma was unexpectedly found to be significantly associated with death due to neoplasia ($p < 0.05$). Surgery alone was significantly associated ($p < 0.05$) with death due to non-neoplastic causes, while no treatment was significantly associated with death due to neoplasia ($p < 0.05$). Tumor location in the gastrointestinal/hepatic system was significantly associated with death due to neoplasia, while tumor location in the gastrointestinal and alimentary parts of the abdominal organ were significantly associated with death due to neoplasia ($p < 0.05$) (Table 8).

Table 8. Predictor variable values of lizards with neoplasia that had significant association with treatment outcomes based on boosted modeling from the ESCRA database.

Predictor Variable	Variable Value	Sample Size	Outcome [a]
Benign vs. malignant	Benign	6	0.13
	Unknown	6	−0.13
Life stage	Adult	34	0.085 [b]
	Unknown	6	−0.11 [b]
Treatment	Surgery only	15	0.18 [b]
	No treatment	23	−0.13 [b]
Tumor location	Abdominal organ (gastrointestinal/alimentary)	5	−0.094 [b]

[a] Outcome effect is relative to positive or negative outcomes with each clinical report scored as +1 or 0, respectively. Negative outcome is death due to neoplasia, and positive outcome is non-neoplastic (survival or death due to non-neoplastic cause). [b] Significance of outcome effect based on $p < 0.05$.

4. Discussion

Exotic animal oncology is still a developing field of study within veterinary medicine with large gaps in the understanding of disease progression, prognosis, and treatment [113]. This lack of understanding of exotic animal oncology is partly due to a lack of research focused on characterizing disease in these animals. This issue results in a large amount of extrapolation from the understanding of similar neoplastic diseases in cat, dog, and human models. The purpose of this study was to increase the knowledge of neoplasia in lizards specifically and to evaluate the effects of neoplasia on individual animals by exploring the relationships between different variables and patient outcomes. This information could be used to aid in the management of similar clinical cases by providing insight into factors that may positively or negatively affect a patient's ultimate outcome.

Through this study, we sought to evaluate neoplasia in lizards by investigating the literature as well as a database mainly containing unpublished data on tumor cases. The combination of these two data sources produced a broader range of predictions than either data source analyzed alone. Our approach in using both ESCRA and published datasets for the analyses therefore provided a more informative evaluation of the clinical treatment of neoplasia in lizards.

Consensus between predictions stemming from each separate data source were limited and were specific to negative prognoses for cases having no treatment attempted and positive prognoses for cases involving having surgery only. ESCRA uses a standardized approach to collecting relevant clinical information. Unfortunately, a similar guideline for published data in the literature does not exist and would have improved comparisons. The literature was also substantially affected by the publication bias of unique cases, first-reported treatments, or first-reported deaths. Another issue impacting consensus may relate to how it was determined if a patient died due to underlying neoplasia or another cause, which is reliant on the interpretation of the treating veterinarian, pathologist, or researcher.

It does appear that the measurement of survival times is useful in interpreting the association of treatment with neoplasia prognoses. Further investigation would ideally collect information on tumor response (size over time) and other time series data, which would lead to better information regarding clinical intervention strategies for neoplasia in lizards.

While evaluating the species represented in this study, two species (Komodo dragons and savannah monitors) were identified as being statistically more likely to die due to neoplasia than non-neoplastic outcomes. Conversely, panther chameleons and round island skinks were identified as being statistically more likely to die due to non-neoplastic outcomes than neoplasia. These species were not among the most represented species within the literature data, with representation being at $n = 5$ or less, so additional prevalence or incidence investigation is needed to determine if each finding is a true species difference, influenced by limited sample size, or a publication bias.

Of the neoplasms identified in this study, islet cell carcinoma and leukemia cases within the combined data of the literature and ESCRA data sources were associated with death due to neoplasia. Papilloma cases were also associated with death due to neoplasia, but only within the literature data source. Chondrosarcoma, myxosarcoma, osteosarcoma, and squamous cell carcinoma cases from the combined data of the ESCRA and literature data sources were found to be significantly associated with non-neoplastic outcomes of death. The findings concerning papillomas were unexpected due to the typical behavior of the neoplasms. Possible explanations for these findings may be advanced life stage, concurrent systemic disease, owner decisions to pursue euthanasia, excessive size of the neoplasm that inhibited movement, eating or normal life actions, or other complications that lead to death due to other reasons. Additional investigation should be undertaken to confirm these findings, as a better understanding of the general progression and severity of a disease and the effectiveness of treatment is of great clinical importance. Compared to this, squamous cell carcinoma cases were found to have the greatest association with non-neoplastic outcomes. This neoplasia, while typically significantly locally aggressive, can often be managed with adjunctive therapies that allow an animal to survive long enough to die from non-neoplastic causes [114] and may represent a true association.

Of the combined set of 274 cases, metastasis was reported for only 6.6% of these cases. The presence of metastasis was unknown in the majority of cases evaluated (73%). The reason for having a majority of cases where metastasis data is not reported is most likely due to a lack of thorough post-mortem testing on all tissues in these retrospective cases to confirm or deny the presence of metastasis. The encouragement of full necropsies of representative tissues will improve the future evaluation of lizard neoplasia cases to assess metastasis. Benign neoplasia was expectedly associated significantly with outcomes of survival or death due to non-neoplastic causes in the ESCRA data set. Malignant neoplasia was significantly associated with death due to neoplasia in the literature data set. Generally, treatment was not sought in the majority of all cases (81%). A lack of treatment was found to significantly negatively impact patient outcomes. Multiple factors can play into the decision to seek treatment or not, including the type of neoplasm present, the current condition of the patient, the availability of different treatments, and personal factors for the owner. It is not unexpected that the lack of treatment would lead to euthanasia or the death of the animal due to the underlying neoplasia. Without addressing the underlying disease process, most neoplasms will negatively affect the quality of life and lifespan of an individual over time.

Sex as a variable was not found to have a significant association with case outcome, but cases where sex was not reported were found to be significantly associated with death due to non-neoplastic causes. In reviewing the cases of individuals with unknown sex included in this study, it was observed that the majority of cases also experienced a lack of treatment by the owners, which, as discussed above, was shown to be significantly associated with neoplasia-related death. This may reflect a lack of investment in the animals by the owners or could be a reflection of the significant effect of not seeking treatment for these conditions. Regardless, these animals appeared to survive or to die of non-neoplastic causes, and ongoing investigation and statistical analysis is indicated to confirm this association. The surgical removal of neoplasia had a significant association with averting death due to neoplasia and had a greater effect in this regard than treatment with radiation therapy alone. This is consistent with previous publications investigating neoplasia in lizards [1,113]. Surgical excision is the basis of most oncological treatment procedures, and even if no further treatment is performed, it is known to prolong the duration and quality of life in veterinary patients. Surgical excision gives a chance for the complete removal of the neoplasia, through which the best prognosis can be achieved [115]. Treatment protocols remain a developing aspect of exotic animal oncology, and precise dosing and choice of medications are still being established [14,113]. An even greater range of positive outcomes in patients receiving treatment of different modalities may result once more refined protocols are created and used with greater frequency.

5. Conclusions

Multiple predictor variables were significantly associated with death or euthanasia due to neoplasia. Some of the greatest effects in this association were for lizards not receiving treatment, tumors of the liver, leukemia, and tumors for which the statuses of life stage and of being benign versus malignant were unknown. Conversely, the greatest effects of predictor variable values for lizards that did not die due to neoplasia were due to those receiving radiation alone, those with benign tumors, squamous cell carcinoma tumor types, and tumors of the skin/scales head region. In general, treatment for neoplasia contributed to longer survival times in contrast to the rapid rate of death associated with animals dying of neoplasia that did not receive treatment.

Author Contributions: Conceptualization, T.M.H., E.G.D., F.W. and F.J.T.; validation, S.H.H. and B.T.; formal analysis, S.H.H.; data curation, F.J.T., F.W., E.G.D. and T.M.H.; writing—original draft preparation, F.W., F.J.T., T.M.H. and B.T.; writing—F.W., F.J.T., E.G.D., T.M.H., B.T., S.H.H., L.M.A. and A.M.B.; supervision, T.M.H. and E.G.D.; project administration, T.M.H. All authors have read and agreed to the published version of the manuscript.

Funding: This study was partially funded by the National Cancer Institute of the National Institutes of Health under award number U54CA217376.

Institutional Review Board Statement: Not applicable.

Informed Consent Statement: Not applicable.

Data Availability Statement: Deidentified data are available through an approved research request to the Exotic Species Cancer Research Alliance (https://escra.cvm.ncsu.edu/, accessed on 4 May 2024).

Acknowledgments: The authors would like to thank all of the institutions that contributed data to this manuscript, including the Audubon Nature Institute, Audubon Nature Zoo, Avian and Exotic Animal Center, Avian and Exotic Animal Clinic of Indianapolis, Berkeley Dog and Cat Hospital, Caroline Veterinary Specialists, Central Florida Zoo, Denver Zoo, Detroit Zoo, Good Zoo, HerpVet Services, Holly House Veterinary Hospital, Indianapolis Zoo, London Zoo, National Zoo, North Carolina State University, North Carolina Zoo, Oklahoma City Zoo, Parris Creek Veterinary Clinic, Potawatomi Zoo, Prospect Park Zoo, Roger Williams Park Zoo, Santa Barbara Zoo, Tufts University, Veterinary Center for Birds and Exotics, and Wright Veterinary Medical Center. We also would like to thank those who are publishing their cases on the treatment of cancer in reptiles to continue to help to improve outcomes for these animals.

Conflicts of Interest: LMA is a shareholder of and consultant for Peel Therapeutics, Inc.

References

1. Hernandez-Divers, S.M.; Garner, M.M. Neoplasia of reptiles with an emphasis on lizards. *Vet. Clin. N. Am. Exot. Anim. Pr.* **2003**, *6*, 251–273. [CrossRef]
2. Cooper, J.E. Tumours (neoplasms) of Reptiles: Some significant cases from Jersey Zoo. *Dodo J. Durrell Wildl. Conserv. Trust.* **2000**, *36*, 82–86.
3. Eleni, C.; Corteggio, A.; Altamura, G.; Meoli, R.; Cocumelli, C.; Rossi, G.; Friedrich, K.G.; Cerbo, P.d.; Borzacchiello, G. Detection of papillomavirus DNA in cutaneous squamous cell carcinoma and multiple papillomas in captive reptiles. *J. Comp. Pathol.* **2017**, *157*, 23–26. [CrossRef] [PubMed]
4. Kadekaru, S.; Suzuki, T.; Une, Y. Gastric carcinoid in three bearded dragons. *J. Jpn. Vet. Med. Assoc.* **2010**, *63*, 945–949. [CrossRef]
5. Gyimesi, Z.S.; Garner, M.M.; Burns, R.B., III; Nichols, D.K.; Brannian, R.E.; Raymond, J.T.; Poonacha, K.B.; Kennedy, M.; Wojcieszyn, J.W.; Nordhausen, R. High incidence of lymphoid neoplasia in a colony of Egyptian spiny-tailed lizards (*Uromastyx aegyptius*). *J. Zoo. Wildl. Med.* **2005**, *36*, 103–110. [CrossRef] [PubMed]
6. Cooper, J.E.; Gschmeissner, S.; Holt, P.E. Viral particles in a papilloma from green lizard (*Lacerta viridis*). *Lab. Anim.* **1982**, *16*, 12–13. [CrossRef] [PubMed]
7. Ahmad, M.; Taqawi, I.H. Radiation induced leukemia and leukopenia in the lizard Uromastix hardwickii. (Preliminary information). *Radiobiol. Radiother.* **1978**, *19*, 353–360.

8. Eustace, R.; Garner, M.M.; Cook, K.; Miller, C.; Kiupel, M. Multihormonal islet cell carcinomas in three komodo dragons (*Varanus komodoensis*). *J. Zoo. Wildl. Med.* **2017**, *48*, 241–244. [CrossRef] [PubMed]
9. Jankowski, G.; Sirninger, J.; Borne, J.; Nevarez, J.G. Chemotherapeutic treatment for leukemia in a bearded dragon (*Pogona vitticeps*). *J. Zoo. Wildl. Med.* **2011**, *42*, 322–325. [CrossRef] [PubMed]
10. Bel, L.; Tecilla, M.; Borza, G.; Pestean, C.; Purdoiu, R.; Ober, C.; Oana, L.; Taulescu, M. Diagnosis and surgical management of malignant ovarian teratoma in a green iguana (*Iguana iguana*). *BMC Vet. Res.* **2016**, *12*, 144. [CrossRef]
11. Gardhouse, S.; Eshar, D.; Lee-Chow, B.; Foster, R.A.; Ingrao, J.C.; Poirier, V.J. Diagnosis and treatment of a periocular myxosarcoma in a bearded dragon (*Pogona vitticeps*). *Can. Vet. J.* **2014**, *55*, 663–666. [PubMed]
12. Newkirk KM, B.E. Kusewitte DF. Chapter 6: Neoplasia and Tumor Biology. In *Pathologic Basis of Veterinary Disease*, 6th ed.; Zachary, J.F., Ed.; Elsevier: St. Louis, MI, USA, 2017; pp. 286–321.
13. Mayr, A.; Hofner, B.; Waldmann, E.; Hepp, T.; Meyer, S.; Gefeller, O. An Update on Statistical Boosting in Biomedicine. *Comput. Math. Methods Med.* **2017**, *2017*, 6083072. [CrossRef] [PubMed]
14. Hepps Keeney, C.M.; Intile, J.L.; Sims, C.S.; Harrison, T.M. Lymphoid leukemia in five bearded dragons (*Pogona vitticeps*). *J. Am. Vet. Med. Assoc.* **2021**, *258*, 748–757. [CrossRef] [PubMed]
15. Abou-Madi, N.; Kern, T.J. Squamous cell carcinoma associated with a periorbital mass in a veiled chameleon (*Chamaeleo calyptratus*). *Vet. Ophthalmol.* **2002**, *5*, 217–220. [CrossRef] [PubMed]
16. Anderson, N.L.; Williams, J.; Sagartz, J.E.; Barnewall, R. Ovarian teratoma in a green iguana (*Iguana iguana*). *J. Zoo. Wildl. Med.* **1996**, *27*, 90–95.
17. Behncke, H.; Heckers, K. Hemangiosarcoma in a Hispaniolan-curly-tailed-lizard (*Leiocephalus schreibersii*, Cope 1866): Diagnosis and Pathogenesis of a Rare Tumor in a Small Iguanid. In *Proceedings of the 7th International Symposium on Pathology and Medicine in Reptiles and Amphibians (Berlin 2004)*, 7th ed.; Edition Chimaira: Berlin, Germany, 2004.
18. Bronson, E.; Pereira, M.; Sanchez, C.; Murray, S. Iridophoroma in a veiled chameleon, *Chamaeleo calyptratus*. *J. Herpetol. Med. Surg.* **2006**, *16*, 58–60. [CrossRef]
19. Brot, S.d.; Sydler, T.; Nufer, L.; Ruetten, M. Histologic, immunohistochemical, and electron microscopic characterization of a malignant iridophoroma in a dwarf bearded dragon (*Pogona henrylawsoni*). *J. Zoo. Wildl. Med.* **2015**, *46*, 583–587. [CrossRef] [PubMed]
20. Burt, D.G.; Chrisp, C.E.; Gillett, C.S.; Rush, H.G. Two cases of renal neoplasia in a colony of desert iguanas. *J. Am. Vet. Med. Assoc.* **1984**, *185*, 1423–1425.
21. Cardona, J.A.C.; Conley, K.J.; Wellehan, J.F.X., Jr.; Farina, L.L.; Origgi, F.C.; Wamsley, H.L. Incomplete ovariosalpingectomy and subsequent malignant granulosa cell tumor in a female green iguana (*Iguana iguana*). *J. Am. Vet. Med. Assoc.* **2011**, *239*, 237–242. [CrossRef] [PubMed]
22. Cooper, R.H. Melanoma in *Heloderma suspectum* Cope. *Proc. Indiana Acad. Sci.* **1969**, *78*, 466–467.
23. Cruz-Cardona, J.A.; Wamsley, H.L.; Myers, D.A.; Wellehan, J.F.X., Jr.; Origgi, F.C.; Farina, L.L. Malignant granulosa cell tumor in a spayed female green iguana (*Iguana iguana*). In Proceedings of the Association of Reptilian and Amphibian Veterinarians, 15th Annual Conference, Los Angeles, CA, USA, 12–15 October 2008; Association of Reptilian and Amphibian Veterinarians: Chester Heights, PA, USA, 2008.
24. Cushing, A.C.; Ossiboff, R.J.; Knafo, S.E.; Priest, H.; Kraus, M.S.; Kollias, G.V. Coelomic and Pericardial Effusion Associated with Mesothelioma in a Savannah Monitor (*Varanus exanthematicus*). *J. Herpetol. Med. Surg.* **2014**, *24*, 66–71. [CrossRef]
25. Darrow, B.G.; McLean, N.S.J.; Russman, S.E.; Schiller, C.A. Periorbital adenocarcinoma in a bearded dragon (*Pogona vitticeps*). *Vet. Ophthalmol.* **2013**, *16*, 177–182. [CrossRef]
26. Dietz, J.; Heckers, K.O.; Pees, M.; Aupperle, H. Bone tumours in lizards and snakes. A rare clinical finding/Knochentumoren bei Echsen und Schlangen: Ein seltener klinischer Befund in der Reptilienpraxis. *Tierärztliche Praxis. Ausg. K. Kleintiere/Heimtiere* **2015**, *43*, 31–39.
27. Dumonceaux, G.A.; Smith, A.J.; Garner, M.M. Anterior coelomic rhabdomyosarcoma in a central American banded gecko, Coleonyx mitratus. *Bull. Assoc. Reptil. Amphib. Vet.* **1999**, *9*, 23–25. [CrossRef]
28. Emerson, J.A.; Walling, B.E.; Whittington, J.K.; Zarfoss, M.K. Pathology in practice. Squamous cell carcinoma (SCC) of the skin of the upper right eyelid. *J. Am. Vet. Med. Assoc.* **2012**, *240*, 1175–1177. [CrossRef] [PubMed]
29. Folland, D.W.; Johnson, M.S.; Thamm, D.H.; Reavill, D. Management of lymphoma in a green iguana. In Proceedings of the Association of Reptilian and Amphibian Veterinarians, 16th Annual Conference, Milwaukee, WI, USA, 8–15 August 2009; Baer, C.K., Ed.; Association of Reptilian and Amphibian Veterinarians: Chester Heights, PA, USA, 2009; pp. 65–66.
30. Folland, D.W.; Johnston, M.S.; Thamm, D.H.; Reavill, D. Diagnosis and management of lymphoma in a green iguana (*Iguana iguana*). *J. Am. Vet. Med. Assoc.* **2011**, *239*, 985–991. [CrossRef] [PubMed]
31. Frye, F.L. Diagnosis and surgical treatment of reptilian neoplasms with a compilation of cases 1966–1993. *In Vivo* **1994**, *8*, 885–892. [PubMed]
32. Frye, F.L.; Dutra, F.R. Reticulum cell sarcoma in an American anole. *Vet. Med. Small Anim. Clin.* **1974**, *69*, 897–899. [PubMed]
33. Frye, F.L.; Rodger, B.; Nevill, H. Testicular and ovarian tumors in a hermaphroditic savannah monitor lizard, Varanus exanthematicus. *Proc. Annu. Conf. Assoc. Reptil. Amphib. Vet.* **1999**, *6*, 59–62.
34. Gál, J.; Jakab, C.; Balogh, B.; Tóth, T.; Farkas, B. First occurrence of periosteal chondroma (juxtacortical chondroma) in *Uromastyx maliensis* (Reptilia: Sauria: Agamidae). *Acta Vet. Hung.* **2007**, *55*, 327–331. [CrossRef] [PubMed]

35. Garner, M.; Johnson, C.; Funk, R. Liposarcoma in a shingleback skink (*Trachydosaurus rugosus*). *J. Zoo. Wildl. Med.* **1994**, *25*, 150–153.
36. Geczy, C.; Jakab, C. Oral fibrosarcoma in a bearded dragon (*Pogona vitticeps*). Case report. *Magy. Allatorv. Lapja* **2013**, *135*, 413–419.
37. Georoff, T.A.; Stacy, N.I.; Newton, A.N.; McAloose, D.; Post, G.S.; Raskin, R.E. Diagnosis and treatment of chronic T-lymphocytic leukemia in a green tree monitor (*Varanus prasinus*). *J. Herpetol. Med. Surg.* **2009**, *19*, 106–114. [CrossRef]
38. Goe, A.M.; Heard, D.J.; Abbott, J.R.; de Mello Souza, C.H.; Taylor, K.R.; Sthay, J.N.; Wellehan, J.F.X. Surgical management of an odontogenic tumor in a banded Gila monster (*Heloderma suspectum cinctum*) with a novel herpesvirus. *Vet. Q.* **2016**, *36*, 109–114. [CrossRef] [PubMed]
39. Goldberg, S.R.; Holshuh, H.J. A case of leukemia in the desert spiny lizard (*Sceloporus magister*). *J. Wildl. Dis.* **1991**, *27*, 521–525. [CrossRef]
40. Gregory, C.; Latimer, K.S.; Fontenot, D.K.; Lamberski, N.; Campagnoli, R.P. Chronic monocytic leukemia in an inland bearded dragon, *Pogona vitticeps*. *J. Herpetol. Med. Surg.* **2004**, *14*, 12–16. [CrossRef]
41. Hadfield, C.A.; Clayton, L.A.; Clancy, M.M.; Beck, S.E.; Mangus, L.M.; Montali, R.J. Proliferative thyroid lesions in three diplodactylid geckos: *Nephrurus amyae*, *Nephrurus levis*, and *Oedura marmorata*. *J. Zoo. Wildl. Med.* **2012**, *43*, 131–140. [CrossRef]
42. Hannon, D.E.; Garner, M.M.; Reavill, D.R. Squamous cell carcinomas in inland bearded dragons (*Pogona Vitticeps*). *J. Herpetol. Med. Surg.* **2011**, *21*, 101–106. [CrossRef]
43. Heckers, K.O.; Aupperle, H.; Schmidt, V.; Pees, M. Melanophoromas and iridophoromas in reptiles. *J. Comp. Pathol.* **2012**, *146*, 258–268. [CrossRef] [PubMed]
44. Hernandez-Divers, S.J.; Knott, C.D.; MacDonald, J. Diagnosis and surgical treatment of thyroid adenoma-induced hyperthyroidism in a green iguana (*Iguana iguana*). *J. Zoo. Wildl. Med.* **2001**, *32*, 465–475. [PubMed]
45. Hernandez-Divers, S.M.; Stahl, S.J.; Keating, J.H.; Orcutt, C.J.; Garner, M.M.; Schiller, C.A.; Wojcieszyn, J.W. Lymphoma in lizards: Three case reports. *J. Herpetol. Med. Surg.* **2003**, *13*, 14–22. [CrossRef]
46. Irizarry Rovira, A.R.; Holzer, T.R.; Credille, K.M. Systemic mastocytosis in an African fat-tail gecko (*Hemitheconyx caudicinctus*). *J. Comp. Pathol.* **2014**, *151*, 130–134. [CrossRef] [PubMed]
47. Irizarry-Rovira, A.R.; Wolf, A.; Ramos-Vara, J.A. Cutaneous melanophoroma in a green iguana (*Iguana iguana*). *Vet. Clin. Pathol.* **2006**, *35*, 101–105. [CrossRef] [PubMed]
48. Jacobson, E.R.; Reese, D.J.; Berry, C.R.; Brock, P.; Agnew, D.W.; Toplon, D.E.; Abbott, J.R.; Kridel, H.A.; Alleman, A.R.; Dunbar, M.D. What is your diagnosis? Granulosa cell tumor. *J. Am. Vet. Med. Assoc.* **2013**, *243*, 1533–1535. [CrossRef] [PubMed]
49. Jakab, C.; Rusvai, M.; Szabo, Z.; Galfi, P.; Marosan, M.; Kulka, J.; Gal, J. Claudin-7-positive synchronous spontaneous intrahepatic cholangiocarcinoma, adenocarcinoma and adenomas of the gallbladder in a Bearded dragon (*Pogona vitticeps*). *Acta Vet. Hung.* **2011**, *59*, 99–112. [CrossRef] [PubMed]
50. Jayathangaraj, M.G.; Muralimanohar, B.; John, M.C. Reticulum cell sarcoma in a monitor lizard. *Intas Polivet* **2001**, *2*, 288–289.
51. Johnson, J.G., III; Naples, L.M.; Chu, C.; Kinsel, M.J.; Flower, J.E.; Bonn, W.G.v. Cutaneous squamous cell carcinoma in a panther chameleon (*Furcifer pardalis*) and treatment with carboplatin implantable beads. *J. Zoo. Wildl. Med.* **2016**, *47*, 931–934. [CrossRef] [PubMed]
52. Kehoe, S.P.; Guzman, D.S.M.; Sokoloff, A.M.; Grosset, C.; Weber, E.S., III; Murphy, B.; Culp, W.T.N. Partial glossectomy in a blue-tongued skink (*Tiliqua scincoides*) with lingual squamous cell carcinoma. *J. Herpetol. Med. Surg.* **2016**, *26*, 36–41. [CrossRef]
53. Kent, M.S. The use of chemotherapy in exotic animals. *Vet. Clin. North. Am. Exot. Anim. Pract.* **2004**, *7*, 807–820. [CrossRef]
54. Knotek, Z.; Dorrestein, G.M.; Hrda, A.; Tomek, A.; Proks, P.; Knotkova, Z.; Jekl, V.; Lewis, W. Hepatocellular carcinoma in a green iguana—A case study. *Acta Vet. Brno* **2011**, *80*, 243–247. [CrossRef]
55. Kubiak, M.; Denk, D.; Stidworthy, M.F. Retrospective review of neoplasms of captive lizards in the United Kingdom. *Vet. Rec.* **2020**, *186*, 28. [CrossRef]
56. Lawson, R. A malignant neoplasm with metastases in the lizard *Lacerta sicula cetti* Cara. *Br. J. Herpetol.* **1962**, *3*, 22–24.
57. Lemberger, K.Y.; Manharth, A.; Pessier, A.P. Multicentric benign peripheral nerve sheath tumors in two related bearded dragons, *Pogona vitticeps*. *Vet. Pathol.* **2005**, *42*, 507–510. [CrossRef] [PubMed]
58. Levine, B. Visceral hemangiosarcoma and egg yolk peritonitis in a common green iguana, Iguana iguana. In Proceedings of the Annual Conference of the Association of Reptilian and Amphibian Veterinarians, Reno, NV, USA, 9–12 October 2002; Volume 9, pp. 69–71.
59. Levine, B.S. Treatment of a malignant ovarian teratoma in a green iguana. *Exot. DVM* **2004**, *6*, 12–14.
60. Lewis, N.; Martinson, S.; Wadowska, D.; Desmarchelier, M. Malignant mixed chromatophoroma with cutaneous, pulmonary, and testicular metastases in a veiled chameleon (*Chamaeleo calyptratus*). *J. Herpetol. Med. Surg.* **2015**, *25*, 16–20. [CrossRef]
61. Lindemann, D.M.; Eshar, D.; Lin, D.; Narayanan, S.K. Cholangiocarcinoma with concurrent ovarian adenocarcinoma in a green iguana (*Iguana iguana*). *Companion Anim.* **2017**, *22*, 162–168. [CrossRef]
62. Lojszczyk-Szczepaniak, A.; Śmiech, A.; Chlebicka, N.; Szczepaniak, K.O.; Klimiuk, P. First case of intestinal leiomyosarcoma in a bearded dragon: Ultrasonographic findings. *Med. Weter.* **2016**, *72*, 303–306. [CrossRef]
63. Lopez, A.; Bons, J. Studies on a case of papillomatosis in the eyed lizard *Lacerta lepida*. *Br. J. Herpetol.* **1981**, *6*, 123–125.
64. Lyons, J.A.; Newman, S.J.; Greenacre, C.B.; Dunlap, J. A gastric neuroendocrine carcinoma expressing somatostatin in a bearded dragon (*Pogona vitticeps*). *J. Vet. Diagn. Investig.* **2010**, *22*, 316–320. [CrossRef]
65. Martorell, J.; Ramis, A.; Espada, Y. Use of ultrasonography in the diagnosis of hepatic spindle-cell sarcoma in a savannah monitor (*Varanus exanthematicus*). *Vet. Rec.* **2002**, *150*, 282–284. [CrossRef]

66. Mikaelian, I.; Levine, B.S.; Smith, S.G.; Harshbarger, J.C.; Wong, V.J. Malignant peripheral nerve sheath tumor in a bearded dragon, *Pogona vitticeps*. *J. Herpetol. Med. Surg.* **2001**, *11*, 9–12. [CrossRef]
67. Mikaelian, I.; Lynch, S.; Harshbarger, J.C.; Reavill, D.R. Malignant chromatophoroma in a day gecko. *Exot. Pet. Pract.* **2000**, *5*, 73–74.
68. Mooij, T.S.; Martel, A.; Bosseler, L.; Chiers, K.; Pasmans, F.; Hellebuyck, T. Atypical clinical presentation of a metastatic gastric neuroendocrine carcinoma in a bearded dragon (*Pogona vitticeps*). [Atypische klinische presentatie van een metastatisch gastrisch neuro-endocrien carcinoom bij een baardagame (*Pogona vitticeps*)]. *Vlaams Diergeneeskd. Tijdschr.* **2014**, *83*, 293–298.
69. Morales Salinas, E.; Aguilar Arriaga, B.O.; Ramirez Lezama, J.; Mendez Bernal, A.; Lopez Garrido, S.J. Oral Fibrosarcoma in a Black Iguana (*Ctenosaura Pectinata*). *J. Zoo. Wildl. Med.* **2013**, *44*, 513–516. [CrossRef] [PubMed]
70. Naples, L.M.; Langan, J.N.; Mylniczenko, N.D.; Kagan, R.; Colegrove, K. Islet cell tumor in a Savannah monitor (*Varanus exanthematicus*). *J. Herpetol. Med. Surg.* **2009**, *19*, 97–105. [CrossRef]
71. Needle, D.; McKnight, C.A.; Kiupel, M. Chondroblastic Osteosarcoma In Two Related Spiny-Tailed Monitor Lizards (*Varanus acanthurus*). *J. Exot. Pet. Med.* **2013**, *22*, 265–269. [CrossRef]
72. Patterson-Kane, J.C.; Redrobe, S.P. Colonic adenocarcinoma in a leopard gecko (*Eublepharis macularius*). *Vet. Rec.* **2005**, *157*, 294–295. [CrossRef] [PubMed]
73. Pellett, S. Squamous cell carcinoma in an inland bearded dragon (*Pogona vitticeps*). In Proceedings of the British Veterinary Zoological Society Spring Meeting 2014, Marwell Zoo, UK, 26–27 April 2014; Focus on Exotic Pets (with expert-led debate "Exotic Pets: Welfare and Trade"). Molenaar, F., Stidworthy, M., Roberts, V., Eds.; British Veterinary Zoological Society: Romford, UK, 2014; p. 54.
74. Pellett, S.; Pinborough, M. Squamous cell carcinoma in a central bearded dragon. *Companion Anim.* **2014**, *19*, 379–384. [CrossRef]
75. Pellett, S.; Pinborough, M.; Fiddes, M.; Cope, I. Ovarian teratoma in a bearded dragon, *Pogona vitticeps*. In Proceedings of the BSAVA Congress 2014, Birmingham, UK, 3–6 April 2014; Scientific Proceedings Veterinary Programme; British Small Animal Veterinary Association: Quedgeley, UK, 2014; p. 643.
76. Pryor, S.G.; Cutler, D.; Yau, W.; Diehl, K.A. Adnexal cystadenoma in a bearded dragon (*Pogona vitticeps*). *J. Exot. Pet. Med.* **2018**, *27*, 85–89. [CrossRef]
77. Raiti, P. Husbandry, diseases, and veterinary care of the bearded dragon (*Pogona vitticeps*). *J. Herpetol. Med. Surg.* **2012**, *22*, 117–131. [CrossRef]
78. Reichenbach-Klinke, H.H.; Schroder, J.H.; Zeller, I. Attempted treatment of a maxillary tumour of a red-throated anolis, *Anolis carolensis* (Iguanidae; Reptilia). [Versuch der Therapierung einer Kiefergeschwulst beim Rotkehlanolis *Anolis carolinensis* Voigt]. *Vet.-Med. Nachrichten* **1972**, *2*, 159–163.
79. Ritter, J.M.; Garner, M.M.; Chilton, J.A.; Jacobson, E.R.; Kiupel, M. Gastric neuroendocrine carcinomas in bearded dragons (*Pogona vitticeps*). *Vet. Pathol.* **2009**, *46*, 1109–1116. [CrossRef] [PubMed]
80. Rivera, S.; Crane, M.M.; McManamon, R.; Gregory, C.R. Surgical treatment of pulmonary melanophoroma in a beaded lizard (*Heloderma horridum exasperatum*). *J. Zoo. Wildl. Med.* **2015**, *46*, 397–399. [CrossRef]
81. Romagnano, A.; Jacobson, E.R.; Boon, G.D.; Broeder, A.; Ivan, L.; Homer, B.L. Lymphosarcoma in a green iguana (*Iguana iguana*). *J. Zoo. Wildl. Med.* **1996**, *27*, 83–89.
82. Rousselet, E.; Souza, C.H.d.M.; Wellehan, J.F.X., Jr.; Epperson, E.D.; Dark, M.J.; Wamsley, H.L. Cutaneous iridophoroma in a Green iguana (*Iguana iguana*). *Vet. Clin. Pathol.* **2017**, *46*, 625–628. [CrossRef] [PubMed]
83. Rowland, M.N. Granulosa-thecal cell tumor in an Asian water dragon (*Physignathus cocincinus*). In Proceedings of the Association of Reptilian and Amphibian Veterinarians, 16th Annual Conference, Milwaukee, WI, USA, 8–15 August 2009; Baer, C.K., Ed.; Association of Reptilian and Amphibian Veterinarians: Chester Heights, WI, USA, 2009; pp. 141–143.
84. Sander, S.J.; Ossiboff, R.J.; Stokol, T.; Steeil, J.C.; Neiffer, D.L. Endolymphatic Sac Carcinoma In Situ in a Tokay Gecko (*Gekko gecko*). *J. Herpetol. Med. Surg.* **2015**, *25*, 82–86. [CrossRef]
85. Sasi, N.N. Adenocarcinoma in Varanus monitor. *Indian. Vet. J.* **1993**, *70*, 775.
86. Savageau, N.R.; Gamble, K.C. Clinical challenge: Renal adenocarcinoma in a beaded lizard (*Heloderma horridum horridum*). *J. Zoo. Wildl. Med.* **2016**, *47*, 945–947. [CrossRef] [PubMed]
87. Schilliger, L.; Paillusseau, C.; Gandar, F.; De Fornel, P. Iridium 192 ((192)-Ir) High dose rate brachytherapy in a central bearded dragon (*Pogona vitticeps*) with rostral squamous cell carcinoma. *J. Zoo. Wildl. Med.* **2020**, *51*, 241–244. [CrossRef] [PubMed]
88. Schilliger, L.; Selleri, P.; Gandar, F.; Rival, F.; Bonwitt, J.; Frye, F.L. Adenoid Hepatocellular Carcinoma Accompanied by Uncharacterized Eosinophilic Intracytoplasmic Inclusions in a Green Iguana (*Iguana iguana*). *J. Herpetol. Med. Surg.* **2012**, *22*, 70–75. [CrossRef]
89. Schmidt, R.E. Plasma cell tumor in an East Indian water lizard (*Hydrosaurus amboinensis*). *J. Wildl. Dis.* **1977**, *13*, 47–48. [CrossRef] [PubMed]
90. Schönbauer, M.; Loupal, G.; Schönbauer-Längle, A. Osteoid chondrosarcoma in a desert monitor (Varanus griseus). *Berl. Munch. Tierarztl. Wochenschr.* **1982**, *95*, 193–194.
91. Schultze, A.E.; Mason, G.L.; Clyde, V.L. Lymphosarcoma with leukemic blood profile in a savannah monitor lizard (*Varanus exanthematicus*). *J. Zoo. Wildl. Med.* **1999**, *30*, 158–164. [PubMed]

92. Sonntag, F.D.; Schroff, C.; Dietz, J.; Heckers, K.O. Metastatic leiomyosarcoma of the ovary in an Inland bearded dragon (*Pogona vitticeps*)—A case report. [Metastasierendes Leiomyosarkom des Ovars bei einer Bartagame (*Pogona vitticeps*)—Ein Fallbericht]. *Prakt. Tierarzt* **2014**, *95*, 518–523.
93. Stacy, B.A.; Vidal, J.D.; Osofsky, A.; Terio, K.; Koski, M.; Cock, H.E.V.d. Ovarian papillary cystadenocarcinomas in a green iguana (*Iguana iguana*). *J. Comp. Pathol.* **2004**, *130*, 223–228. [CrossRef]
94. Stegeman, N.; Aloisio, F.; Edwards, J.; Belcher, C.; Hoppes, S. Disseminated spindle cell sarcoma in a Savannah monitor (*Varanus exanthematicus*). In Proceedings of the Association of Reptilian and Amphibian Veterinarians, 16th Annual Conference, Milwaukee, WI, USA, 8–15 August 2009; Baer, C.K., Ed.; Association of Reptilian and Amphibian Veterinarians: Chester Heights, WI, USA, 2009; p. 175.
95. Stolk, A. Tumours of reptiles. 4. Multiple osteomas in the lizard *Lacerta viridis*. *Beaufortia* **1958**, *7*, 1–9.
96. Strunk, A.; Haney, S.; Reavill, D. Malignant chromatophoroma in a bearded dragon (*Pogona vitticeps*). In Proceedings of the Association of Reptilian and Amphibian Veterinarians, 16th Annual Conference, Milwaukee, WI, USA, 8–15 August 2009; Baer, C.K., Ed.; Association of Reptilian and Amphibian Veterinarians: Chester Heights, WI, USA, 2009; pp. 73–76.
97. Suedmeyer, W.K.; Turk, J.R. Lymphoblastic leukemia in an inland bearded dragon, *Pogona vitticeps*. *Bull. Assoc. Reptil. Amphib. Vet.* **1996**, *6*, 10–12. [CrossRef]
98. Sykes, J.M.t.; Trupkiewicz, J.G. Reptile neoplasia at the Philadelphia Zoological Garden, 1901–2002. *J. Zoo. Wildl. Med.* **2006**, *37*, 11–19. [CrossRef]
99. Tocidlowski, M.E.; McNamara, P.L.; Wojcieszyn, J.W. Myelogenous leukemia in a bearded dragon (*Acanthodraco vitticeps*). *J. Zoo. Wildl. Med.* **2001**, *32*, 90–95. [PubMed]
100. Tocidlowski, M.E.; Merrill, C.L.; Loomis, M.R.; Wright, J.F. Teratoma in desert grassland whiptail lizards (*Cnemidophorus uniparens*). *J. Zoo. Wildl. Med.* **2001**, *32*, 257–259.
101. Tsuruno, S.; Yamanouchi, A.; Kajigaya, H.; Saito, T. A case of hepatocellular carcinoma in green iguana (*Iguana iguana*). *Jpn. J. Zoo. Wildl. Med.* **2006**, *11*, 87–92. [CrossRef]
102. Tsuruno, S.; Yamashita, Y.; Yamanouchi, A.; Kajigaya, H.; Saito, T. Two cases of adrenal cortical adenocarcinoma in female green iguana (*Iguana iguana*). *Jpn. J. Zoo. Wildl. Med.* **2010**, *15*, 115–119. [CrossRef]
103. Tynes, V.V. Cholangiocarcinoma in a green iguana. *Exot. Pet. Pract.* **1997**, *2*, 87.
104. Vasil'ev, D.B.; Solov'ev, I.N. Ganglioneuroblastoma in Yemenite chameleon (*Chamaeleo calyptratus*): The first recorded case of a tumor of neuroectodermal histogenesis in reptiles. *Arkh Patol.* **2006**, *68*, 45–47. [PubMed]
105. Welle, M.; Rödiger, K.S. Cholangioma in a green iguana. [Cholangiom bei einem Grünen Leguan (*Iguana iguana*)]. *Kleintierpraxis* **1992**, *37*, 415–417.
106. Wenger, S.; Simova-Curd, S.; Grest, P.; Wteinmetz, H.W.; Hatt, J.-M. Ovarian teratoma in a Fiji Island banded iguana (*Brachylophus fasciatus*) and a green iguana (*Iguana iguana*). *J. Herpetol. Med. Surg.* **2010**, *20*, 20–23. [CrossRef]
107. Whiteside, D.P.; Garner, M.M. Thyroid adenocarcinoma in a crocodile lizard, *Shinisaurus crocodilurus*. *J. Herpetol. Med. Surg.* **2001**, *11*, 13–16. [CrossRef]
108. Williams, M.J.; Wong, H.E.; Priestnall, S.L.; Szladovits, B.; Stapleton, N.; Hedley, J. Anaplastic Sarcoma and Sertoli Cell Tumor in a Central Bearded Dragon (*Pogona vitticeps*). *J. Herpetol. Med. Surg.* **2020**, *30*, 68–73. [CrossRef] [PubMed]
109. Wilson, G.H.; Fontenot, D.K.; Brown, C.A.; Kling, M.A.; Stedman, N.; Greenacre, C.B. Pseudocarcinomatous biliary hyperplasia in two green iguanas, *Iguana iguana*. *J. Herpetol. Med. Surg.* **2004**, *14*, 12–18. [CrossRef]
110. Zhelev, V. Disseminated squamous epithelial keratogenous carcinoma in a lizard (*Lacerta viridis*). *Obs. I Sravn. Patol.* **1978**, 117–120.
111. Zschiesche, W. Cardiac rhabdomyoma in a small lizard, *Anolis equestris*. [Rabdomioma cardiaco in una piccola lucertola] *Anolis equestris*. *Summa* **1986**, *3*, 105–106.
112. Zwart, P.; Harshbarger, J.C. Hematopoietic neoplasms in lizards: Report of a typical case in *Hydrosaurus amboinensis* and of a probable case in *Varanus salvator*. *Int. J. Cancer* **1972**, *9*, 548–553. [CrossRef] [PubMed]
113. Zehnder, A.; Graham, J.; Antonissen, G. Update on Cancer Treatment in Exotics. *Vet. Clin. N. Am. Exot. Anim. Pr.* **2018**, *21*, 465–509. [CrossRef] [PubMed]
114. Duke, E.G.; Harrison, S.H.; Moresco, A.; Trout, T.; Troan, B.V.; Garner, M.M.; Smith, M.; Smith, S.; Harrison, T.M. A Multi-Institutional Collaboration to Understand Neoplasia, Treatment and Survival of Snakes. *Animals* **2022**, *12*, 258. [CrossRef] [PubMed]
115. Christman, J.; Devau, M.; Wilson-Robles, H.; Hoppes, S.; Rech, R.; Russell, K.E.; Heatley, J.J. Oncology of Reptiles: Diseases, Diagnosis, and Treatment. *Vet. Clin. North. Am. Exot. Anim. Pr.* **2017**, *20*, 87–110. [CrossRef] [PubMed]

animals

MDPI

Case Report

Successful Treatment of an Acinar Pancreatic Carcinoma in an Inland Bearded Dragon (*Pogona vitticeps*): A Case Report

Johannes Hetterich [1,*], Marion Hewicker-Trautwein [2], Wencke Reineking [2], Lisa Allnoch [2] and Michael Pees [1]

[1] Department of Small Mammal, Reptile and Avian Medicine and Surgery, University of Veterinary Medicine Hannover, Foundation, 30559 Hannover, Germany
[2] Department of Pathology, University of Veterinary Medicine Hannover, Foundation, 30559 Hannover, Germany
* Correspondence: johannes.hetterich@tiho-hannover.de

Simple Summary: This report describes the diagnostic evaluation and therapy of a bearded dragon with an irregular mass within its body cavity. Initially, the eight-year-old lizard was presented with unspecific clinical symptoms (reduced overall condition and reduced forage intake). A round-shaped mass was determined by physical examination. The irregular structure subsequently was visualized by ultrasonography. After two days of stabilization therapy in the clinic, the mass was removed surgically under general anesthesia. The histological examination of the irregular tissue revealed a pancreatic tumor. Neoplasms of the pancreas are rare in lizards, and successful treatment and long-term survival are not commonly reported. The lizard recovered slowly but gradually after the surgical procedure and regained regular forage intake and behavior within three weeks. Long-term survival was confirmed by follow-ups within two years after the surgery.

Abstract: An adult, 362 g, male, intact inland bearded dragon (*Pogona vitticeps*) was admitted to a veterinary clinic due to a temporary cloacal prolapse and a two-week history of reduced overall condition and forage intake. Physical examination revealed an approximately 2 × 1 cm round-shaped, rigid intracoelomic tissue mass. Multiple sand deposits were present on the cloacal mucous membranes, though no signs of cloacal prolapse were present. The lizard was otherwise responsive but showed reduced body tension and movement behavior. Initial fecal examination revealed a high-grade oxyuriasis. A 2 × 1.5 cm sized intracoelomic, well-vascularized, round-shaped mass was subsequently visualized by ultrasonography. After a two-day stabilization therapy, the intracoelomic mass was removed by performing a standard ventral coeliotomy under general anesthesia. Histopathological examination of the excised mass revealed an acinar pancreatic adenocarcinoma with infiltration of the peritumorous connective soft tissue. The lizard remained at the clinic for a further seven days. Its postsurgical condition improved slowly. However, the lizard started regular forage intake 10 days after surgery, and general behavior enhanced constantly within the following three weeks. The animal was presented for a follow-up six weeks after surgery, showing bright and alert behavior with no signs of disease or illness. The lizard was re-examined 20 months after the initial presentation due to a reduced overall condition and reduced food intake. Blood chemistry evaluation revealed markedly decreased protein parameters, and moderate ascites was identified ultrasonographically. A distinct association with the preceding neoplastic disease could not be made, and the lizard returned to its regular condition under supportive therapy within three weeks. To the authors' knowledge, this is the first report of successful treatment of a pancreatic carcinoma in a bearded dragon.

Keywords: lizard; neoplasia; pancreatic disease; ultrasonography; histopathology; coeliotomy

Citation: Hetterich, J.; Hewicker-Trautwein, M.; Reineking, W.; Allnoch, L.; Pees, M. Successful Treatment of an Acinar Pancreatic Carcinoma in an Inland Bearded Dragon (*Pogona vitticeps*): A Case Report. *Animals* **2024**, *14*, 1976. https://doi.org/10.3390/ani14131976

Academic Editors: Josep Pastor and Juan Carlos Illera del Portal

Received: 22 March 2024
Revised: 17 June 2024
Accepted: 1 July 2024
Published: 4 July 2024

1. Introduction

Diagnosis of pancreatic disease in reptiles is challenging, as clinical signs are often nonspecific and indicative signs, like pronounced changes in blood parameters, are lacking compared to mammalian medicine [1]. Accurate diagnoses of pancreatic diseases in reptiles are oftentimes based on postmortem examinations.

Pancreatic neoplasia reports in reptiles are rare [2]. For lizards, multihormonal islet cell carcinomas have been described for Komodo dragons (*Varanus komodoensis*) [3]. A pancreatic glucagonoma was found in a rhinoceros iguana (*Cyclura c. figgensi*) [4]. A retrospective study of neoplasms of captive lizards in the United Kingdom evaluated 158 tumors, revealing a prevalence of pancreatic neoplasia in 1.3% of all tumor diseases [5]. Survey-based data on long-term outcome of pancreatic neoplastic disease in lizards are not available. Disease is usually found to be in an advanced stage by the time of presentation [1]. Clinical symptoms like lethargy, anorexia and weakness are pronounced but nonspecific [1]. Treatment attempts should include nutritional support, fluid therapy, and analgesia [1,2]. However, most animals are euthanized due to severe clinical conditions and poor prognosis [1].

2. Clinical Case

2.1. History and Clinical Examination

An eight-year-old, 362 g, male inland bearded dragon (*Pogona vitticeps*) raised under human care was presented to the veterinary clinic due to a temporary cloacal prolapse. Also, the animal owner reported a three-week history of progressing lethargy and reduced forage intake. Husbandry conditions included a 180 × 60 × 60 cm glass terrarium with a temperature gradient of 24–29 °C (75–85 °F) and a basking area of 35 °C (95 °F). The animal owner provided an ultraviolet light source for 10 h/day and maintained a 40–50% humidity within the terrarium. The daily diet included various herbs, salads, and vegetables. Insects (crickets, cockroaches) were given twice a week and were constantly dusted with supplemental calcium powder. The animal owner did not report any pre-existing disease or medical problem in the lizard.

Clinical examination revealed a good body condition. However, the lizard showed reduced body tension and movement behavior. An approximately 2 × 1 cm round-shaped, rigid intracoelomic tissue mass in the middle third of the medial right coelom was identified by coelomic palpation. Minor sand deposits were stuck at cloacal mucous membranes, though no signs of cloacal prolapse were present. Based on the general examination, no other abnormalities than those mentioned above were present. Upon case history and general examination, neoplastic disease of different coelomic organs (intestinals, spleen, liver, gallbladder, pancreas) and an intracoelomic abscess were the main differential diagnoses.

2.2. Diagnostic Procedures

Initial diagnostic methods included parasitological fecal examination, radiography, and ultrasonography. The animal owner declined a blood examination due to financial restrictions. Microscopic fecal examination at 100× magnification revealed a high infestation of oxyurids (>20 eggs per field of view) and a low-grade flagellate infestation. Under manual restraint, dorsoventral and horizontal beam lateral radiographic projections were performed (Figure 1; digital X-ray; detector system: Fujifilm Console Advance DR-ID 300 CL, Fujifilm Europe GmbH, Düsseldorf, Germany; tube system: Gierth X-ray International GmbH, Riesa, Germany; film focus distance FFD 60 cm, 50 kV, 5 mAs). The lateral radiograph was insufficient for any evaluation due to a technical issue. On the dorsoventral view, the medial right coelom was filled with soft-tissue material and hypodense areas (moderately filled gastrointestinal tract and gastrointestinal gas). Mineral opacities were visible in the left coelom, indicating deposits of foreign material (sand or similar anorganic substrate) in the large intestine. Both lung fields were visible on the dorsoventral radiograph, though superimpositions with the gastrointestinal tract did not allow a clear evaluation of the entire lung field size. No distinct intracoelomic lesion was determinable upon radiography. Ultrasonographic examination under manual restraint and in dorsal recumbency (Micro curved array transducer, 5–9 MHz; Vivid 7 Dimension; GE Healthcare GmbH, Solingen, Germany) with transverse and longitudinal views from the ventral coupling site showed an approximately 2 × 1.5 cm round-shaped, intracoelomic mass located in the middle third of the medial right coelom. The structure was surrounded by intestinal and fat body tissue (Figure 2). However, the mass showed good vascularization, and demarcation to surrounding tissue was clearly feasible (Figure 3). No other abnormalities of coelomic organs were visualized during the ultrasound examination.

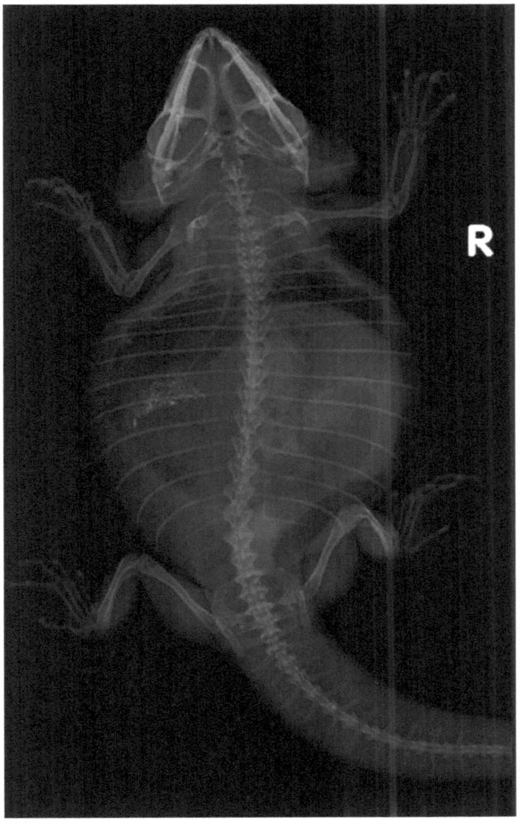

Figure 1. Dorsoventral radiograph of an eight-year-old bearded dragon (*Pogona vitticeps*). No indicative coelomic alterations are visible.

Based on the ultrasonographic findings and the reduced clinical condition, the presumptive diagnosis for the intracoelomic mass was a neoplastic disease involving the gastrointestinal tract. Distinct organ affiliation could not be interpreted at this stage. Further diagnostic imaging, including a coelioscopy (and, if possible, a biopsy of the affected tissue), might have been reasonable. However, the animal owner declined any further diagnostic testing. An explorative coeliotomy as a treatment attempt was conducted. The animal preoperatively was treated with fluid administration (20 mL/kg SC SID; Sterofundin, ISO 1/1 E, B. Braun Melsungen AG, Melsungen, Germany), analgetic therapy (0.3 mg/kg SC SID; Meloxicam; Metacam 2 mg/mL, Boehringer Ingelheim GmbH, Ingelheim, Germany), and housing in the species' preferred optimal temperature zone (POTZ) over a period of two days.

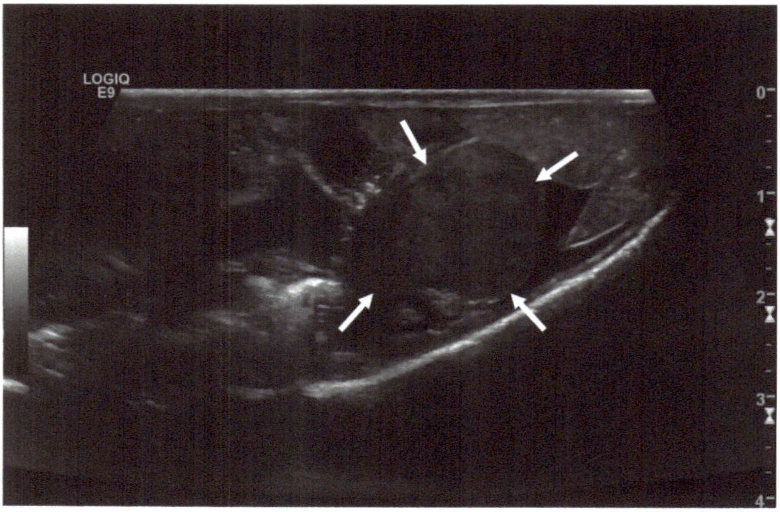

Figure 2. Transverse ultrasonographic image of the medial right coelom in a bearded dragon acquired with a 5 to 9 MHz micro curved array transducer (penetration depth 3 cm; frequency 15 MHz) showing an approximately 2 × 1.5 cm round-shaped, irregular intracoelomic mass (arrows). Additionally, the gastrointestinal tract (medial, here: to the left of the mass) and fat body (lateral, here: to the right of the mass) are visible and closely related to the mass.

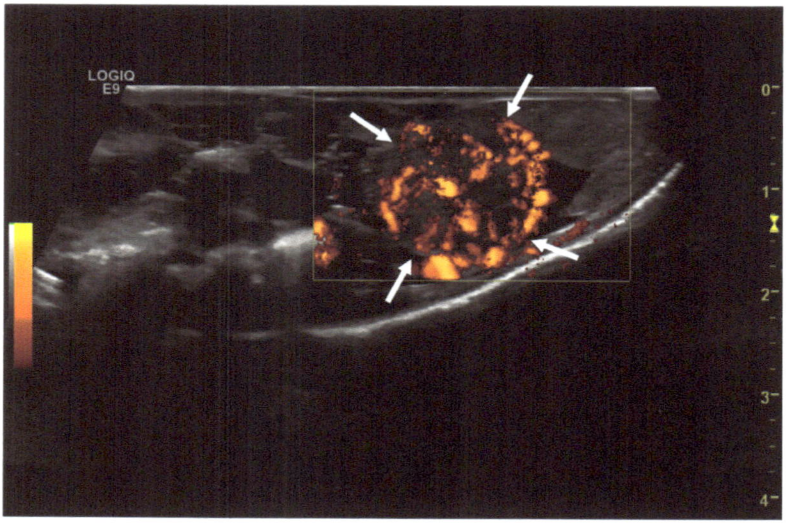

Figure 3. Transverse ultrasonographic image of the medial right coelom in a bearded dragon acquired with a 5 to 9 MHz micro curved array transducer and Color Doppler evaluation (penetration depth 2 cm; frequency 15 MHz). Note the distinct vascularization of the round-shaped intracoelomic mass (arrows). Additionally, gastrointestinal tract (medial, here: to the left of the mass) and fat body (lateral, here: to the right of the mass) are visible and closely related to the mass.

General anesthesia was induced with hydromorphone (1 mg/kg IM; Hydromorphon 2 mg/mL, hameln pharma GmbH, Hameln, Germany), ketamine (10 mg/kg IV; Ketamin 100 mg/mL, CP-Pharma Handelsgesellschaft mbH, Burgdorf, Germany) and medetomidine (0.1 mg/kg IV; Cepetor 1 mg/mL, CP-Pharma Handelsgesellschaft mbH). The bearded dragon was intubated with a modified 14-gauge IV catheter. Anesthesia was maintained using inhalant isoflurane (1–3%) and oxygen (0.8 L/min) via intermittent positive-pressure ventilation. Anesthesia monitoring included monitoring the heart and respiratory rate as well as cloacal temperature. Thermal support was provided throughout the surgical procedure. The lizard was placed in dorsal recumbency and, after a standardized aseptical preparation using an iodine solution (Jodosept®, Vetoquinol GmbH, Ismaning, Germany) and chlorhexidine, a coeliotomy using a ventral paramedian approach was performed. After a skin incision with a scalpel blade (#11), the ventral abdominal muscles and the peritoneal membrane were incised stepwise using Metzenbaum scissors. A prominent, round-shaped, well-vascularized mass in the medial right mid-coelom was identified. The tan-colored bulging tissue with an uneven but smooth surface was closely related to the small intestines and even more closely adjacent to pancreatic tissue. However, no capsule-like demarcations were visibly connected to further organs and the mass was resected completely. An extensive hematoma was present in between the pancreatic tissue, small intestines, and the irregular mass lesion. It was removed partly using Metzenbaum scissors, as its inseparable connection with the surrounding small intestine and pancreatic tissue did not allow for a complete extirpation. No resection or incision of regular small intestine and pancreatic tissue was performed. The surgical procedure was proceeded by a thorough coelomic inspection and lavage with sterile warm fluids (Sterofundine ISO 1/1 E, 15 mL; B. Braun AG, Melsungen, Germany). Conclusively, stepwise standard closure of the coelom with a continuous over-and-over one-layer suture for the peritoneal membrane and ventral abdominal muscles (Polyglycolide, coated, absorbable), as well as simple interrupted skin sutures (Polyglycolide, coated, absorbable), was performed.

2.3. Postoperative Care and Follow-Up

The lizard recovered slowly from anesthesia. Despite administering a reversal agent of atipamezole (0.1 mg/kg IM; atipamezole 5 mg/mL, CP-Pharma Handelsgesellschaft mbH, Burgdorf, Germany) and fluids (20 mL/kg SC BID, Sterofundin, ISO 1/1 E) post-operatively, the lizard remained unresponsive, with markedly decreased reflexes and no signs of independent movement behavior for the first 18 h postoperatively. However, fluids (20 mL/kg SC BID; Sterofundin, ISO 1/1 E) and analgetic therapy (Meloxicam 0.3 mg/kg SC SID; Tramadol, 10 mg/kg every 48 h IM; Tramadolhydrochloride 50 mg/mL, Dechra Veterinary Products GmbH, Aulendorf, Germany) were proceeded and an antibiotic treatment (10 mg/kg SID SC; Enrofloxacin, Baytril 25 mg/mL, Elanco GmbH, Cuxhaven, Germany) was started. The patient's behavior improved continuously on day one post surgical procedure.

On the second day after surgery, the animal's body tension and reflexes returned to their full extent. The animal also regained independent movement behaviour. Postsurgical therapy for three days included the fluid, analgetic and antibiotic treatment mentioned above as well as supportive care with sucralfate (50 mg/kg SID PO; Sucrabest®, Combustin GmbH, Hailtingen, Germany), forced feeding (15 mL/kg SID PO; Oxbow Critical Care Herbivore, Oxbow Pet Products, Murdock, NE, USA) and UV-radiation therapy for 15 min daily. The lizard was released from the clinic three days after surgery. The owner continued oral therapy with enrofloxacin (prescribed for further ten days), meloxicam (seven days), tramadol (five days), sucralfate (seven days) and forced feeding (salad and herbs every 48 h, and crickets twice per week). Three days later, the owner presented the bearded dragon with markedly reduced movement behavior and body tension and anorexia. Blood sampling and a follow-up ultrasonography were recommended to the animal owner but rejected due to financial restrictions. The patient was treated in the clinic for a period of another three days, continuing the therapy mentioned, which included enrofloxacin (10 mg/kg SID SC), Meloxicam (0.3 mg/kg SC SID), tramadol (10 mg/kg every 48 h IM), fluids (20 mL/kg SC BID; Sterofundin, ISO 1/1 E), sucralfate (50 mg/kg SID PO) and forced feeding (daily). Its general condition improved, and one week after release from the clinic the owner reported slow but constant progress in the animal's condition. The lizard regained regular food intake after 16 days. Its overall condition was evaluated as unremarkable in a follow-up six weeks after surgery. The healing of the incision site on the ventral coelom was unremarkable, and all skin sutures were removed.

The lizard was re-examined 20 months after the initial presentation due to a reduced overall condition and appetite. Clinically, it showed nonspecific symptoms, including a reduced body tension and moderately weakened overall condition. Based on the general examination, no other abnormalities than those mentioned above were present. Ultrasonography of the coelom under manual restraint in dorsal recumbency identified moderate ascites, but no other distinct findings relating to any of the coelomic organs were found. Also, the coelomic area of the previous surgery revealed no abnormalities. Blood collection from the ventral tail vein was performed using a 22G cannula (0.5 mL, lithium-heparine). Blood chemistry evaluation revealed markedly decreased protein parameters (total protein 1.95 g/dL (3.0–8.1 g/dL), albumin 1.02 g/dL (1.2–4.0 g/dL), references according to Exotic Animal Formulary, J. Carpenter, fifth edition, 2018). All other blood parameters were evaluated to be within the species-specific references. Differential diagnoses included various liver and gastrointestinal diseases. Any further plausible diagnostic procedures (complete blood cell count, ascites sampling, coelioscopy, magnetic resonance imaging) including stationary therapy were declined by the animal owner. A supportive therapy for the lizard included furosemide (5 mg/kg SID PO; furosemide 50 mg/mL, cp-pharma, Handelsgesellschaft mbH, Burgdorf, Germany), a liver-supporting product (50 mg/kg SID PO; Legaphyton®, Vetoquinol S.A., Lure, France), and forced feeding of a high-protein diet (raising the number of cockroaches and crickets from 2–3 up to 7–8 weekly). The lizard's condition improved, and the animal regained regular food intake three weeks later. The animal owner neither presented the lizard for the follow-up appointment four weeks later nor any later follow-up.

2.4. Histological, Immunohistochemical and Electron Microscopical Evaluation

The surgically removed mass was fixed in 10% neutral buffered formalin, routinely embedded in paraffin, and 2–3 µm sections were stained with hematoxylin and eosin (H&E) for histologic examination.

Histological examination of the completely excised tissue revealed a well-differentiated moderately cell-dense demarcated multinodular neoplastic mass with an acinar growth pattern that extended multifocally into the organ capsule (Figure 4A,B). The mass replaced normal pancreatic architecture. Upon multiple cross sections, no endocrine islets were detectable. The multifocally dilated acinar structures consisted of cuboidal to columnar tumor cells. Throughout the tumor, a strong eosinophilic, slightly granular staining of the apical cytoplasm was seen, indicating the presence of cytoplasmic zymogen granules (white arrows, Figure 4C). The medium-sized or large nuclei of the tumor cells were located in the periphery of the cells and had prominent nucleoli (black arrows, Figure 4C). Sometimes tumor cells with vacuolated cytoplasm and vesicular appearing nuclei were found (black arrowheads, Figure 4C). There was moderate anisocytosis and anisokaryosis with nucleolar pleomorphism. Mitotic figures were not seen. Between the tumor cells, a scant amount of fibrovascular stromal tissue was present. The mass was surrounded by a rim of connective tissue (interpreted as organ capsule) which had multiple hyperemic blood vessels, a mild focal hemorrhage, and multifocal infiltration by tumor cells (Figure 4D). Based on the histomorphology, an acinar pancreatic carcinoma was suspected.

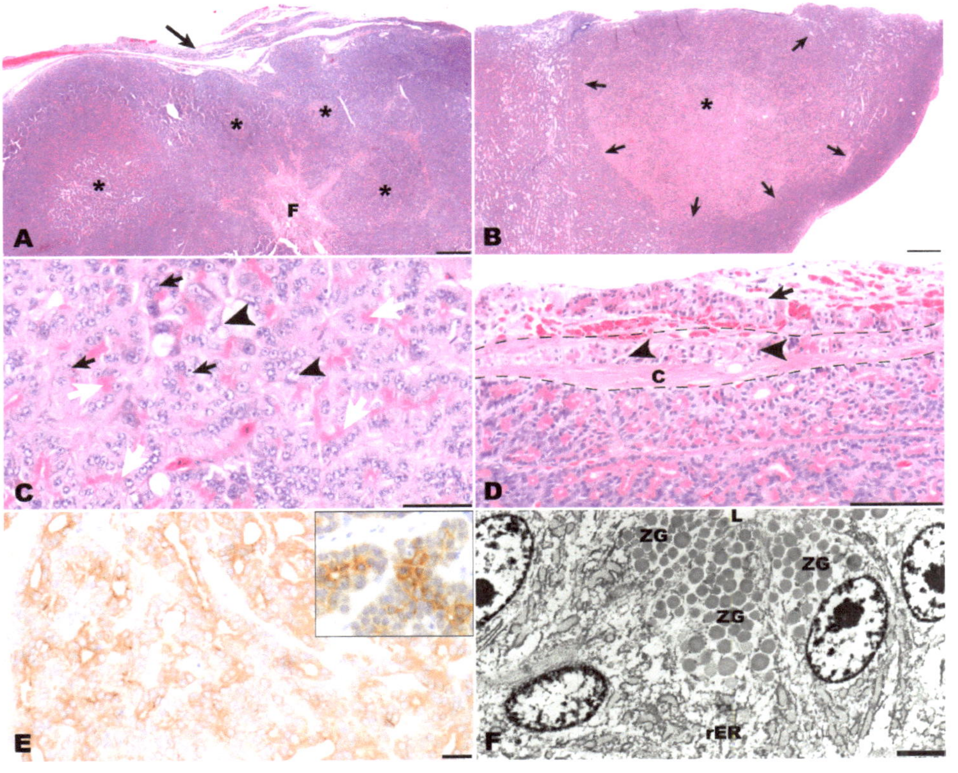

Figure 4. Pathomorphological findings in the surgically excised mass, bearded dragon. (**A**) Overview of the multinodular pancreatic mass (asterisks) with capsular infiltration (arrow) and central fibrosis (F). Hematoxylin-Eosin (HE). Bar = 500 µm. (**B**) Low magnification of one nodule (asterisk) highlighting that nodules are unencapsulated but well demarcated (arrows) and paler than surrounding tissue. HE. Bar = 500 µm. (**C**) Tumor cells are arranged in tubulo−acinar structures with bright pink granular material in the apical cytoplasm (white arrows). The cells exhibit prominent nucleoli (black arrows) and the partly vesicular appearance of nuclei (black arrowheads). HE. Bar = 50 µm. (**D**) Multifocally tumor cells were present within the capsule (arrowheads) as well as extracapsular adjacent to blood vessels (arrow). The dashed line indicates the thickness of the fibrous capsule (C). HE. Bar = 100 µm (**E**) A diffuse centroacinar immunolabeling of tumor for pan-cytokeratin was detected. The immunolabeling represents the reaction that can be observed in normal exocrine pancreas of bearded dragons (insert). IHC for pan-cytokeratin. Bar = 20 µm. (**F**) Ultrastructurally, epithelial tumor cells surrounding an empty lumen (L) were detected. Within the cytoplasm, large amounts of rough endoplasmic reticulum (rER) were present in the perinuclear areas. In addition, moderately electron-dense, well-demarcated granules were seen in the apical cytoplasm of the tumor cells resembling zymogen granules (ZG). Transmission electron microscopy. Bar = 2500 nm.

To confirm the diagnosis, additional testings have been applied to the samples.

Periodic acid-Schiff (PAS) reaction and PAS reaction with diastase digestion were used for demonstration of zymogen granules. Immunohistochemistry (IHC) on paraffin sections was carried out using the avidin−biotin complex (ABC) technique and primary mouse monoclonal antibodies to pan-cytokeratin (panCK, clone AE1/AE3: 1:500), vimentin (clone V9: 1:100), and neuroendocrine markers, i.e., synaptophysin (clone DAK-SYNAP; 1:500), and neuron-specific enolase (NSE, clone BBS/NC/VI-H14: 1:100). In addition, a rabbit polyclonal antibody against bovine chromogranin A (bovine pituitary secretory protein chromogranin, SP1 CrgA: 1:2000) was applied. All antibodies were purchased from Dako Deutschland GmbH, Hamburg, Germany (now Agilent Technologies GmbH, Waldbronn, Germany). As secondary antibodies, biotinylated goat anti-mouse and goat anti-rabbit antibodies (each 1:200) were used. Signal enhancement was achieved with the avidin–biotin–peroxidase complex kit (VECTASTAIN® Elite® ABC-Kit, Vector Laboratories, Inc., Burlingame, CA, USA). Peroxidase activity was detected using 0.1% H_2O_2 with 3,3′-diaminobenzidine solution as the chromogen. Sections were counterstained with Mayer's hematoxylin. All immunohistochemical reactions included positive control archival tissue of bearded dragon intestine and pancreas, as well as appropriate canine tissues. As negative controls, respective dilutions of immunoglobulin isotypes or rabbit serum replaced the primary antibodies (Supplementary Table S1).

All H&E and IHC slides were digitized using the Olympus VS200 digital slide scanner (Olympus Deutschland GmbH, Hamburg, Germany). Representative images of slides were exported with the OlyVIA software (Olympus Deutschland GmbH, Hamburg, Germany; www.olympus-lifescience.com/de, accessed on 2 March 2024).

Transmission electron microscopy was performed by re-embedding H&E-stained paraffin sections in epoxy resin using the pop-off technique described by Lehmbecker et al. [6]. Slides were evaluated with an EM906-transmission electron microscope (Carl Zeiss CMP GmbH, Göttingen, Germany). Pictures were recorded on a CCD-digital camera (TRS Tröndle Restlichverstärkersysteme, Moorenweis, Germany).

The PAS reaction showed that PAS-positive zymogen granules were present in the cytoplasm of tumor cells and that their numbers varied from numerous to frequent between different areas of the tumor. Also, the number of cells with diastase-resistant cytoplasmic zymogen granules varied from frequent to uncommon between different areas of the tumor. Immunohistochemically, tumor cells showed a positive diffuse centroacinar immunolabeling for panCK. (Figure 4E). In addition, multifocally single to few acinar tumor cells showed a homogenous cytoplasmic staining. No labeling was observed with all other applied markers (vimentin, NSE, CrgA, synaptophysin). Immunolabeling for pan cytokeratin resembled the pancreatic pattern observed in the control tissue of another bearded dragon (insert, Figure 4E). Altogether, the findings of the immunohistochemical examination mentioned above confirmed the diagnosis of an acinar pancreatic carcinoma.

Transmission electron microscopical examination revealed the presence of zymogen granules in the apical cytoplasm of tumor cells. Zymogen granules were distinct, round, and had a homogenous and sometimes fine granular appearance. They showed a slight to medium electron density, and their number varied between different tumor cells (ZG, Figure 4F). In addition, high amounts of rough endoplasmic reticulum are present in the tumor cells (rER, Figure 4F).

3. Discussion

Diseases of the reptile pancreas are not commonly reported. However, several differential diagnoses should be considered relating to the case reports in the literature [1]. Herpes viral inclusion bodies, as well as arenavirus and ferlavirus related inclusion bodies, have been found to affect the pancreas in snakes [7,8]. Also, differential diagnoses of pancreatic neoplasia include pancreatitis, which has also been described for reptile species [2]. Acute necrotizing pancreatitis, as in mammals, can be a result of the escape and activation of pancreatic enzymes, leading to severe inflammation processes [1]. Depending on the location and severity of the inflammation, both exocrine and endocrine portions of the pancreas may be involved. Also, pancreatitis can be described as being associated with endoparasitic infection [2]. Migratory nematodes and trematodes might cause damage to pancreatic tissue. They might also trigger secondary bacterial inflammation processes. Validated blood chemistry values might be highly beneficial as elevated enzyme parameters could serve as pancreatitis markers. However, validation of these enzymes has not yet been established for reptiles. Therefore, diagnostic imaging methods, mainly ultrasonography

3. Eustace, R.; Garner, M.M.; Cook, K.; Miller, C.; Kiupel, M. Multihormonal islet cell carcinomas in three komodo dragons (*Varanus komodoensis*). *J. Zoo Wildl. Med.* **2017**, *48*, 241–244. [CrossRef] [PubMed]
4. Frye, F.L.; McNeely, H.E.; Corcoran, J.H. Functional pancreatic glucagonoma in the rhinoceros iguana *Cyclura c. figgensi*, characterized by immunocytochemistry. In Proceedings of the Association of Reptile and Amphibian Veterinarians, Columbus, OH, USA, 5–9 October 1999; p. 315.
5. Kubiak, M.; Denk, D.; Stidworthy, M.F. Retrospective review of neoplasms of captive lizards in the United Kingdom. *Vet. Rec.* **2020**, *186*, 28. [CrossRef] [PubMed]
6. Lehmbecker, A.; Rittinghausen, s.; Rohn, K.; Baumgärtner, W.; Schaudien, D. Nanoparticles and pop-off technique for electron microscopy: A known technique for a new purpose. *Toxicol. Pathol.* **2014**, *42*, 1041–1046. [CrossRef] [PubMed]
7. Jacobson, E.R. *Infectious Diseases and Pathology of Reptiles*; CRC Press: Boca Raton, FL, USA, 2007.
8. Pees, M.; Neul, A.; Müller, K.; Schmidt, V.; Truyen, U.; Leinecker, N.; Marschang, R. Virus distribution and detection in corn snakes (*Pantherophis guttatus*) after experimental infection with three different ferlavirus strains. *Vet. Microbiol.* **2016**, *182*, 213–222. [CrossRef] [PubMed]
9. Bucy, D.S.; Guzman, D.S.-M.; Zwingenberger, A.L. Ultrasonographic anatomy of bearded dragons (*Pogona vitticeps*). *J. Am. Vet. Med. Assoc.* **2015**, *246*, 868–876. [CrossRef] [PubMed]
10. Miller, M.R.; Lagios, M. *Biology of the Reptilia*; Academic Press: New York, NY, USA, 1970.
11. Moscona, A.A. Anatomy of the pancreas and Langerhans islets in snakes and lizards. *Anat. Rec.* **1990**, *227*, 232–244. [CrossRef] [PubMed]
12. Calimano-Ramirez, L.F.; Daoud, T.; Gopireddy, D.R.; Morani, A.C.; Waters, R.; Gumus, K.; Klekers, A.R.; Bhosale, P.R.; Virarkar, M.K. Pancreatic acinar cell carcinoma: A comprehensive review. *World J. Gastroenterol.* **2022**, *28*, 5827–5844. [CrossRef] [PubMed]
13. Aupperle-Lellbach, H.; Törner, K.; Staudacher, M.; Müller, E.; Steiger, K.; Klopffleisch, R. Characterization of 22 canine pancreatic carcinomas and review of literature. *J. Comp. Pathol.* **2019**, *173*, 71–82. [CrossRef] [PubMed]
14. Cony, F.G.; Slaviero, M.; Santos, I.R.; Cecco, B.S.; Bandinelli, M.B.; Panziera, W.; Sonne, L.; Pavarini, S.P.; Driemeier, D. Pathological and immunohistochemical characterization of pancreatic carcinoma in cats. *J. Comp. Pathol.* **2023**, *201*, 123–129. [CrossRef] [PubMed]
15. Chaudhary, P. Acinar Cell Carcinoma of the Pancreas: A Literature Review and Update. *Indian J. Surg.* **2015**, *77*, 226–231. [CrossRef] [PubMed]
16. Kiupel, M.; Capen, C.; Miller, M.; Smedley, R. Histological classification of tumors of the endocrine system of domestic animals. In *World Health Organization International Histological Classification of Tumors of Domestic Animals*; Armed Forces Institute of Pathology: Washington, DC, USA, 2008.
17. Pavone, S.; Manuali, E.; Eleni, C.; Ferrari, A.; Bonanno, E.; Ciorba, A. Canine pancreatic clear acinar cell carcinoma showing an unusual mucinous differentiation. *J. Comp. Pathol.* **2011**, *145*, 355–358. [CrossRef] [PubMed]
18. Ordónez, N.G. Pancreatic acinar cell carcinoma. *Adv. Anat. Pathol.* **2001**, *8*, 144–159. [PubMed]
19. Ritter, J.M.; Garner, M.M.; Chilton, J.A.; Jacobsen, E.R.; Kiupel, M. Gastric neuroendocrine carcinomas in bearded dragons (*Pogona vitticeps*). *Vet. Pathol.* **2009**, *46*, 1109–1116. [CrossRef] [PubMed]
20. Lyons, J.A.; Newman, S.J.; Greenacre, C.B.; Dunlap, J. A gastric neuroendocrine carcinoma expressing somatostatin in a bearded dragon (*Pogona vitticeps*). *J. Vet. Diagn. Investig.* **2010**, *22*, 316–320. [CrossRef] [PubMed]
21. LaDouceur, E.E.B.; Argue, A.; GARNER, M.M. Alimentary Tract Neoplasia in Captive Bearded Dragons (*Pogona* spp.). *J. Comp. Pathol.* **2022**, *194*, 28–33. [CrossRef] [PubMed]
22. Jakab, C.; Rusvai, M.; Szabó, Z.; Gálfi, P.; Marosán, M.; Kulka, J.; Gál, J. Claudin-7-positive synchronous spontaneous intrahepatic cholangiocarcinoma, adenocarcinoma and adenomas of the gallbladder in a bearded dragon (*Pogona vitticeps*). *Acta Vet. Hung.* **2011**, *59*, 99–112. [CrossRef] [PubMed]
23. Crouch, E.E.V.; McAloose, D.; McEntire, M.S.; Morrisey, J.K.; Miller, A.D. Pathology of the Bearded Dragon (*Pogona vitticeps*): A Retrospective Analysis of 36 Cases. *J. Comp. Pathol.* **2021**, *186*, 51–61. [CrossRef] [PubMed]
24. Page, M.J.; McKenzie, J.E.; Bossuyt, P.M.; Boutron, I.; Hoffmann, T.C.; Mulrow, C.D.; Shamseer, L.; Tetzlaff, J.M.; Akl, E.A.; Brennan, S.E.; et al. The PRISMA 2020 statement: An updated guideline for reporting systematic reviews. *BMJ* **2021**, *372*, n71. [CrossRef] [PubMed]

animals

Article

The Use of Prefemoral Endoscope-Assisted Surgery and Transplastron Coeliotomy in Chelonian Reproductive Disorders

Tom Hellebuyck * and Ferran Solanes Vilanova

Department of Pathobiology, Pharmacology and Zoological Medicine, Faculty of Veterinary Medicine, Ghent University, Salisburylaan 133, B-9820 Merelbeke, Belgium
* Correspondence: tom.hellebuyck@ugent.be

Simple Summary: Tortoises and turtles (chelonians) are routinely presented in veterinary practice because of their reproductive disorders. Although egg binding in chelonians can often be resolved with conventional therapy, the diagnosis and treatment of complicated cases of egg binding and various other disorders of the reproductive tract in chelonians often require a more advanced therapeutic approach. If surgical intervention is required, endoscope-assisted techniques comprise the least invasive and thus preferred surgical approach. In cases where the use of endoscope-assisted procedures is not feasible, the reproductive tract needs to be accessed through the plastron (transplastron coeliotomy). The present report describes the diagnostic and surgical approach applied in seven cases of female chelonians with reproductive disorders. The therapeutic efficacy largely relies on the choice of minimally invasive endoscope-assisted surgery versus transplastron coeliotomy.

Abstract: Throughout the last decades, the increased popularity of the keeping of reptiles has led to a better understanding of the captive needs of a wide variety of species. Although this is reflected by the successful captive reproduction in many of those species, reproductive disorders such as preovulatory follicular stasis, postovulatory dystocia, secondary yolk coelomitis, and prolapse of the oviduct and male copulatory organ are commonly encountered in veterinary practice. In comparison to squamates, chelonians with postovulatory dystocia seem to be more responsive to oxytocin treatment, even in cases of chronic dystocia. There are various conditions, however, that necessitate the use of surgical procedures for the treatment of dystocia and other reproductive disorders in chelonians. Although restrictions may be encountered, the endoscope-assisted prefemoral approach is the least invasive and thus preferred technique instead of the ventral transplastron coeliotomy. The present report describes the diagnostic and surgical approach applied in seven cases of female chelonians with reproductive disorders. The therapeutic efficacy largely relied on the choice of minimally invasive endoscope-assisted surgery versus transplastron coeliotomy that was primarily dictated by the involved species, etiology, and associated pathology.

Keywords: chelonians; coeliotomy; endoscopy; oophorectomy; transplastronectomy

Citation: Hellebuyck, T.; Solanes Vilanova, F. The Use of Prefemoral Endoscope-Assisted Surgery and Transplastron Coeliotomy in Chelonian Reproductive Disorders. *Animals* **2022**, *12*, 3439. https://doi.org/10.3390/ani12233439

Academic Editor: Volker Schmidt

Received: 6 November 2022
Accepted: 5 December 2022
Published: 6 December 2022

1. Introduction

The increased popularity of reptile pets has led to continued advances in husbandry and nutrition and the successful breeding of a wide variety of species. Nevertheless, reproductive disorders such as preovulatory follicular stasis (PFS) and dystocia are still some of the most frequently observed disorders in the reptile patient [1,2]. Follicular activity and ovulation can occur in solitary female reptiles, mostly resulting in the production of infertile ova. Although PFS and dystocia are mostly seasonal problems, environmental control in captivity may result in reproductive activity throughout the year, especially in nontemperate species [1–3].

Dystocia can be classified into obstructive or nonobstructive dystocia. Obstructive dystocia is caused by a barrier that prevents the normal passage of eggs or feti through the oviduct or cloaca and may be related to a maternal or fetal abnormality including deformation of the pelvis, space-occupying processes in the coelomic cavity, oviductal strictures, adhesive salpingitis and eggs or feti with an abnormal shape, size, or mineralization as well as broken or damaged eggs [2]. Nonobstructive dystocia can be attributed to a wide variety of primary etiologies, including inappropriate environmental factors and malnutrition, mainly those that result in calcium deficiency, or any disorder that results in a poor physical condition of the female. The etiology is mostly multifactorial, and often it is challenging to obtain a proven causation [1,2].

True nonobstructive dystocia is routinely treated with the administration of oxytocin or, if available, arginine vasotocin, but therapeutic efficacy is variable and seems to be higher in chelonians than in squamates [4,5]. If medical treatment fails or if a permanent solution is desirable, surgical intervention is required. In squamates, reproductive surgery, mainly consisting of oophorectomy, salpingotomy, and/or salpingectomy, is routinely and relatively easily performed through coeliotomy. Surgical intervention for the treatment of dystocia and other reproductive disorders in chelonians, on the other hand, is more challenging due to the constraints of the chelonian shell [6,7]. Whenever possible, the minimally invasive endoscope-assisted prefemoral approach is the preferred technique instead of the ventral transplastron coeliotomy. The latter is largely dictated by the anatomical characteristics of the involved species, and especially in cases of obstructive dystocia, the primary and secondary pathologies [6–9].

The present report describes the diagnostic and therapeutic approach of reproductive disorders in seven cases of female chelonians with an emphasis on the deciding factors towards selecting a transplastron coeliotomy or endoscope-assisted surgical approach.

2. Materials and Methods

2.1. Animals and Diagnostic Procedures

The cases in this study involve female chelonians that were presented at a veterinary teaching hospital with signs that were attributed to a reproductive disorder. Table 1 provides an overview of the species, the age, and bodyweight at initial presentation for each case, as well as the provided housing and feeding regimen.

Table 1. Overview of chelonian cases with reproductive disorders that were treated with transplastron coeliotomy (TC) or endoscope-assisted coeliotomy. Tx: treatment; Dx: diagnosis; CT: cystotomy; OE: oophorectomy; EAO: endoscope-assisted oophorectomy; ST: salpingotomy.

Case nr.	Species	Age, Body Weight	Management and Nutrition	Clinical Signs	Dx	Tx	Follow-Up Period
1	Red-eared slider (*Trachemys scripta elegans*)	29 years, 0.85 kg	Housing: glass tank of 200 L. Heat bulb installed in the dry part (35 × 50 cm) of the enclosure creating a local hotspot of 28 °C. UV irradiation not provided. Diet: cat food pellets on a daily basis. Co-housed with a female (20 yrs) and male (18 yrs) red-eared slider.	Apathy, anorexia	Scirrhous ovarian carcinoma, chronic egg binding	TC, OE	24 months
2	Yellow-bellied slider (*Trachemys scripta scripta*)	14 years, 1.185 kg	Housing: aquarium containing 120 L of water and dry area (50 × 50 cm) covered with sand as a substrate. Ultraviolet irradiation provided with an average temperature at the basking spot of 27 °C. Diet: commercial cat food pellets combined and gammarus on a daily basis. Co-housed with a female yellow-bellied slider of the same age.	Apathy, anorexia	Ectopic eggs, coelomitis, chronic egg binding	EAO	48 months

Table 1. *Cont.*

Case nr.	Species	Age, Body Weight	Management and Nutrition	Clinical Signs	Dx	Tx	Follow Up Period
3	Common snapping turtle (*Chelydra serpentina*)	32 years, 11.2 kg	Housing: outside pond (6 × 3 × 1.2 m) where the animals hibernated from early September to late April. Diet: adult mice and liver twice a week. Co-housed with a male snapping turtle of the same age.	Post-hibernation anorexia	Granulomatous oophoritis	EAO	18 months
4	Black marsh turtle (*Siebenrockiella crassicollis*)	19 years, 0.34 kg	Housing: inside pond of 80 L and a dry area of 80 × 80 cm with cocopeat as substrate. UV irradiation provided with hotspot of 29 °C. Diet: commercial turtle pellets, liver once every 2 days. Co-housed with two females and a male black marsh turtle of the same age. UV irradiation was provided in the dry part and a water part of 80 L was present.	Apathy, anorexia	Bacterial salpingitis	ST	12 months
5	Red-eared slider (*Trachemys scripta elegans*)	22 years, 1.1 kg	Glass tank of 150 L with a dry part (40 × 30 cm) provided with local hot spot of on average 27 °C created by an UV irradiation heat bulb. Diet: commercial turtle pellets on a daily basis. Co-housed with a male and female red-eared slider of the same age.	Tenesmus, buoyancy disorder	Obstructive dystocia	TC, ST, OE	18 months
6	Hermann's tortoise (*Testudo hermanni*)	51 years, 0.98 kg	Housing: outside enclosure of 6 × 3 m with glass house of 80 × 50 × 80 cm. Individually housed. Diet: herbs and vegetables on a daily basis.	Apathy, anorexia, dehydration	Chronic egg binding, oviductal invagination	TC, ST, OE	8 months
7	Hermann's tortoise (*Testudo hermanni*)	5 years, 0.630 kg	Housing: terrarium of 150 × 60 × 40 cm with wood chips as a substrate. A heat lamp creating a local hotspot of 27 °C on average. Neither UV irradiation, calcium, nor vitamin supplementation were provided. Co-housed with a male Hermann's tortoise of same age. Diet: vegetables on a daily basis.	Anorexia, apathy, dyspnea	Dystocia, ectopic eggs within urinary bladder	TC, CT, OE	6 months

If a radiographic examination was performed, radiographic projections comprised a laterolateral and craniocaudal projection (horizontal beam) as well as a dorsoventral projection (vertical beam). An ultrasonographic examination was performed using an 8–15 MHz microconvex transducer using the prefemoral coupling window. Unless stated otherwise, ultrasonographic abnormalities were confined to the reproductive tract.

2.2. Anesthetic Procedures

In all cases, an identical anesthetic and analgesic protocol was used consisting of intravenous (IV) induction of anesthesia with alfaxalone (10 mg/kg, Alfaxan Multidose, Jurox Limited, Crawley, UK) administered via the right jugular vein and intubation with an uncuffed endotracheal tube. Anesthesia was maintained with 1.5–2% isoflurane (Isoflo, Abbott Logistics B.V., Breda, The Netherlands) in 1 L medical oxygen per minute with intermittent positive-pressure ventilation. In case of performing a transplastron coeliotomy, morphine was administered perioperatively at 2 mg/kg (IM, Morphine HCL Sterop 10 mg/mL, Laboratoires Sterop NV, Brussels, Belgium) and tramadol (5 mg/kg, PO, Tramadol EG, Eurogenerics NV, Brussels, Belgium) postoperatively. For the endoscope-assisted prefemoral approach, local infiltration with 2 mg/kg lidocaine (Xylocaine 1%, Aspen Pharma Trading Limited, Dublin, Ireland) of the prefemoral incision site was performed prior to the start of the surgery, and meloxicam was administered perioperatively (0.3–0.5 mg/kg, IM, Metacam 20 mg/mL, Boehringer Ingelheim, Vetmedica GmbH, Ingelheim, Germany).

2.3. Surgical Procedures

Routine considerations for preanesthetic evaluation and surgical preparation were performed, and fasting times varied between 24 and 36 h. The chelonians were encouraged to defecate and urinate before surgery by stimulation of the cloaca using a cotton tip applicator or shallow bathing in the turtles and testudinid species, respectively.

For the endoscope-assisted prefemoral procedures, a routine approach as previously described by Innes and Hernandez-Divers [10] and Proença and Divers [11] was used to enter the coelomic cavity with the animal placed in lateral recumbency. Depending on the species and/or the surgical aim, a left or bilateral prefemoral approach was applied.

Transplastron coeliotomies were performed as described by Divers and Wüst [6] and special preoperative considerations included assessing the plastron thickness, including its expansion towards the plastrocarpacial bridge, and locating the hinge region and the degree of access required to perform the procedure. An oscillating sagittal saw was used to incise the plastron, and the caudal cut was generally made incomplete, leaving a few millimeters of bone thickness and allowing the segment to break spontaneously when lifting the segment with a periostal elevator. The latter was considered to help stabilizing the loose segment during closure and increased the chance of maintaining blood supply and thus primary healing. Entry into the coelomic cavity was achieved by making a midline incision of the coelomic membrane between the abdominal veins in the Testudo species and a unilateral paramedian incision in the (semi-)aquatic species. For closure of the plastron incision, veterinary acrylic (Technovit 6091, Kulzer GmbH, Hanau, Germany) was used.

3. Results

3.1. Case 1

A 29-year-old female red-eared slider (*Trachemys scripta elegans*) was presented with anorexia and apathy after 2 weeks. During the past 22 years, the turtle annually produced 4 to 6 eggs and had no clinical history. Radiographic projections (Figure 1A,B) revealed the presence of 5 eggs located at the right side of the caudal coelomic cavity. Three eggs had an irregular shape, and one egg had an unusual small size and seemed to be fused with one of the other eggs. All eggs showed abnormally thickened shells, and a lamellar appearance was noticed in the most cranial egg. A generalized soft tissue opacity occupying the entire coelomic cavity was noted, causing obvious compression of the lung field on the latero-lateral (Figure 1B) and craniocaudal projection. An ultrasound confirmed the presence of retained eggs and the presence of a large soft tissue mass measuring approximately 11.5 × 6.1 cm (Figure 1C) at the left-mid- to caudal coelomic cavity. The mass showed a heterogenic appearance, and a color doppler examination revealed pronounced vascularization. The latter findings yielded obstructive dystocia and chronic egg retention caused by a space-occupying mass.

Taking into account the size of the mass and the presence of multiple retained eggs with an abnormal appearance, the transplastron coeliotomy was chosen as the most appropriate surgical procedure for this case. The mass appeared to originate from the right ovary. Following ligation and resection of the mass, static displacement of the heart, liver, gastrointestinal tract, and left oviduct to the right coelomic cavity was noted. Based on a visual inspection, no signs of metastasis could be observed. Next, a bilateral salpingotomy was performed to remove two eggs from each oviduct followed by the oophorectomy of the right ovary that showed an inactive and normal appearance. Recovery from anesthesia was uneventful, and the turtle showed a good appetite and normal behavior the day after surgery. The histological examination of the ovarian mass revealed a cell-rich tissue that was growing infiltrative in dense collagen stroma (Figure 1D). The neoplastic cells were organized in islets to multi-layered tubules and showed a moderate amount of clear, polygonal cytoplasm. The nuclei were pale with a small nucleolus, and the number of mitosis per high power field was less than one. A histological diagnosis of a scirrhous ovarian carcinoma was made. During a 2-year follow-up period, the turtle showed a good general condition, and based on the ultrasonography and radiographic examination performed 6 and 18 months after surgery, no indications of neoplastic disease could be noted.

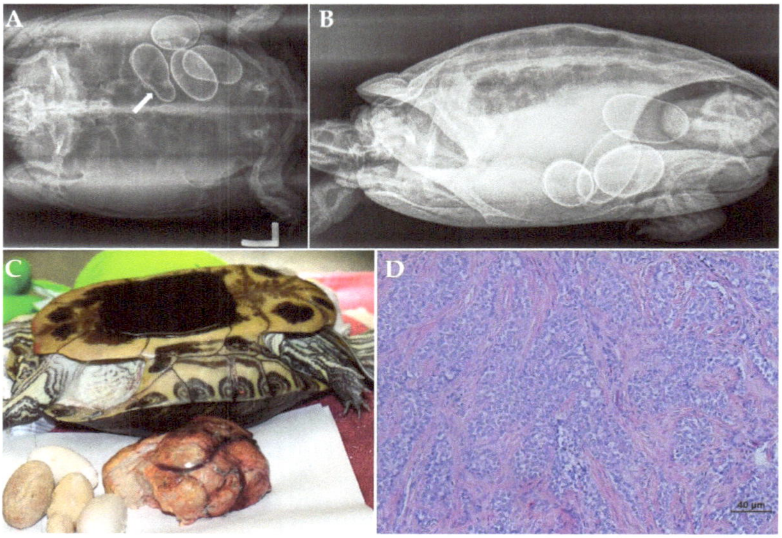

Figure 1. Chronic egg binding associated with a scirrhous ovarian carcinoma in a red-bellied slider (*Trachemys cripta elegans*). (**A**) Dorsoventral radiographic projection: 5 eggs located at the right side of the caudal coelomic cavity. An irregular shape is noted in 3 eggs and all eggs show abnormally thickened shells with a lamellar appearance in the most cranial egg. One egg has an unusual small size and is fused with one of the other eggs (arrow). (**B**) Generalized soft tissue opacity of the coelomic cavity with obvious compression of the lung field on the left laterolateral projection. (**C**) Postoperative view after removal of the ovarian neoplasia and retained eggs. (**D**) Histological section of a scirrhous ovarian carcinoma (Hematoxylin and eosin stain) composed of a cell-rich tissue that is growing infiltrative in dense collagen stroma. Neoplastic cells are organized in islets to multi-layered tubules.

3.2. Case 2

A 19-year-old female yellow-bellied slider (*Trachemys scripta scripta*) was presented because of anorexia and having displayed overactive behavior for two weeks. During the past 10 years, the turtle had produced two clutches of eggs per year with, on average, 8 eggs per clutch. During the past 5 years, recurrent episodes of dystocia were resolved following the administration of oxytocin by a local vet. Two days before initial presentation, the turtle had received 3 administrations of oxytocin (IM, Oxytocine Kela 10 IU/mL, Kela Veterinaria NV, Antwerpen, Belgium) at 15 IU/kg without effect. An ultrasound revealed the presence of at least 2 oval-shaped eggs, and a radiographic examination (Figure 2A,B) confirmed the presence of 4 eggs with a well-mineralized but thin shell. Three eggs had an abnormal shape, and one egg showed an abnormally small size. The most caudal, large egg showed a vertical position and was located cranial to the pelvis inlet.

Based on these findings, and especially because of the positioning of the most caudal egg, obstructive dystocia was presumed, and an endoscope-assisted left coeliotomy was performed. All eggs appeared to be free in the coelomic cavity. Despite the large size of the eggs relative to the size of the prefemoral incision, an endoscope-assisted localization and manipulation of the eggs followed by bilateral oophorectomy was attempted instead of performing a transplastron coeliotomy. Taking into account the diameter of the eggs in relation to the size of the prefemoral incision, ovocentesis needed to be performed using a 19 gauge needle. After aspiration of the content of the eggs, the eggshells were collapsed and exteriorized. Once the eggs were removed, a bilateral oophorectomy was performed via the left prefemoral fossa. No indications of oviductal rupture were found. Recovery was uneventful, and one week postoperatively the turtle displayed a normal and active behavior, and the appetite was restored. The turtle remained healthy during a 4-year follow-up period.

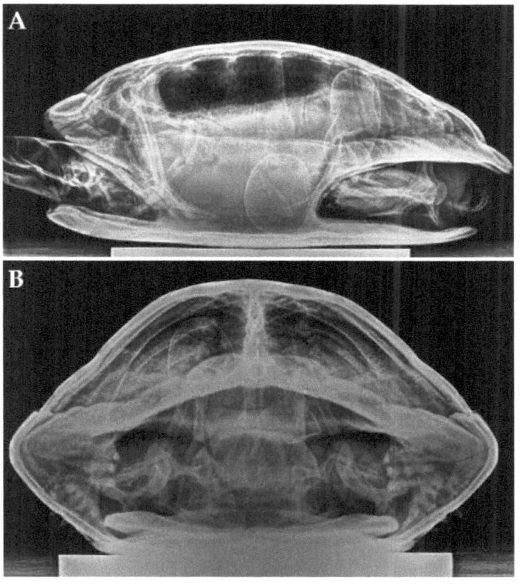

Figure 2. Laterolateral (**A**) and craniocaudal (**B**) radiograph (horizontal beam) of a yellow-bellied slider (*Trachemys scripta scripta*) revealing the presence of 4 ectopic eggs with 3 eggs that have a well-mineralized but thin shell and an abnormal shape. The fourth egg has an abnormally small size. The long axis of the most caudal, large egg showed a vertical position and is located in the right caudal coelomic cavity cranial to the pelvis inlet.

3.3. Case 3

A 32-year-old female common snapping turtle (*Chelydra serpentina*) was presented because of post-hibernation anorexia. The turtle was housed together with a male of the same age in an outside pond where the animals hibernated from early September to late April. Although mating behavior had been noticed, the female never produced eggs.

A radiographic examination did not reveal abnormalities, but based on the ultrasound, both ovaries contained previtellogenic follicles as well as a large number of heterogeneous vittelogenic follicles with an average diameter of 1.2 cm showing an anechogenic central core and a peripheral anechoic rim. Blood was collected from the jugular vein for a serum biochemistry profile and hematological evaluation. Besides hypercalcemia (14.9 mg/dL), the serum biochemistry did not reveal abnormalities in comparison to physiological reference intervals [12], and a complete blood cell count revealed pronounced heterophilia and monocytosis. Based on these findings, a presumptive diagnosis of oophoritis was made, and an endoscope-assisted oophorectomy was planned, taking into account the typically nonrestrictive prefemoral fossa in this mature snapping turtle. Following a routine left prefemoral approach to the coelom, the ipsilateral ovary was located, and a 3 mm atraumatic forceps was used to grasp the interfollicular tissue and retract the ovary to the prefemoral incision. Next, the whole ovary was gently exteriorized and the mesovarium dissected using radiosurgery. Following the same procedure, the contralateral ovary was removed.

The macroscopic appearance of both ovaries (Figure 3A,B) complied to the ultrasonographic findings with adhesions of multiple follicles to the coelomic wall, and a bilateral oophorectomy was performed. A moderate amount of free coelomic fluid with a blurry appearance and diphtheroid plaques at the serosal surface of the liver were noticed. As the snapping turtle showed persistent anorexia following surgery, force-feeding was performed every 3 days. A histological examination of ovarian tissue revealed the presence of a small number of previtellogenic follicles and multiple vitellogenic follicles that contained protein globules surrounded by large numbers of foamy macrophages and giant cells (Figure 3C). Occasionally, aggregates of lymphocytes, plasma cells, and heterophils were noticed. In the ovarian parenchyma, the multifocal presence of cholesterol crystals that were surrounded by macrophages and giant cells could be observed (Figure 3D). Ziehl Neelsen staining was negative. A histological diagnosis of granulomatous inflammation with the formation of cholesterol granulomas, presumably caused as a reaction to the leakage of yolk, was made. A microbiological examination of ovarian tissue yielded negative results. Four weeks postoperatively, the turtle regained her appetite and was sent home. During the first months of an 18-month follow-up period, the owner stated that the appetite largely exceeded the appetite that was seen preoperatively, and 6 months postoperatively the bodyweight had increased to 12.3 kg. After adapting the feeding schedule, the body weight was reduced to the preoperative body weight 3 months later and remained stable until the end of the 18 month follow-up period.

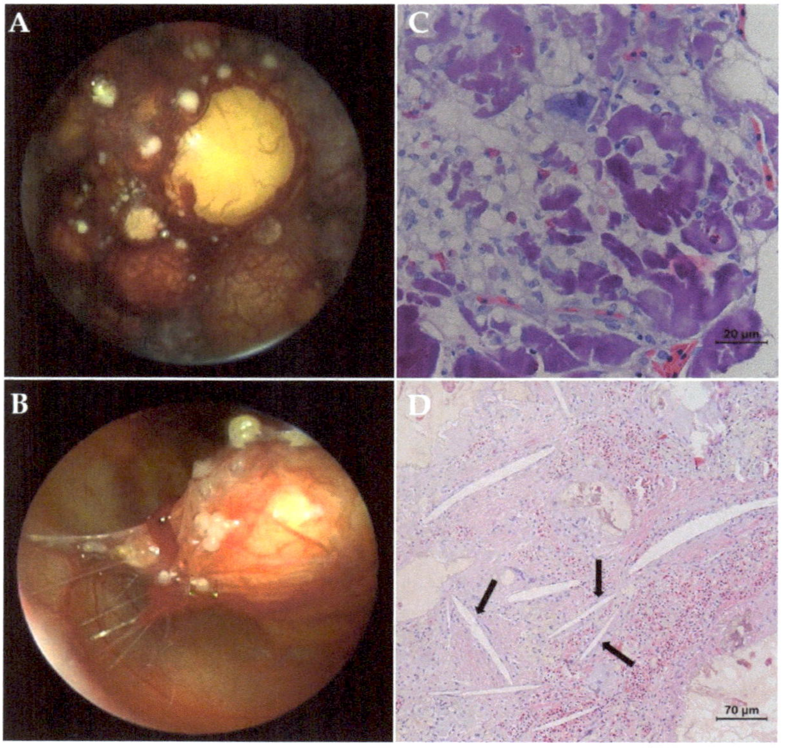

Figure 3. (**A**) Left prefemoral coelioscopic view of the left ovary in a common snapping turtle (*Chelydra serpentina*) with oophoritis. Previtellogenic and a large number of excessively vascularized early vitellogenic follicles can be observed. (**B**) Several follicles show adhesions to the coelomic membrane. (**C**) Histological section of the ovarian tissue (Hematoxylin and eosin stain). Note the presence of multiple follicles surrounded by large numbers of foamy macrophages and giant cells. (**D**) In the ovarian parenchyma, the multifocal presence of cholesterol crystals (arrows) surrounded by macrophages and giant cells can be observed.

3.4. Case 4

A 19-year-old female black marsh turtle (*Siebenrockiella crassicollis*) was presented with pronounced apathy and anorexia after one week. During the previous year, the turtle had produced 2 to 4 fertilized eggs each year. Based on a radiographic examination, a mild but decreased mineralization of the skeleton and a large, excessively mineralized egg were seen in the left caudal coelomic cavity, while in the right caudal coelomic cavity, the remnants of an eggshell could be noticed (Figure 4A,B). An ultrasound confirmed the presence of a well-developed egg in the left oviduct and a hyperechoic structure surrounded by a moderate amount of mildly hyperechoic fluid in the right oviduct. As the turtle did not respond to 3 injections with oxytocin at increasing doses of 5, 10, and 15 IU/kg administered at 2 h intervals, an endoscope-assisted bilateral salpingotomy was planned. The prefemoral fossa was deemed nonrestrictive, and it was considered that both oviducts could be reached using a unilateral approach. Via a routine left prefemoral approach to the coelom, the ipsilateral oviduct was identified and grasped using a 3 mm atraumatic forceps, and a left salpingotomy was performed. Obvious inflammation of the oviductal mucosa was noticed, and a decayed and collapsed eggshell that contained thickened yolk as well as free fluid was removed from the oviduct and sampled for microbiological examination. Next, the contralateral oviduct was located and exteriorized via the left prefemoral incision, and the egg was removed through salpingotomy. Postoperatively, antimicrobial treatment with amoxicillin-clavulanic acid (20 mg/kg, PO, once daily, 10 days, Synulox 50 mg, Zoetis Belgium SA, Louvain-la-Neuve, Belgium) was initiated, and the appetite was restored after four days. A microbiological examination revealed a pure culture of *Citrobacter freundii* that was sensitive to the used antimicrobial treatment. The bacterium was presumed to have established salpingitis following ascending infection, secondary to egg retention and decay. During a one-year follow-up period, the turtle did not show recurrence of salpingitis. Despite several mating attempts, the turtle did not ovulate or produce eggs during this period.

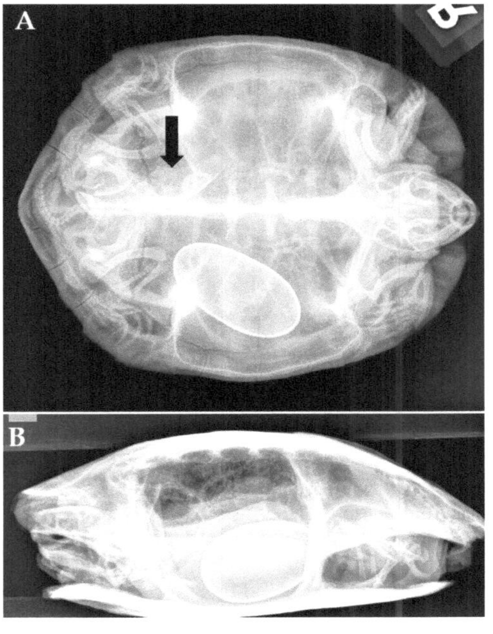

Figure 4. Ventrodorsal (**A**) and laterolateral (**B**) radiograph of a black marsh turtle (*Siebenrockiella crassicollis*) showing mild generalized decreased mineralization of the skeleton and a large, excessively calcified egg in the left caudal coelomic cavity as well as the remnants of an eggshell (arrow) in the right coelomic cavity.

3.5. Case 5

A 22-year-old female red-eared slider was presented with buoyancy disorder and tenesmus after one week. During the past decade, the turtle produced a clutch of 5 to 8 eggs each year. A radiographic examination revealed the presence of 7 eggs with a well-mineralized shell, but the two most cranial eggs with a normal size as well as an excessively small egg were collapsed. The most caudal egg had a relatively large size and was considered to obstruct the pelvis inlet (Figure 5A,B). Based on these findings, a diagnosis of obstructive dystocia was made, and a transplastron coeliotomy was performed, taking into account the diagnosis of obstructive dystocia, the large number of eggs, as well as the abnormal appearance of several eggs and the variable size of the eggs.

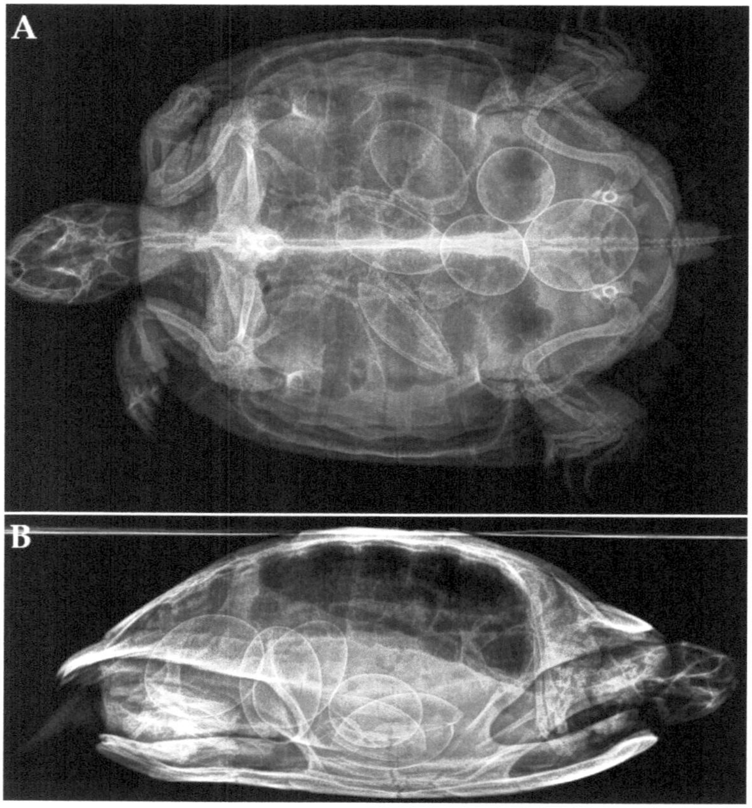

Figure 5. Ventrodorsal (**A**) and laterolateral (**B**) radiograph of a red-eared slider (*Trachemys scripta elegans*) with obstructive dystocia revealing the presence of 7 eggs with a well-mineralized shell. The 2 most cranial eggs with a normal size as well as an excessively small egg are collapsed, and the most caudal egg has a relatively large size and obstructs the pelvis inlet.

Following bilateral salpingotomy, 3 eggs were removed from each oviduct. The most caudal egg needed to be moved in retrograde using gentle digital manipulation via the cloaca before it could be removed through the left salpingotomy incision. Next, bilateral oophorectomy was performed. The day following surgery, the turtle showed good appetite, and buoyancy was restored one week after the surgery. During an 18-month follow-up period, an increased food intake was noticed according to the owner which was reflected by a substantial gain in body weight to 1.35 kg. The transplastronectomy segment was removed as a bony sequestrum after 6.5 months, and bony growth could be noticed below.

3.6. Case 6

A 51-year-old female Hermann's tortoise (*Testudo hermanni*) was presented because of chronic egg binding. Based on a ventrodorsal radiograph, the local vet diagnosed the presence of a presumed egg with an abnormal shape 2 years prior to initial presentation. Treatment with oxytocin by the local vet did not result in oviposition, and the owner declined further treatment and did not seek further veterinary advice as the tortoise did not show clinical signs at that time. Two years after the initial diagnosis, the animal was irresponsive and anorectic immediately after awakening from hibernation and was presented 3 weeks later. Based on the clinical examination as well as the biochemistry test results and hematological evaluation, severe dehydration, moderate hypercalcemia, and a marked rise in creatinine kinase, in addition to a relatively low PCV (17%) and mild heterophilia were demonstrated. A radiographic examination revealed a large radiopaque structure in the left caudal coelomic cavity (Figure 6A). The size, shape, and position of the structure were identical to the radiographic findings of the local vet.

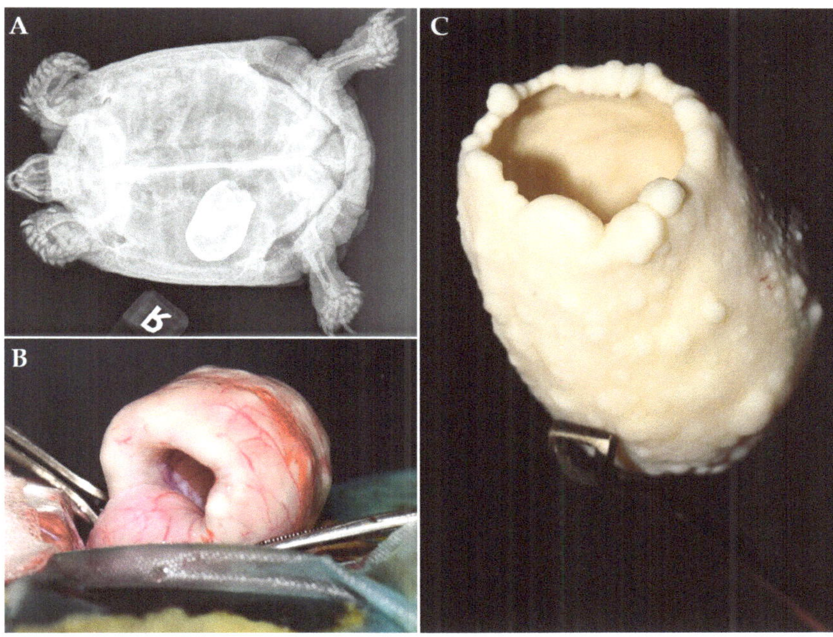

Figure 6. (**A**) Dorsoventral radiograph of a Hermann's tortoise (*Testudo hermanni*) showing a large, retained egg with an abnormal shape in the left caudal coelomic cavity. (**B**) Intraoperative view during transplastron coeliotomy exteriorization of the oviduct containing the retained egg. (**C**) The egg after surgical removal from the oviduct.

Following 5 days of rehydration and nutritional therapy, the tortoise showed a marked improvement of the general condition, and a transplastron coeliotomy was performed to remove the retained egg from the left oviduct through a salpingotomy. The latter surgical approach was preferred, taking into account the large size of the retained structure and the chronicity of the condition. The cranial part of the oviduct seemed to be invaginated into this opening and could be reposed using digital manipulation (Figure 6B). The structure seemed to consist of a severely thickened eggshell with a central lumen (Figure 6C). Next, a bilateral oophorectomy was performed. The tortoise showed an obvious improvement in general condition and a restored appetite one week after surgery, respectively. During an 8-month follow-up period, the tortoise remained without clinical signs.

3.7. Case 7

A 5-year-old female Hermann's tortoise was presented due to showing anorexia for one week, and apathy as well as open-mouth breathing for two days. Prior to showing anorexia, the tortoise had shown overactive and nesting behavior for 5 days. The tortoise showed prominent flattening and pyramiding of the carapace and decreased mineralization of the shell as well as bilateral distention of the prefemoral fossae. Based on an ultrasound, an overfilled urinary bladder was noticed as well as well-mineralized eggs that were considered to be present within the urinary bladder, and a percutaneous cystocentesis was performed. As the vertical distance between the caudal edge of the carapace and plastron was considered to be abnormally limited, it was presumed that the eggs had entered the urinary bladder from the cloacal vestibule through the urodeum and urethral opening as they could not pass through the vent. A radiographic examination was advised but declined by the owner. Taking into account the abnormal development of the tortoise as well as the presence of 4 eggs within the urinary bladder, a transplastron coeliotomy was performed, and following entry of the coelom, the presence of the eggs within the urinary bladder was confirmed. Next, a cystotomy was performed as previously described by Divers [13] to remove the eggs after intraoperative cystocentesis followed by bilateral oophorectomy.

Recovery from the surgery was smooth, and following a week of supportive care and force-feeding, the appetite was restored. Both management and nutritional advice were provided to the owner. During a 6-month follow-up period, the tortoise appeared to show a normal condition.

4. Discussion

The cases in the present report illustrate the successful use of endoscope-assisted prefemoral surgery as well as transplastron coeliotomy for various reproductive disorders in several chelonian species. Although most signs at initial presentation could be related to egg retention in 6 out of 7 cases and chronic retention was demonstrated in 3 cases, primary conditions such as neoplastic disease as well as secondary complications, such as salpingitis, ectopic eggs in the coelomic cavity or urinary bladder, and coelomitis need to be considered in cases with obstructive or nonobstructive dystocia [1,2,14].

Although transplastron coeliotomy is considered as an invasive procedure, it has been documented as an effective and safe procedure for the treatment of reproductive and other disorders (e.g., cystic calculus removal) in various chelonian species [8]. In the present report, four out of seven cases required the use of transplastron coeliotomy. The use of this technique versus an endoscopy-assisted prefemoral approach should be well-considered, and in chelonians this choice is primarily dictated by the involved species. In chelonian species with nonrestrictive prefemoral fossae (e.g., *C. serpentina*, *Sternotherus* spp., *Terrapene* spp., softshell turtles) [15], the endoscope-assisted surgery of reproductive disorders is often possible. Depending on the size of the animal, the size and number of retained eggs, or other pathology of the reproductive tract as well as complicating factors, a uni- or bilateral approach can be used [6]. As previously documented, endoscope-assisted oophorectomy usually can be performed through a single fossa in *Trachemys* sp. [6,10], as demonstrated for the cases involving terrapins as well as the snapping turtle in the present study. In the terrapins with the ovarian carcinoma (case 1) and the obstructive dystocia (case 5), a prefemoral approach would not have allowed successful treatment, while in the yellow-bellied slider (case 2) and the black marsh turtle (case 4), the prefemoral approach was adequate and a single point of entry proved to be sufficient for the removal of retained eggs and performing of a bilateral oophorectomy in the slider. In the latter case, however, intraoperative ovocentesis was necessary in order to remove multiple eggs from the coelomic cavity.

Taking into account the signalment of the described cases, it is noteworthy that 6 out of 7 cases had an average to relatively old age [15]. Several of the described cases had successfully produced clutches of eggs prior to developing dystocia and other reproductive disorders. Although several reproductive disorders in chelonians may be expected to be seen more frequently in aged animals (e.g., neoplastic disease) and solitary females [2], we do not consider the age as a fundamental risk factor in the development of reproductive disorders, and five out of seven cases in the present study were co-housed with a conspecific male.

Specifically, nutritional secondary hyperparathyroidism (NSHP) and other disorders that cause calcium deficiency are considered to predispose the development of dystocia and PFS in reptiles [2]. In the present study, mild to severe signs of skeletal demineralization were demonstrated in the black marsh turtle (case 4) and the Hermann's tortoises (cases 6 and 7). Especially in case 7, the retrograde migration of eggs from the cloaca to the urinary bladder through the urodeum was undoubtedly related to developmental abnormalities of the shell that were primarily related to NSHP. Although relatively uncommon, the authors observed several cases of ectopic eggs in the coelomic cavity in saurian and chelonian species as seen in the yellow-bellied slider (case 2) and in a Hermann's tortoise (case 7). In most cases, the repeated administration of usually high doses of oxytocin seemed to have a causal relationship with this pathology, especially in cases where obstructive dystocia was not recognized prior to the start of the treatment [14,16].

Several cases in the present study illustrate that chronic egg retention can be left unnoticed during long periods of time until secondary complications occur or the primary cause leads to overt signs. Radiographic signs of chronic egg retention may include the excessive mineralization of the eggshells that sometimes develops into a lamellar appearance after a prolonged period of time as observed in several cases in this study. In addition, abnormally shaped or collapsed eggs and the rupture of the eggshell with the subsequent development of salpingitis as observed in the black marsh turtle (case 4) constitute other pathologies that can be related to chronic egg binding.

While postovulatory dystocia, especially obstructive cases, can be considered as one of the most common reproductive disorders that requires surgical treatment, the prevalence of true PFS in chelonians seems to be low, and to the authors' knowledge there are no literature data that unambiguously demonstrate cases of PFS in chelonians. Although other disorders, such as ovarian and oviductal neoplasia as described in the first case of this study, are also relatively rare in chelonians [17], oophoritis and salpingitis as diagnosed in the snapping turtle (case 3) and the black marsh turtle (case 4), respectively, may be underdiagnosed in chelonians in comparison to other reptile taxa. This may be largely attributed to the chelonian shell that hampers easy visualization and sampling of the reproductive tract.

5. Conclusions

There are various reproductive disorders in chelonians that necessitate the use of surgical procedures. We consider the series of disorders described in the present report as representative of the reproductive conditions in chelonians that may be encountered in veterinary practice and that require advanced surgical treatment. The findings of this study may aid the practitioner in establishing an etiological diagnosis, differentiating between conditions that need a conservative or surgical approach, and selecting the most suitable surgical treatment for these and comparable reproductive disorders. Although restrictions may be encountered, the endoscope-assisted prefemoral approach is the least invasive, and thus preferred, technique instead of the ventral transplastron coeliotomy. In conclusion, the therapeutic efficacy for the described cases largely relied on the choice of minimally invasive endoscopic-assisted surgery versus transplastron coeliotomy that was primarily dictated by the involved species, etiology, and associated pathology.

Author Contributions: Conceptualization, T.H. and F.S.V.; methodology, T.H.; investigation, T.H. and F.S.V.; writing—original draft preparation, T.H.; writing—review and editing, T.H. and F.S.V.; visualization, T.H.; supervision, T.H. All authors have read and agreed to the published version of the manuscript.

Funding: This research received no external funding.

Institutional Review Board Statement: Not applicable.

Informed Consent Statement: Informed consent was obtained from all subjects involved in the study.

Data Availability Statement: The data presented in this study are available on request from the corresponding author.

Acknowledgments: The authors would like to express their gratitude to Annemie van Caelenberg and Merle Toom of the Department of Morphology, Imaging, Orthopedics, Rehabilitation and Nutrition of the Faculty of Veterinary Medicine, Ghent University for their kind assistance with the interpretation of radiographic images.

Conflicts of Interest: The authors declare no conflict of interest.

References

1. Innes, C.J.; Boyer, T.H. Chelonian reproductive disorders. *Vet. Clin. Exot. Anim. Pract.* **2002**, *5*, 555–578. [CrossRef] [PubMed]
2. Stahl, S.J.; DeNardo, D.F. Section 9 Medicine. Chapter 80 Theriogenology. In *Mader's Reptile and Amphibian Medicine and Surgery*, 3rd ed.; Divers, S.J., Stahl, S.J., Eds.; Saunders Elsevier: St. Louis, MO, USA, 2019; pp. 849–893.
3. Chitty, J.; Raftery, A. Egg retention/dystocia. In *Essentials of Tortoise Medicine and Surgery*, 1st ed.; Chitty, J., Raftery, A., Eds.; John Wiley & Sons: West Sussex, UK, 2013; pp. 195–197.
4. Di Ianni, F.; Parmigiani, E.; Pelizzone, I.; Bresciani, C.; Gnudi, G.; Volta, A.; Manfredi, S.; Bigliardi, E. Comparison between intramuscular and intravenous administration of oxytocin in captive-bred red-eared sliders (*Trachemys scripta elegans*) with nonobstructive egg retention. *J. Exot. Pet. Med.* **2014**, *23*, 79–84. [CrossRef]
5. Tucker, J.K.; Thomas, D.I.; Rose, J. Oxytocin dosage in turtles. *Chelonian Conserv. Biol.* **2007**, *6*, 321–324. [CrossRef] [PubMed]
6. Divers, S.J. Section 8 Endoscopy. Chapter 65 Endoscope-assisted and endoscopic surgery. In *Mader's Reptile and Amphibian Medicine and Surgery*, 3rd ed.; Divers, S.J., Stahl, S.J., Eds.; Saunders Elsevier: St. Louis, MO, USA, 2019; pp. 615–623.
7. Wüst, E.; Divers, S.J. Section 10 Surgery. Chapter 99 Chelonian prefemoral coeliotomy. In *Mader's Reptile and Amphibian Medicine and Surgery*, 3rd ed.; Divers, S.J., Stahl, S.J., Eds.; Saunders Elsevier: St. Louis, MO, USA, 2019; pp. 1054–1056.
8. Divers, S.J.; Wüst, E. Section 10 Surgery. Chapter 100 Chelonian transplastron coeliotomy. In *Mader's Reptile and Amphibian Medicine and Surgery*, 3rd ed.; Divers, S.J., Stahl, S.J., Eds.; Saunders Elsevier: St. Louis, MO, USA, 2019; pp. 1057–1061.
9. Innes, C.J. Endoscopy and endosurgery of the chelonian reproductive tract. *Vet. Clin. Exot. Anim. Pract.* **2010**, *13*, 243–254. [CrossRef] [PubMed]
10. Innes, C.J.; Hernandez-Divers, S.J. Coelioscopic-assisted prefemoral oophorectotomy in chelonians. *J. Am. Vet. Med. Assoc.* **2007**, *230*, 1049–1052. [CrossRef] [PubMed]
11. Proença, L.M.; Kleine, S.; Quandt, J.; Mullen, C.O.; Divers, S.J. Coelioscopic-assisted sterilization of female Mojave desert tortoises (*Gopherus agassizii*). *J. Herpetol. Med. Surg.* **2014**, *24*, 95–100. [CrossRef]
12. Gibbons, P.M.; Whitaker, B.R.; Carpenter, J.W.; McDermott, C.T.; Klaphake, E.; Sladky, K.K. Section 4 Infectious diseases and laboratory sciences. Chapter 35 Hematology and biochemistry tables. In *Mader's Reptile and Amphibian Medicine and Surgery*, 3rd ed.; Divers, S.J., Stahl, S.J., Eds.; Saunders Elsevier: St. Louis, MO, USA, 2019; pp. 333–350.
13. Divers, S.J. Section 10 Surgery. Chapter 104 Urinary tract. In *Mader's Reptile and Amphibian Medicine and Surgery*, 3rd ed.; Divers, S.J., Stahl, S.J., Eds.; Saunders Elsevier: St. Louis, MO, USA, 2019; pp. 1071–1076.
14. Mans, C.; Sladky, K.K. Diagnosis and management of oviductal disease in three red-eared slider turtles (*Trachemys scripta elegans*). *J. Small Anim. Pract.* **2012**, *53*, 234–239. [CrossRef] [PubMed]
15. Boyer, T.H.; Innes, C.J. Section 2 Biology (Taxonomy, Anatomy, Physiology, and Behavior). Chapter 7 Chelonian taxonomy, anatomy, and physiology. In *Mader's Reptile and Amphibian Medicine and Surgery*, 3rd ed.; Divers, S.J., Stahl, S.J., Eds.; Saunders Elsevier: St. Louis, MO, USA, 2019; pp. 31–49.
16. Knotek, Z.; Jekl, V.; Knotkova, Z.; Grabensteiner, E. Eggs in chelonian urinary bladder: Is coeliotomy necessary? In Proceedings of the Proceedings of the Association of Reptilian and Amphibian Veterinarians, 16th Annual Conference, Milwaukee, WI, USA, 8–15 August 2009; pp. 118–121.
17. Simard, J.; Ducatelle, R.; Van Caelenberg, A.; Hellebuyck, T. Oviductal prolapse associated with a leiomyoma in a Hermann's tortoise (*Testudo hermanni*). *Flem. Vet. J.* **2021**, *90*, 71–77. [CrossRef]

animals

MDPI

Case Report

Egg Removal via Cloacoscopy in Three Dystocic Leopard Geckos (*Eublepharis macularius*)

Alessandro Vetere [1,*], Enrico Bigliardi [1], Marco Masi [2], Matteo Rizzi [1], Elisa Leandrin [3] and Francesco Di Ianni [1]

1 Department of Veterinary Science, University of Parma, Strada del Taglio 10, 43126 Parma, Italy
2 Centro Veterinario Specialistico, Via Sandro Giovannini, 51/53, 00137 Roma, Italy
3 Clinica Veterinaria Madonna Di Rosa, Via Rosa 1, 33078 San Vito Al Tagliamento, Italy
* Correspondence: alessandro.vetere88@gmail.com

Simple Summary: Three adult female leopard geckos (*Eublepharis macularius*) belonging to three different owners were referred to for coelomic distention, anorexia, and weight loss. X-rays showed the presence of a macrosomic egg set in the third caudal of the coelomic cavity, and the diagnosis of dystocia was made in all three geckos. A cloacal endoscopic examination was performed on all three animals. A voluminous egg protruding through the urogenital papilla to the cloaca was visible. All the eggs were removed easily using endoscopic forceps. In two geckos, the eggshell was torn, and the content was aspirated to reduce the egg volume. After the procedure, a subcutaneous deslorelin implant was implanted. All geckos recovered rapidly after surgery. Two of the three geckos were healthy at the follow-up visit (respectively one and two years after the surgery) and did not show any signs of dystocia. Only in the third gecko, the dystocia recurred again 6 months later, and bilateral ovariosalpingectomy was performed. After surgery, the gecko recovered rapidly, resumed eating, and was discharged after one week of hospitalization in good condition.

Abstract: Dystocia is a multifactorial, life-threatening condition commonly affecting pet reptiles. Treatment for dystocia can be either medical or surgical. Medical treatment usually involves the administration of oxytocin, but in some species or, in some cases, this treatment does not work as expected. Surgical treatments such as ovariectomy or ovariosalpingectomy are resolutive, but invasive in small-sized reptiles. In this paper, we describe three cases of post ovulatory egg retention in three leopard geckos (*Eublepharis macularius*) successfully treated through a cloacoscopic removal of the retained eggs, after a non resolutive medical treatment. The intervention was fast, non-invasive, and no procedure-related adverse effects were noted. The problem relapsed six months later in one animal, and a successful bilateral ovariosalpingectomy was performed. Cloacoscopy should be considered a valuable, non-invasive tool for egg removal in dystocic leopard geckos when the egg is accessible to manipulation. Recrudescence or complications such as adhesions, oviductal rupture, or the presence of ectopic eggs should recommend surgical intervention.

Keywords: *Eublepharis macularius*; dystocia; endoscopy; reproductive disorders

Citation: Vetere, A.; Bigliardi, E.; Masi, M.; Rizzi, M.; Leandrin, E.; Di Ianni, F. Egg Removal via Cloacoscopy in Three Dystocic Leopard Geckos (*Eublepharis macularius*). *Animals* **2023**, *13*, 924. https://doi.org/10.3390/ani13050924

Academic Editor: Jean-Marie Exbrayat

Received: 23 January 2023
Revised: 17 February 2023
Accepted: 23 February 2023
Published: 3 March 2023

1. Introduction

Dystocia (egg binding) in reptiles is a common, multifactorial disease [1]. Different factors of captivity are often related to the occurrence of dystocia, such as inadequate husbandry (improper temperature gradients or humidity levels, inadequate nesting sites, overcrowding), poor physical conditions (illness, dehydration, malnutrition), reproductive apparatus disorders (infectious diseases, traumatic injuries, misshapen or large eggs), and metabolic diseases (hypocalcemia secondary to nutritional or renal hyperparathyroidism) [1,2]. Two common forms of reproductive disorders in reptiles are recognizable: preovulatory follicular stasis and postovulatory stasis or egg binding. The first form occurs when egg development stops prior to ovulation after vitellogenesis, resulting in persistent follicles and leading to inflammation and eventually rupture, coelomitis, and death [2–4]. The second form, the postovulatory stasis, occurs when the eggs are not laid and retained inside the oviduct for an indefinite period of time, or ectopic [1,3–5]. In both conditions, the animal can refuse to eat and its conditions can deteriorate rapidly, leading it to death [4,5]. The diagnosis of postovulatory stasis is usually made by X-ray [4,6,7] due to the radiopacity of the eggshell. Retained eggs can become overcalcified and appear more radiopaque on radiographic images [6]. Egg presence can also be assessed through careful palpation of the mid- or caudal coelom. Vitellogenic follicles during preovulatory stasis are better visualized and even measured through ultrasound or CT scan [8–11]. Treatment for dystocia can be either medical or surgical. Medical treatment involves the administration of oxytocin (1 to 10 UI/kg) intramuscularly (IM) or the administration of arginine vasotocine [12,13]. Oxytocin seems to work better in chelonians than in snakes and lizards [12]. It is generally used one hour after IM calcium gluconate administration, especially if the blood calcium level is low [12,14]. Arginine vasotocine seems to be more effective in reptiles than oxytocin, but the short storage life, high cost, and unlicensed use make this medical treatment option problematic in clinical practice [12]. Percutaneous ovocentesis is often reported by some authors as a treatment of dystocia in geckos [7]. However, it carries a high risk of organ rupture, with consequent coelomitis. If the eggs can be visualized through the cloacal opening, ovocentesis can be performed to make the eggs collapse and therefore be removed more easily [3]. Hormonal therapy, such as deslorelin implants, seems to be ineffective in suppressing ovarian activity in leopard geckos [15,16] and its use is still under debate in other reptiles [17–19]. In cases where medical treatment fails or is not applicable, surgery remains the elective choice. Ovariectomy or ovariosalpingectomy is resolutive [4,5,20].

2. Clinical Cases

2.1. Histories and Clinical Examination

2.1.1. Case 1

A 2-year-old, 45 g captive bred female leopard gecko (*Eublepharis macularius*) was presented for clinical examination due to one week of anorexia and weight loss. The animal was kept in a 60 (length) × 50 (height) × 30 (depth) cm glass terrarium without substrate, with only paper towels at the bottom that were changed weekly. The temperature inside the cage was 30 °C (86 °F) during the day and 25 °C (77 °F) at night. The diet was consistent and consisted of insects with calcium powder given three times a week. A UVB light bulb with 5.0 spectrum was provided and changed every 6 months. A plastic box with peat moss was present in the terrarium and misted daily. The owner indicated that the gecko was acquired two weeks earlier from another breeder. Once introduced in the new environment, the gecko spent the most of time burrowed in the substrate. At clinical examination, the gecko was lethargic and mildly to moderately dehydrated. The coelomic cavity was distended, and the presence of a large egg was palpable and visible in transparency through the skin in the caudal half of the coelom on the right half of the sagittal plane.

2.1.2. Case 2

A 4-year-old, 50 g captive bred female leopard gecko (*Eublepharis macularius*) was presented for clinical examination due to dysecdysis, lethargy, anorexia, and distention of the coelomic cavity (Figure 1). The animal was housed with a two-year-old male in a 50 (length) × 50 (height) × 50 (depth) cm glass terrarium with paper towels as a substrate and maintained at 30 °C (86 °F) during the day and 24 °C (75.2 °F) at night. The diet was consistent and consisted of insects dusted with calcium powder given twice a week. A UVB light bulb with 5.0 spectrum was provided and changed every 6 months. The owner also reported that a box with a wet mixture of peat moss and coconut fiber was added to the terrarium to facilitate deposition, and the female spent most of the time burrowing inside the box. No eggs were found during the daily check by the owner. At clinical examination, the gecko was lethargic and moderately dehydrated. The abdomen was swollen, and the presence of two voluminous masses was palpable in the caudal half of the coelom on the left and right halves of the sagittal plane.

Figure 1. Case 2: The gecko presented with distenction of the coelomic cavity, dysecdysis, and poor nutritional status.

2.1.3. Case 3

A 4-year-old, 52 g captive bred female leopard gecko (*Eublepharis macularius*) was presented for a veterinary second opinion. The owner indicated that the gecko was kept with an adult male of unknown age; both were kept in an 80 (length) × 40 (height) × 40 (depth) glass terrarium with paper towels as a substrate at 32 °C (89.6 °F) at day and 24 °C (75.2 °F) at night. In the cage, a box with a wet mixture of peat moss and coconut fiber was present as a wet nest to facilitate molting, and after the suggestion of the original breeder, deposition. The female spent most of the time burrowing inside without eating for two weeks. The diet was consistent and consisted of insects dusted with calcium powder given twice a week. No UVB light was provided. The animal was first presented due to lethargy and anorexia, and a diagnosis of egg binding was made by the first veterinarian based on the anamnesis and by the dorsoventral X-ray taken the same day. Oxytocin at 10 UI/kg was administered by the previous veterinarian. IM was performed twice with a one-hour interval without any results. At clinical examination, the gecko was lethargic and mildly dehydrated. The coelomic cavity was distended, and the presence of a large egg was palpable in the caudal half of the coelom on the right half of the sagittal plane.

2.2. Diagnostic Procedures

2.2.1. Case 1

Complete blood work, X-rays, and ultrasounds were performed. The CBC count and PCV were unremarkable. The biochemistry tests showed a moderate increase in AST and the CK X-rays showed the presence of large eggs in the caudal half of the coelomic cavity on the right half of the sagittal plane, and the presence of radiopaque foreign material was interpreted as ingested sand in the intestinal tract (Figure 2). Ultrasound examination confirmed the egg presence and the presence of hyperechoic material inside the intestinal lumen; no oviduct rupture was noted. Oxytocin at 5 UI/kg (10 UI/mL neurofisin, FATRO S.p.a, Ozzano dell'Emilia, Italy) [14] was administered in the right tricep after IM administration of calcium gluconate at 100 mg/kg (200 mg/mL Calcio PH, FATRO S.p.a, Ozzano dell'Emilia, Italy) 1 h before, without success. Given the inefficacy of the treatment, cloacoscopy was planned to examine any potential oviduct abnormalities and determine whether the egg could be removed through endoscopic grasping forceps, avoiding a more invasive celiotomy. The animal was anaesthetized using alfaxalone (10 mg/mL Alfaxan, Dechra Veterinary Products Srl, Torino, Italia) at a dosage of 5 mg/kg delivered intravenously in the right jugular vein [21]. The heart rate and respiratory rate were monitored during all procedures. The animal lost the rightening reflex after approximately 40 s but experienced apnea, so it was intubated and mechanically ventilated (one breath every 10 s). A 2.7 mm × 18 cm, 30° oblique telescope (within a 4.8 mm operative sheath) (Storz Telepack TP100 EN, Karl Storz Endoscopia Italia Srl, Verona, Italy) was used for cloacal inspection, and showed one egg protruding from the right salpinx to the cloaca (Figure 3). During the procedure, the operator held the animal in ventral recumbency with the left hand, while the right hand maneuvered the telescope. The eggshell was broken using grasping forceps, and the content was removed without any difficulties (Figure 4, Video S1).

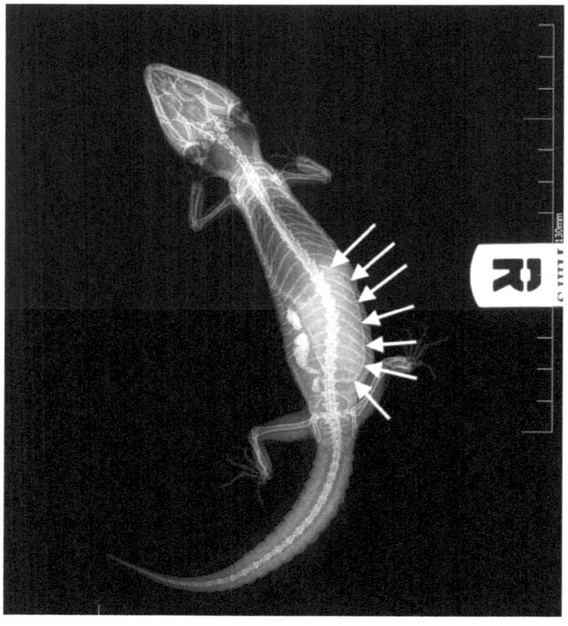

Figure 2. Case 1: Dorso-ventral X-ray. A large egg was visible in the caudal coelom (white arrows).

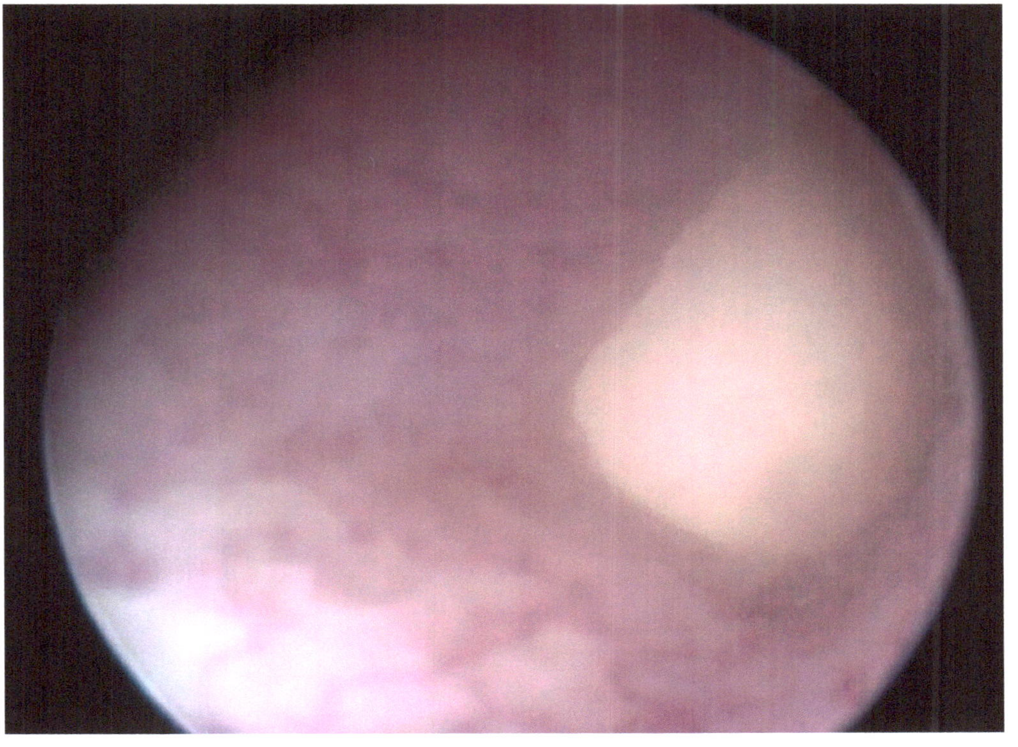

Figure 3. Case 1: Cloacoscopy. Endoscopic view of the right egg protruding from the oviduct to the cloaca.

2.2.2. Case 2

Complete blood work, X-rays, and ultrasounds were performed. The CBC count showed an increased WBC count with moderate heterofilia and monocytosis, mild basophilia, mild eosinopenia, and lymphocytosis (Table 1). The PCV was mildly increased. Biochemistry tests showed a moderate increase in creatinine kinase (Table 2). X-rays were performed in dorsoventral and latero-lateral projections, showing the presence of two large eggs in the caudal coelom (Figure 5). Ultrasound examination confirmed the presence of eggs; no oviduct rupture was noted. The use of oxytocin as medical treatment was not performed due to the severe weakness of the gecko. Due to the mild dehydration, warm fluids (ringer solution) were administered subcutaneously at 10 mL/kg/die and stabilized at 28 °C for 12 h. Cloacoscopy was planned for the day after to verify whether the eggs were visible and removable using endoscopic grasping forceps, avoiding a more invasive celiotomy. The animal was anesthetized using alfaxalone (10 mg/mL Alfaxan, Dechra Veterinary Products Srl, Via Agostino da Montefeltro, 2, Torino, Italia) at a dosage of 5 mg/kg delivered intravenously in the right jugular vein [21]. The animal lost the rightening reflex after approximately 60 s and maintained spontaneous breathing during the entire procedure. A 2.7 mm × 18 cm, 30° oblique telescope (within a 4.8mm operative sheath (Storz Telepack TP100 EN, Karl Storz Endoscopia Italia Srl, Verona, Italy) was used for cloacal inspection and showed the surface of the two eggs protruding from the left and right salpinx to the cloaca (Figure 6). During the procedure, the operator held the animal in ventral recumbency with the left hand, while the right hand maneuvered the telescope, as described in the Case 1. The eggshell was broken using grasping forceps, and the content was removed by performing cloacal lavages with warm sterile NaCl solution. The empty eggshells were removed as described for case 1.

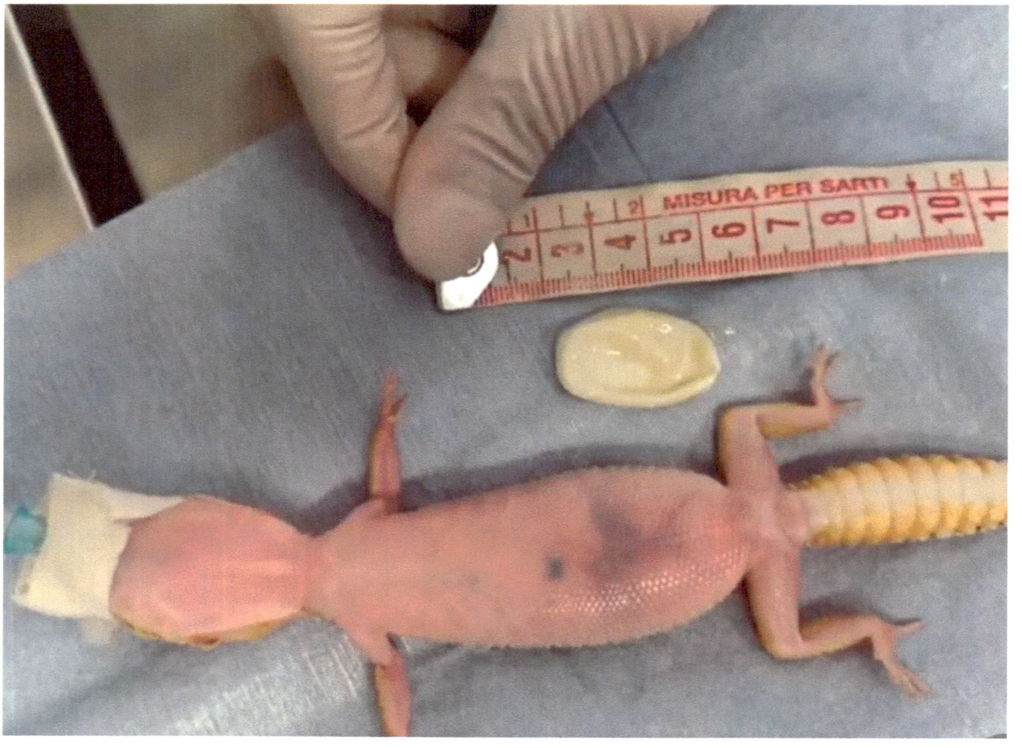

Figure 4. Case 1: End of the procedure. The egg was aspirated and easily extracted through the cloacal opening.

Table 1. Comparison between CBC (complete blood count) parameters of the cases and the reference values. Abnormal values are in bold.

	Case 1	Case 2	Case 3	Normal Values [22,23] (Female)
PCV%	40	**45**	**45**	21–40%
WBC × $10^3/mm^3$	8	**13**	**10**	6–9.4
RBC × $10^6/mm^3$	0.7	**1.3**	**1.1**	0.43–0.89
Heterophils ($10^3/\mu L$)	2.9	**3.5**	2	1.08–2.73
Eosinophils ($10^3/\mu L$)	0.4	**0**	0.8	0.15–1.95
Basophils ($10^3/\mu L$)	1	**3**	**3**	0.00–2.26
Monocytes ($10^3/\mu L$)	1	**4**	2	0.60–2.16
Lymphocytes ($10^3/\mu L$)	3	**4**	3	1.67–5.39

Table 2. Comparison between biochemical values of the cases and the reference values. Abnormal values are in bold.

	Case 1	Case 2	Case 3	Reference Values [22,23]
AST (U/L)	12	**78**		11–65
Total protein (g/dL)	7	**9**	4.2	2.4–8.0
Albumin	18	22	15	13–23
Creatinkinasis (U/L)	**4.897**	**4.000**	1861	0–3.701
Phosphorous (mg/dL)	14.5	8.4	14	1.5–16.4

Table 2. *Cont.*

	Case 1	Case 2	Case 3	Reference Values [22,23]
Calcium (mg/dL)	18	22	31	14–>37
Potassium (mmol/lt)	6.8	5.5	6.1	4.50–7.0
Uric acid (mg/dL)	3.3	**6.7**	4	0.5–6.6 mg/dL
Biliary acids	0.6	3.2	1.1	0.8–21 μmol/L

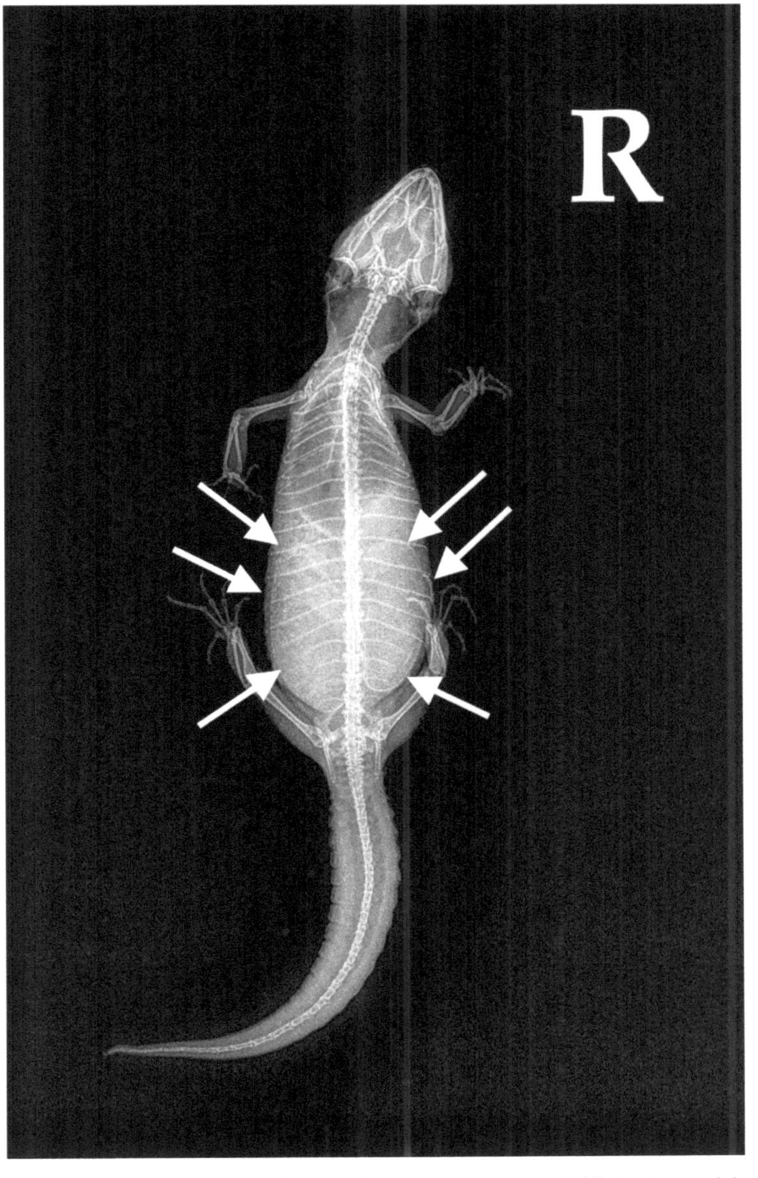

Figure 5. Case 2: Dorsoventral X-ray. Two large eggs were visible in the caudal coelom (white arrows).

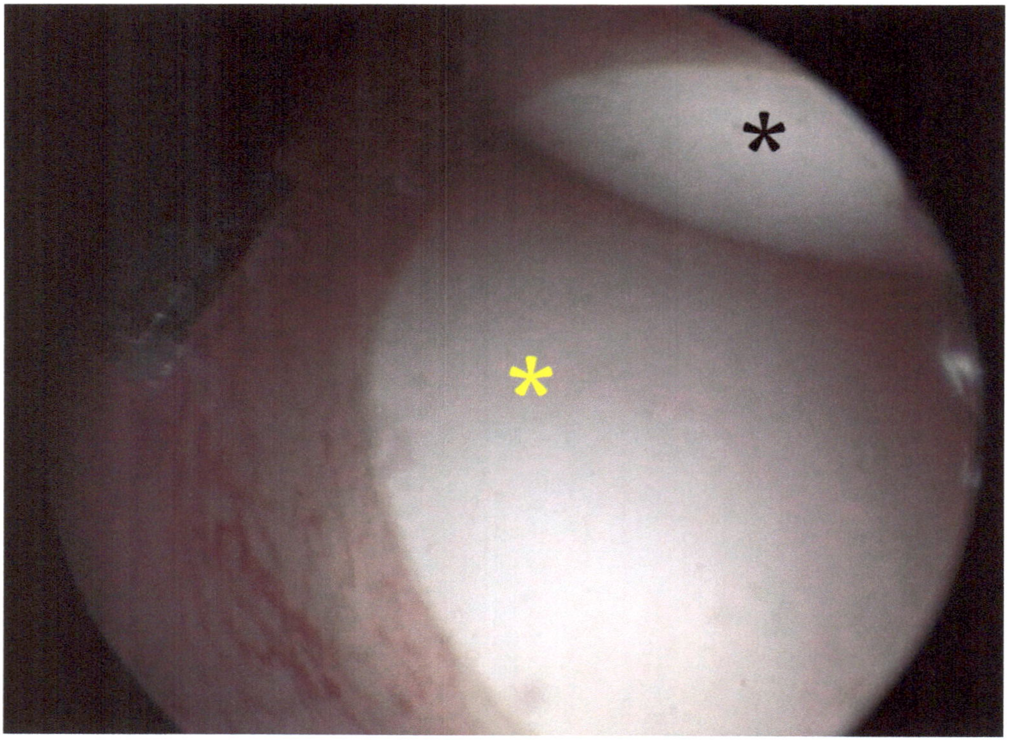

Figure 6. Cloacoscopy. Endoscopic view of the right (black asterisk) and the left (yellow asterisk) egg protruding from the vagina to the cloaca.

2.2.3. Case 3

Complete blood work, X-rays, and ultrasounds were performed. The CBC count showed a mildly increased WBC count with mild basophilia. The PCV was mildly increased. Biochemistry tests were unremarkable. X-rays were repeated and showed the presence of one large hypocalcified egg on the left half of the coelomic cavity (Figure 7) and a high radiopaque area on the left side, measuring approximately 0.5 cm × 1 cm in diameter. Ultrasound examination confirmed the egg presence in the right oviduct and the presence of a hyperechoic foreign body in the left oviduct. No oviduct ruptures were noted. Cloacoscopy was planned for the day after. As described for the case 2, warm fluids (ringer solution) were administered subcutaneously at 10 mL/kg/die and the gecko stabilized at 28 °C for 24 h. The animal was anesthetized using the same protocol described for cases 1 and 2. The animal lost the rightening reflex after approximately 40 s and maintained spontaneous breathing during the entire procedure. A 2.7 mm × 18 cm, 30° oblique telescope (within a 4.8 mm operative sheath) (Storz Telepack TP100 EN, Karl Storz Endoscopia Italia Srl, Verona, Italy) was used for cloacal inspection and showed the surface of the egg protruding from the right salpinx to the cloaca. During the procedure, the operator held the animal as described in case 1 and 2. The eggshell was broken, and the content was removed by performing cloacal lavages with warm sterile NaCl solution. The empty eggshells were removed as described for cases 1 and 2. A subcutaneous deslorelin implant (4.7 mg, Suprelorin®, Milano, Italy) was applied in the neck region upon the owner's request despite the owner having been informed that there is no scientific evidence of its efficacy in leopard geckos.

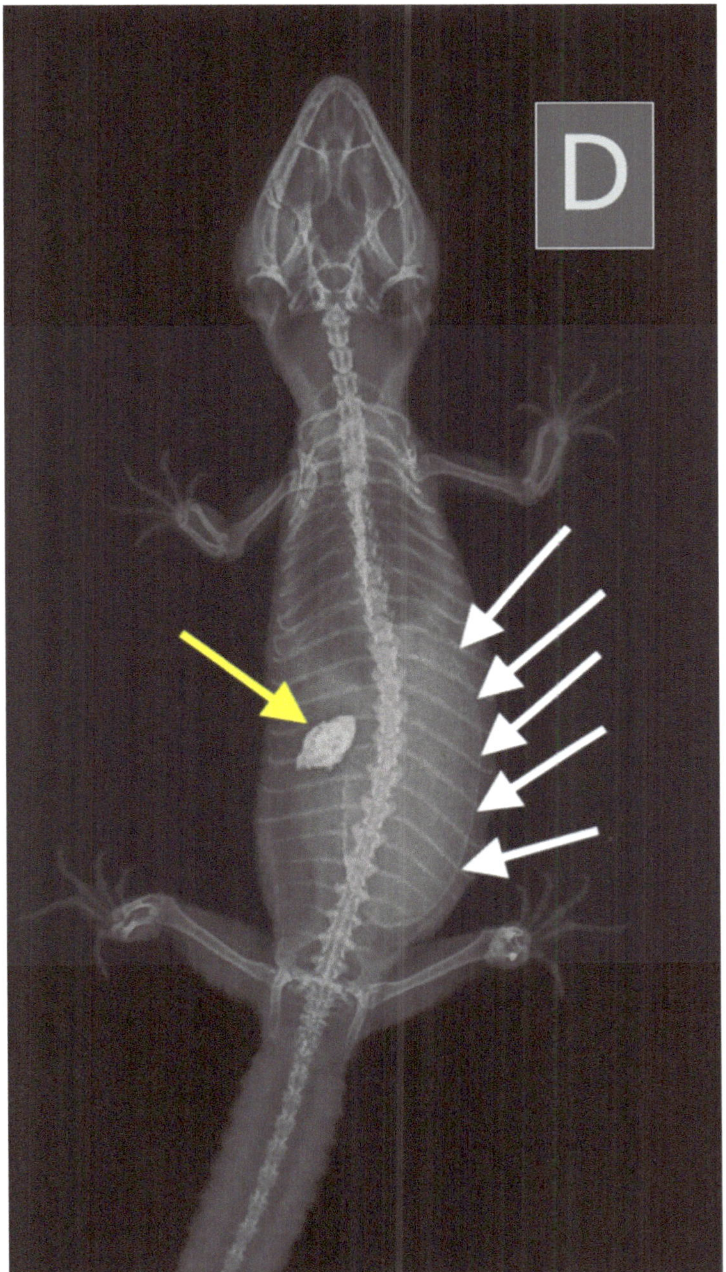

Figure 7. Case 3: Dorso-ventral X-ray. A large egg was visible in the caudal half of the coelomic cavity (white arrows).

2.3. Postoperative Care and Follow-Up

2.3.1. Case 1

The gecko recovered 15 min after the end of the procedure, and no adverse effects were noted. Warm fluids (ringer solution) were administered subcutaneously at 15 mL/kg/die at the end of the endoscopic procedure to support the renal function after the anesthesia. Assisted feeding (Emeraid IC critical care formula, Emeraid LLC, A Division of Lafeber, Cornell, IL, USA) was administered once a day as nutritional support. Five days after the procedure, the gecko started to eat, so it was discharged. One month later, the gecko's weight had increased by 5 g, and she was active and in good nutritional status. One year later, the gecko laid six infertile eggs, but dystocia had not occurred.

2.3.2. Case 2

The gecko recovered 20 min after the end of the procedure, and no adverse effects were noted. Fluid therapy and nutritional support were administered as described in case 1. Two days after the procedure, the gecko started to eat, so it was discharged. The owner was instructed to separate the male gecko from the female gecko and the problem had never occurred at the two-year follow-up.

2.3.3. Case 3

The gecko recovered 18 min after the end of the procedure, and no adverse effects were noted. Warm fluids (ringer solution) were administered subcutaneously at 15 mL/kg/die at the end of the endoscopic procedure. Fluid therapy and nutritional support were administered as described in case 1. Five days after the procedure, the gecko started to eat, so it was discharged. The owner was instructed to separate the male gecko from the female gecko. The leopard gecko did not lay any eggs in the next 6 months, and then a new episode of dystocia occurred; a large egg was evident in the right oviduct again, and the calcification in the left oviduct was still present. A bilateral ovariosalpingectomy was proposed and accepted by the owner. The gecko was premedicated with a mixture of dexmedetomidine at a dosage of 0.1 mg/kg (0.5 mg/mL Dexdomitor®, Vetòquinol Italia S.r.l, Bertinoro, Italy) and ketamine at a dosage of 10 mg/kg (100 mg/mL Ketavet®, Intervet production Srl, Aprilia, Italy) delivered in the right triceps [24]. The animal was intubated, maintained under gaseous anesthesia with 2% isoflurane (IsoFlo®, Zoetis Italia s.r.l, Roma, Italy) and mechanically ventilated (one breath every 15 s). A paramedian incision was performed to avoid the abdominal venous sinus. The cutis, abdominal muscles, and the coelomic membrane were cut, and the right oviduct with a voluminous egg inside was rapidly identified and isolated. A right salpingotomy was performed to remove the egg, making the access to the ovaries easier and improving the visibility of the near organs and blood vessels. Both the right and left ovaries and the respective oviducts were removed after ligature of the main vessels (Figure 8) with a 7/0 absorbable monofilament (Monosyn® Braun Avitum Italy S.p.A. Mirandola, Italy). The hyperechoic structure visible on ultrasound examination revealed a residual hypercalcified eggshell inside the left salpinx. The coelomic membrane was closed with a simple continuous suture through the abdominal muscles with a 6/0 absorbable monofilament (Monosyn® Braun Avitum Italy S.p.A. Mirandola, Italy). The cutis was closed with an everted simple interrupted suture with a 6/0 absorbable monofilament (Monosyn® Braun Avitum Italy S.p.A. Mirandola, Italy). Atipamezole at a dosage of 0.5 mg/kg (Atidorm® Fatro Industria Farmaceutica Veterinaria S.p. A, Ozzano dell'Emilia, Italy) was administered to the left triceps muscles at the end of the surgical procedure. The animal awakened 4 min after administration of the anesthesia reversal agent. A single 5 mg/kg dose of tramadol was administered IM, and 0.2 mg/kg meloxicam was administered SC once a day (SID) [25] postoperatively. The antimicrobial ceftazidime was administered SC every 3 days at a dose of 20 mg/kg [26]. The gecko started to eat after 2 days of therapy and was discharged after one week of hospitalization in good condition. At the 3-month follow-up, the clinical examination was unremarkable.

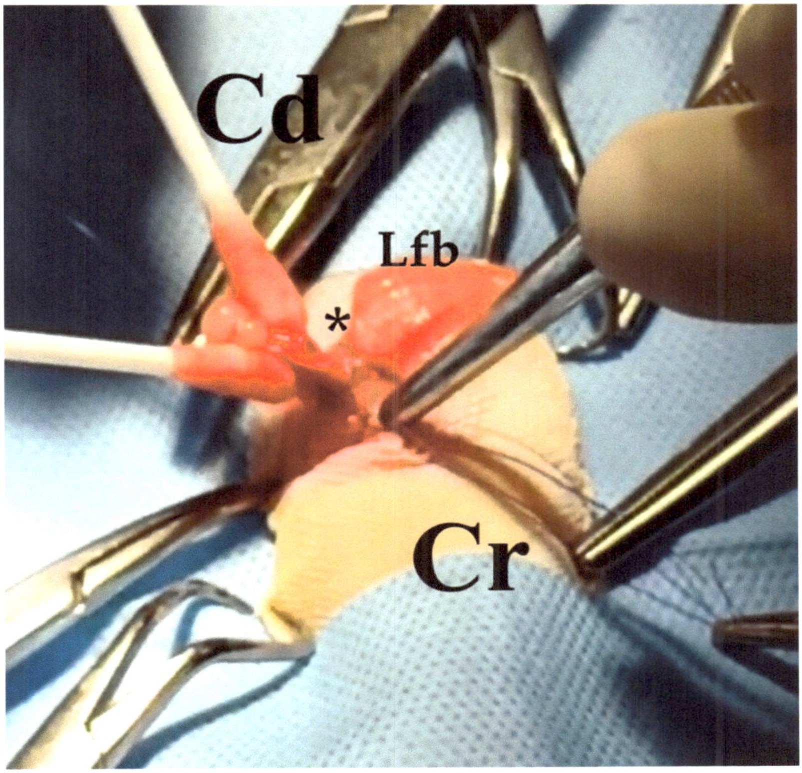

Figure 8. Case 3: Surgery. Left ovarian stalk (black asterisk) ligature during the ovariosalpingectomy intervention. Cr: cranial. Cd: caudal. Lfb: left fat body.

3. Discussion

Rigid endoscopy is a useful tool for the direct and also indirect evaluation of the reproductive tract in reptiles [3,27–29]. Coelioscopy is used as a direct method for the evaluation of the reproductive tract in chelonians, snakes, and lizards [3,28–30]. In tortoises, cystoscopy permits an indirect visualization of the coelomic organs in transparency through the urinary bladder wall [30,31]. This technique also permits the detection of retained eggs in the urinary bladder as well as their removal [32–34]. To the authors' knowledge, this is the first report of endoscopic-assisted egg removal through cloacoscopy in leopard geckos. All three cases presented here were diagnosed as obstructive postovulatory stasis in three female leopard geckos. As stress can be a predisposing factor for dystocia [1,3], it is possible that the already pregnant female, recently acquired and recently housed in a new environment, developed postovulatory stasis due to a high level of stress. As sexually active males can attempt mating persistently, their constant presence in the same cage, together with the female, could have been a predisposing factor to dystocia in case 2 and case 3 [35,36].

In case 1, PCV% and CBC count were unremarkable. In case 2, PCV%, RBC, WBC, heterophils, basophils, monocytes, and lymphocytes were slightly increased (Table 1), and in case 3, PCV%, WBC, RBC and basophils were slightly increased. These alterations were considered to be related to the dehydration status [37] and not related to any reproductive disorder. In cases 1 and 2, CK was mildly augmented (Table 2). In reptiles, CK levels can vary with variations in environmental temperature [30]. Given the above, in case 1 and 2, the increase in CK was considered to not be related to any disease. AST, total protein, and uric acid were slightly increased in case 2 (Table 2). Increases in AST and total protein in reptiles with follicular development have been reported [30]; however, in this case, the mild increases in AST and total protein were thought to be related more to dehydration, as uric acid was also augmented. Oxytocin, as a promoter of muscle contractibility [14], was administered in cases 1 and 3 without success. It is possible that in geckos, oxytocin is less effective than in chelonians, as has been described for snakes and some lizards [12,14].

Deslorelin acetate is a synthetic nonapeptide analogue of the natural gonadotropin-releasing hormone (GnRH). It acts as a contraceptive by temporarily suppressing the hypothalamic–pituitary–gonadal axis (HPG axis), inhibiting the production of pituitary hormones such as follicle-stimulating hormone (FSH) and luteinizing hormone (LH) [16,38,39]. In reptiles, the use of GnRH agonists has been poorly investigated; the use of deslorelin implants in leopard geckos seems to be less effective or ineffective at suppressing gonadal activity [15,16]. In lizards, suppression of gonadic activity was reported only in iguanas [38]. In chelonians, a desloreline implant was successfully used to treat chronic ovodeposition in a Greek tortoises (*Testudo graeca*) up to 24 months after application [40]. In a male *Chelonia mydas*, an annual application of deslorelin implants decreased serum testosterone levels after the fourth treatment [40].

In case 3, the gecko did not lay any eggs in the 6 months after treatment. A mature female leopard gecko can lay four to five clutches of two eggs per season with a one-month interval between clutches [36]. In this case, we do not know whether the gecko was at the end of the reproductive season or whether the ovarian activity was suppressed by the implant. Available scientific data [15] suggest that the latter hypothesis is unlikely. In all three cases, cloacal endoscopy permitted not only egg visualization but also egg removal through the cloacal opening under general anesthesia. Only one gecko (case 1) underwent apnea, and it was intubated and mechanically ventilated. Apnea has been reported in leopard geckos with the protocol used in case 1 [21]. In case 3, the recrudescence of egg binding required ovariosalpingectomy intervention. No procedure-related complications were noted.

The underlying causes of dystocia in these three cases remain unclear: it is likely that the voluminous eggs were too large to pass through the cloaca openings, or the oviduct was hypocontractile. Even the total calcium was considered normal according to the reference values [23], the ionized calcium was not measured. It is possible that an underlying hypocalcemia caused a decreased oviductal smooth muscle contractility [3,14]. The authors find this hypothesis unlikely since calcium supplementation was regularly provided by the owners. Moreover, no clinical signs of hypocalcemia such as seizures or skeletal deformities were noted at the physical examination. No evident fractures or skeletal demineralizations were radiographically detected. Diminishing egg pressure, leaking the shell with grasping forceps, and removal were performed easily. However, this procedure should be performed carefully to avoid the oviduct damage or rupture. A rude egg manipulation with the grasping forceps or a powerful cloacal flush could tear the oviductal or cloacal walls, predisposing to coelomitis.

Even if ultrasonography is a useful tool in diagnosing dystocia in reptiles [3,9,10], we were unable to visualize the distal part of the oviduct in all three geckos. In fact, the distal part of the reproductive tract is often difficult to image in lizards due to its location within the pelvic canal [41]. In these three described cases, the eggs protruding from the salpinx to the cloaca were visible and directly approachable during the cloacoscopy. This procedure should be critically evaluated whenever the eggs are not directly accessible or when adherences between the eggshell and the oviduct is suspected. Cloacoscopic egg removal should be considered before performing more invasive techniques, such as explorative celiotomy, when medical treatment fails. In fact, wound dehiscence and herniation are common post-surgical complication in leopard geckos, in relation to the thinness and the softness of the coelomic membrane and body wall musculature, which causes the muscles to tear more easily when tension is applied during the coelomic closure [33]. Percutaneous ovocentesis is contraindicated in reptiles and is not recommended because it increases the risk of yolk coelomitis and oviduct adherence. [9,28]. Many reptiles that are routinely presented for medical examinations are used for breeding and commercial purposes. In particular, leopard geckos have been bred for many years and today there are selections or "morphs" with a high commercial value on the market [42]. It is therefore of paramount importance to consider the owner's desire to maintain functionality and the integrity of the reproductive system.

4. Conclusions

Considering the predisposition that leopard geckos have to dystocia, the use of the rigid endoscope as a minimally invasive instrument to reduce soft tissue trauma, duration, and post pain surgery, is a valuable option in all subjects affected by dystocia secondary to malformations of the egg, as in the cases described herein. Endoscopic egg removal could be a valuable tool in dystocic leopard geckos when the egg is accessible to manipulation, or in subjects in which it is essential to preserve the possibility of future reproduction. Adhesions, oviductal rupture, or the presence of ectopic eggs should recommend surgical intervention. However, the veterinarian must understand and correct the underlying causes of reproductive disorders to prevent a recurrence.

Supplementary Materials: The following supporting information can be downloaded at: https://www.mdpi.com/article/10.3390/ani13050924/s1, Video S1.

Author Contributions: A.V. was the major contributor to the article's writing. F.D.I., M.M., M.R. and E.L. contributed to the acquisition, analysis and interpretation of the data and contributed to methodology acquisition and validation of the methods. E.B. supervised the procedures. All authors have read and agreed to the published version of the manuscript.

Funding: This research received no external funding.

Institutional Review Board Statement: Ethical review and approval were waived for this study due to the spontaneous pathologies of all reptiles.

Informed Consent Statement: Not applicable.

Data Availability Statement: Data sharing is not applicable. No new data were created or analyzed in this study. Data sharing is not applicable to this article.

Acknowledgments: Not applicable.

Conflicts of Interest: The authors declare no conflict of interest.

References

1. DeNardo, D.; Barten, S.L.; Rosenthal, K.L.; Raiti, P.; Nathan, R. Dystocia. *J. Herpetol. Med. Surg.* **2000**, *10*, 8–17. [CrossRef]
2. Stahl, S.J. Reptile production medicine. *Semin. Avian Exot. Pet Med.* **2001**, *10*, 140–150. [CrossRef]
3. Hedley, J. Reproductive diseases of reptiles. *Practice* **2016**, *38*, 457–462. [CrossRef]
4. Lock, B.A. Reproductive surgery in reptiles. *Vet. Clin. N. Am. Exot. Anim. Pract.* **2000**, *3*, 733–752. [CrossRef]
5. Stahl, S.J. Reptile obstetrics. In Proceedings of the North American Veterinary Conference, Gainesville, FL, USA, 7–11 January 2006; NAVC: Gainesville, FL, USA, 2006; pp. 1680–1683.
6. Di Giuseppe, M.; Silvestre, A.M.; Luparello, M.; Faraci, L. Post-ovulatory dystocia in two small lizards: Leopard gecko (*Eublepharis macularius*) and crested gecko (*Correlophus ciliatus*). *Russ. J. Herpetol.* **2017**, *24*, 128–132 [CrossRef]
7. Hall, A.J.; Lewbart, G.A. Treatment of dystocia in a leopard gecko (*Eublepharis macularius*) by percutaneous ovocentesis. *Vet. Rec.* **2006**, *158*, 737–739. [CrossRef]
8. Bertocchi, M.; Bigliardi, E.; Pelizzone, I.; Vetere, A.; Manfredi, S.; Cattarossi, D.; Rizzi, M.; Di Ianni, F. Monitoring of the reproductive cycle in captive-bred female *Boa constrictor*: Preliminary ultrasound observations. *Animals* **2021**, *11*, 3069. [CrossRef]
9. Rivera, S. Health assessment of the reptilian reproductive tract. *J. Exot. Pet Med.* **2008**, *17*, 259–266. [CrossRef]
10. Isaza, R.; Ackerman, N.; Jacobson, E.R. Ultrasound imaging of the coelomic structures in the *Boa constrictor* (*Boa constrictor*). *Vet. Radiol. Ultrasound* **1993**, *34*, 445–450. [CrossRef]
11. Vetere, A.; Di Ianni, F.; Bertocchi, M.; Castiglioni, V.; Nardini, G. Unilateral ovarian torsion in a Moroccan eyed lizard (*Timon tangitanus*). *J. Exot. Pet Med.* **2022**, *41*, 46–47. [CrossRef]
12. Boyer, T.H. Emergency care of reptiles. *Vet. Clin. N. Am. Exot. Anim. Pract.* **1998**, *1*, 191–206. [CrossRef] [PubMed]

13. Sawyer, W.H.; Munsick, R.A.; Van Dyke, H.B. Evidence for the presence of arginine vasotocin (8-arginine oxytocin) and oxytocin in neurohypophyseal extracts from amphibians and reptiles. *Gen. Comp. Endocrinol.* **1961**, *1*, 30–36. [CrossRef] [PubMed]
14. Jenkins, J.R. Medical management of reptiles. *Compend. Contin. Educ. Pract. Vet.* **1991**, *13*, 980–988.
15. Cermakova, E.; Oliveri, M.; Knotkova, Z.; Knotek, Z. Effect of a GnRH agonist (deslorelin) on ovarian activity in leopard geckos (*Eublepharis macularius*). *Vet. Med.* **2019**, *64*, 228–230. [CrossRef]
16. Korste, M.C. Deslorelin as a Contraceptive in Female Leopard Geckos (*Eublepharis macularius*). Master's Thesis, Utrecht University, Utrecht, The Netherlands, 2019.
17. Bardi, E.; Manfredi, M.; Capitelli, R.; Lubian, E.; Vetere, A.; Montani, A.; Bertoni, T.; Talon, E.; Ratti, G.; Romussi, S. Determination of efficacy of single and double 4.7 mg deslorelin acetate implant on the reproductive activity of female pond sliders (*Trachemys scripta*). *Animals* **2021**, *11*, 660. [CrossRef]
18. Potier, R.; Monge, E.; Loucachevsky, T.; Hermes, R.; Göritz, F.; RGecoochel, D.; Risi, E. Effects of deslorelin acetate on plasma testosterone concentrations in captive yellow-bellied sliders (*Trachemys scripta* sp.). *Acta Vet. Hung.* **2017**, *65*, 440–445. [CrossRef]
19. Rowland, M.N. Use of a deslorelin implant to control aggression in a male bearded dragon (*Pogona vitticeps*). *Vet. Rec.* **2011**, *16*[illegible], 127. [CrossRef]
20. Backues, K.A.; Ramsay, E.C. Ovariectomy for treatment of follicular stasis in lizards. *J. Zoo Wildl. Med.* **1994**, *25*, 111–116.
21. Morici, M.; Di Giuseppe, M.; Spadola, F.; Oliveri, M.; Knotkova, Z.; Knotek, Z. Intravenous alfaxalone anaesthesia in leopard geckos (*Eublepharis macularius*). *J. Exot. Pet Med.* **2018**, *27*, 11–14. [CrossRef]
22. Cojean, O.; Alberton, S.; Froment, R.; Maccolini, E.; Vergneau-Grosset, C. Determination of leopard gecko (*Eublepharis macularius*) packed cell volume and plasma biochemistry reference intervals and reference values. *J. Herpetol. Med. Surg.* **2020**, *30*, 156–16[illegible]. [CrossRef]
23. Knotkova, Z.; Morici, M.; Oliveri, M.; Knotek, Z. Blood profile in captive adult male leopard geckos (*Eublepharis macularius*). *Vet. Med.* **2019**, *64*, 172–177. [CrossRef]
24. Doss, G.A.; Fink, D.M.; Sladky, K.K.; Mans, C. Comparison of subcutaneous dexmedetomidine-midazolam versus alfaxalone-midazolam sedation in leopard geckos (*Eublepharis macularius*). *Vet. Anaesth. Analg.* **2017**, *44*, 1175–1183. [CrossRef] [PubMed]
25. Ting, A.K.Y.; Tay, V.S.Y.; Chng, H.T.; Xie, S. A critical review on the pharmacodynamics and pharmacokinetics of non-steroidal anti-inflammatory drugs and opioid drugs used in reptiles. *Vet. Anim. Sci.* **2022**, *17*, 100267. [CrossRef] [PubMed]
26. Lawrence, K. The use of antibiotics in reptiles: A review. *J. Small Anim. Pract.* **1983**, *24*, 741–752. [CrossRef]
27. Jekl, V.; Knotek, Z. Endoscopic Examination of Snakes by access through an Air SAC. *Vet. Record.* **2006**, *158*, 407–410. [CrossRef]
28. Divers, S.J.; Stahl, S.J. (Eds.) *Mader's Reptile and Amphibian Medicine and Surgery-e-Book*; Elsevier Health Sciences: Amsterdam, The Netherlands, 2018.
29. Schildger, B.; Haefeli, W.; Kuchling, G.; Taylor, M.; Tenhu, H.; Wicker, R. Endoscopic examination of the pleuro-peritoneal cavity in reptiles. In *Seminars in Avian and Exotic Pet Medicine*; WB Saunders: Philadelphia, PA, USA, 1999; Volume 8, pp. 130–138.
30. Divers, S.J. Endoscopy equipment and instrumentation for use in exotic animal medicine. *Vet. Clin. Exot. Anim. Pract.* **2010**, *13*, 171–185. [CrossRef]
31. Di Girolamo, N.; Selleri, P. Clinical applications of cystoscopy in chelonians. *Vet. Clin. Exot. Anim. Pract.* **2015**, *18*, 507–526. [CrossRef]
32. Mans, C.; Foster, J.D. Endoscopy-guided ectopic egg removal from the urinary bladder in a leopard tortoise (*Stigmochelys pardalis*). *Can. Vet. J.* **2014**, *55*, 569.
33. Minter, L.J.; Wood, M.W.; Hill, T.L.; Lewbart, G.A. Cystoscopic guided removal of ectopic eggs from the urinary bladder of the Florida cooter turtle (*Pseudemys floridana* floridana). *J. Zoo Wildl. Med.* **2010**, *41*, 503–509. [CrossRef]
34. Knotek, Z.; Jekl, V.; Knotkova, Z.; Grabensteiner, E. Eggs in chelonian urinary bladder: Is coeliotomy necessary. In Proceedings of the Association of Reptilian and Amphibian Veterinarians, Milwaukee, WI, USA, 8–15 August 2009; pp. 118–121.
35. Khan, M.S. *Natural History and Biology of Hobbyist Choice Leopard Gecko Eublepharis Macularius*; Talim ul Islam College: Rabwah, Pakistan, 2009.
36. Bradley, T.; Nieves, D. Leopard gecko, *Eublepharis macularim*, captive care and breeding. *Bull. Assoc. Reptil. Amphib. Vet.* **1999**, *9*, 36–40. [CrossRef]
37. Stacy, N.I.; Harr, K.E. Hematology of reptiles with a focus on circulating inflammatory cells. *Infect. Dis. Pathol. Reptiles* **2020**, *1*, 267–330. [CrossRef]
38. Grundmann, M.; Möstl, E.; Knotkova, Z.; Knotek, Z. The use of synthetical GnRH agonist implants (deslorelin) for the suppression of reptile endocrine reproductive activity. In Proceedings of the 1st International Congress for Avian, Reptile and Exotic Mammal, Wiesbaden, Germany, 20–26 April 2013; p. 248.
39. Iannaccone, M.; Ulivi, V.; Campolo, M. Use and duration of Deslorelin acetate in a Testudo graeca to solve a chronic re-productive disorder. In Proceedings of the Zoo and Wildlife Health Conference, Berlin, Germany, 24–27 May 2017; pp. 113–116.
40. Graham, K.M.; Mylniczenko, N.D.; Burns, C.M.; Bettinger, T.L.; Wheaton, C.J. Examining factors that may influence accurate measurement of testosterone in sea turtles. *J. Veter. Diagn. Investig.* **2015**, *28*, 12–19. [CrossRef] [PubMed]

41. Di Ianni, F.; Volta, A.; Pelizzone, I.; Manfredi, S.; Gnudi, G.; Parmigiani, E. Diagnostic sensitivity of ultrasound, radiography and computed tomography for gender determination in four species of lizards. *Vet. Radiol. Ultrasound* **2015**, *56*, 40–45. [CrossRef] [PubMed]
42. Agarwal, I.; Bauer, A.M.; Gamble, T.; Giri, V.B.; Jablonski, D.; Khandekar, A.; Mohapatra, P.P.; Masroor, R.; Mishra, A.; Ramakrishnan, U. The evolutionary history of an accidental model organism, the leopard gecko *Eublepharis macularius* (Squamata: Eublepharidae). *Mol. Phylogenetics Evol.* **2022**, *168*, 107414. [CrossRef] [PubMed]

animals

MDPI

Article

Sex Determination in Immature Sierra Nevada Lizard (*Timon nevadensis*)

Alessandro Vetere *, Michela Ablondi, Enrico Bigliardi, Matteo Rizzi and Francesco Di Ianni

Department of Veterinary Science, University of Parma, Strada del Taglio 10, 43126 Parma, Italy
* Correspondence: alessandro.vetere88@gmail.com

Simple Summary: Sex determination in reptiles is frequently requested by reptile breeders, and it is a real challenge in reptiles with little or no sexual dimorphism, such as in immature subjects. Twenty-three clinically healthy young Sierra Nevada lizards (*Timon nevadensis*) aged between 4 and 6 months were included for sex determination using two techniques: cloacal probing and contrast radiography. Results showed that contrast radiography may have major sensitivity for sex determination compared to probing. Given the above, this technique could represent a valid and less invasive aid for sexing young lizards.

Abstract: Sex determination has a fundamental role in a captive breeding context, both for commercial reasons and in relation to animal welfare itself. However, this can be particularly difficult, especially in reptiles with little or no sexual dimorphism. Twenty-three clinically healthy young Sierra Nevada lizards (*Timon nevadensis*) were included in this study for sex determination. The first attempt at sexing was carried out by cloacal probing. A small, buttoned probe was inserted very gently into the hemipenal pouches, and the length of the inserted part was evaluated and measured. Subsequently, for each animal, a contrast medium was administered into the cloaca, and radiography was performed within 5 min. Through probing, 11 males and 8 females were recognized. The test was, however, equivocal in four subjects. In contrast radiography, 14 males and 9 females were identified. All the animals were rechecked after 8 months through an ultrasound examination, confirming 15 of the 14 previously male sexed animals based on contrast radiography. All the animals identified as female ($n = 9$) by contrast radiography were confirmed. From these results, it seems that contrast radiography may have major sensitivity in sex determination compared to probing. This technique could represent a valid and less invasive aid for sexing young lizards.

Keywords: sex determination; reptiles; radiology; lizards

Citation: Vetere, A.; Ablondi, M.; Bigliardi, E.; Rizzi, M.; Di Ianni, F. Sex Determination in Immature Sierra Nevada Lizard (*Timon nevadensis*). *Animals* **2022**, *12*, 2144. https://doi.org/10.3390/ani12162144

Academic Editor: Tom Hellebuyck

Received: 23 July 2022
Accepted: 17 August 2022
Published: 21 August 2022

Publisher's Note: MDPI stays neutral with regard to jurisdictional claims in published maps and institutional affiliations.

1. Introduction

The genus *Timon* currently comprises six species found on three different continents [1]. The Asian species *Timon princeps* and *Timon kurdistanicus* inhabit regions in Turkey, Iran and Iraq; and *Timon tangitanus* and *Timon pater* are found in Morocco, Tunisia and Algeria. *Timon nevadensis* and *Timon lepidus* are restricted to Europe [1]. *Timon nevadensis* are present in the south-eastern part of Spain [1]. Adults can be over 130 mm in snout–vent length (SVL) [2] and they are sexually dimorphic: the male is big compared to the female [2]. Mature males display evident prefermoral pores, whereas in females, these pores are not well developed [3]. Juveniles lack evident sexual dimorphism [2–4]. While *Timon lepidus* has most recently been assessed for the IUCN Red List of Threatened Species in 2008 as near threatened (NT), there are no data available for *Timon nevadensis*.

Sex determination is frequently required in reptiles, especially for species with no evident sexual dimorphism [5]. Sex identification in lizards is usually visually performed by evaluating secondary sexual characteristics, such as the presence of well-developed femoral pores in males or the presence of brilliant colors of livery. Knowing the sex of individual subjects is essential for zoological facilities or for private breeders to make reproductive pairs. In species where intersex aggression is frequent, a male to female ratio favoring females is typically desired [6,7]. The male and female reproductive systems have different disorders, and knowing the sex of a single animal is very useful for the diagnosis of diseases associated with the reproductive organs [8,9]. Male lizards have a pair of inverted hemipenes [10] placed inside two pouches at the base of the tail; only one hemipene at a time is used for copulation [10], whereas in females, hemipenes are lacking. During copulation, semen is carried from the urogenital opening in the roof of the female vent cloaca along a groove called "sulcus spermaticus." Hemipene shape varies among reptile species; sometimes they bear spikes or corneal hooks to adhere better to the cloacal mucosa [11]. However, many species lack evident sexual dimorphism [12]. Different techniques for sexing lizards include the following. In probing, a lubricated, stainless-steel probe of the proper diameter is carefully and gently inserted into the side of the vent of the lizard (or even snake) and then directed towards the tail tip along the interior of the side of the tail. If the lizard is a male, the probe will slip inside the hemipenal pouch for approximately one quarter of the tail length, depending on the species [13]. If the probe advances only a short way to a depth of one to six scales [13,14], the lizard is considered a female, as there are no hemipenes for the probe to advance into [15]. Probing is considered quite invasive and should be carefully performed to avoid tissue damage [12,14]. Manual eversion of the hemipenes involves placing a small amount of pressure at the tail base to evert the hemipenes (in males) out of the hemipenal pouches [16]. This practice may be traumatic if performed with excessive force [12,16]. Ultrasonography is also used in lizards as a useful tool to determine follicular development or for sex determination [17,18]. However, in juvenile animals or in very small specimens, the gonads may be difficult or impossible to visualize [12,14]. In addition, ultrasound examination may be difficult to perform in species with thick scales or in subjects where the gut is filled with gas [12]. Contrast radiography is another useful technique involving the use of contrast media introduced into the hemipenal pouches to highlight the presence of the inverted hemipenes inside [12]. The presence of a certain quantity of cellular debris or fluid inside the precloacal pouches seems to interfere with a proper contrast distribution, leading to a false-negative result [19]. Genotipic sex determination via karyotyping and diagnostic techniques such as computed tomography (CT) scanning are useful tools for sex determination in lizards, but they are still expensive [20]. Regarding CT, the image resolution can be inadequate in very small specimens, leading to inconclusive results [12]. The aim of this study was to compare and evaluate the relative diagnostic sensitivity of two different techniques, probing and contrast radiography, for sex determination in immature specimens of Sierra Nevada lizards (*Timon nevadensis*).

2. Materials and Methods

Ethical approval for the study was given by the University of Parma. The owners gave informed consent to allow participation of their animals in the study. Twenty-three clinically healthy, young Sierra Nevada lizards (*Timon nevadensis*) aged between 4 and 6 months and owned by a private reptile breeder were presented to the Department of Veterinary Science, University of Parma, Italy, for sex determination. The animals weighed, according to an analytical balance (VEVOR® Digital Precision Scale 5000 g × 0.01 g, Taicang Vevor Industry Co., Ltd. 9448 Richmond Pl, #e, Rancho Cucamonga, CA, 91730, USA) from 10 to 18 g (0.022 to 0.039 lb; median, 13.7 g (0.030 lb) and measured from 6 to 12 cm SVL (average 9.8 cm). All the animals were divided individually into different plastic cages and identified with progressive identification numbers. Two different operators performed the imaging, and they were unaware of the sexes of the animals and of the prior results for every performed technique. A first attempt at sex determination was carried out by cloacal probing. The animals were placed under gaseous anesthesia using 2% isoflurane (Isoflo® Zoetis Italia S.r.l, Via Andrea Doria, 41 M, 00,192 Roma RM) for 15 min inside the induction chamber until the rightin reflex was completely lost. A small, buttoned, straight,

1.0 × 60 mm lacrimal duct cannula (E-VET A/S, Ole Rømersvej 26A, 6100 Haderslev, Denmark) was introduced gently into the proctodeum, and then turned backward toward the tail, entering the hemipenal pouch. The length of the inserted part was evaluated and measured. If the probe slipped through the vent inside the hemipenal pouches for more than 5 or 6 caudal scales, the animal was identified as a male. In contrast, if the probe did not go through the vent for more than 4 scales or if any signs of resistance were recorded in both pouches, the animal was classified as female. In every case of dubious results, e.g., different lengths measured for the left and the right pouch in the same subject, the animal was classified as not determined (ND). Subsequently, and blindly, for each animal, 0.1 mL iodinated contrast medium (Ioversol Optiray® 350, Guebert S.p. A, Via Albricci 9, Segrate, Italy) was administered into the postcloacal region by another operator through a buttoned, straight, 1.0 × 60 mm lacrimal duct cannula (E-VET A/S, Ole Rømersvej 26A, 6100 Haderslev, Denmark), and ventro-dorsal (VD) total body radiography was performed within 5 min (Regius 110S, Konica Minolta Health care, Tokyo, Japan). The excess contrast media outlining the cloacal rim was gently cleaned with a piece of a paper towel to avoid artefacts in the radiograms. If one or both of the hemipenes could be visualized, the lizard was classified as male, whereas if the hemipenes were not visualized, the lizard was classified as female. All the animals were rechecked at 8 months (T2) through an ultrasound examination (Esaote Class C, Esaote S.p.a. Via di Caciolle 15, 50127, Florence, Italy) using a 12 MHz linear probe (Esaote Mylab 30 Vet Gold, Esaote S.p.a. Via di Caciolle 15, 50127, Florence, Italy), searching and identifying the gonads. All the animals were placed in dorsal recumbency, and the probe was gently slid cranio-caudally in the third of the coelomic cavity, until the gonads were detected between the spine and the longissimus dorsi in both transversal and longitudinal plane (Figure 1A,B). All the results were compared for statistical analysis.

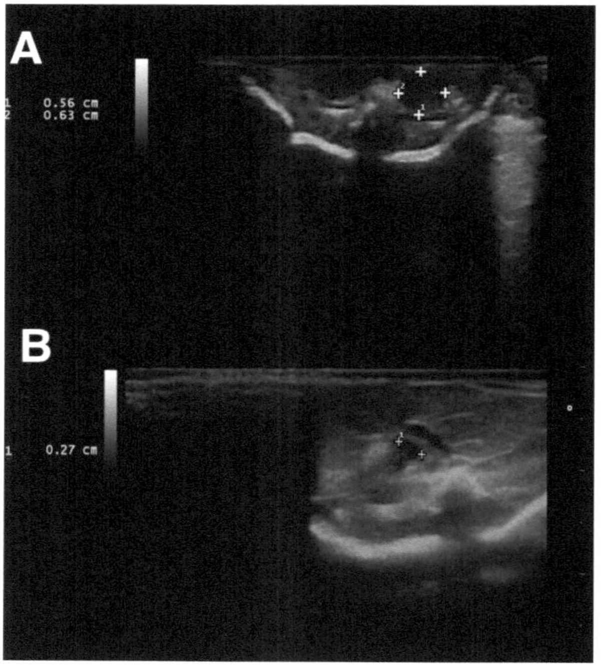

Figure 1. Ultrasonographic examination was performed 8 months after the contrast radiographic study (T3). 1: longitudinal diameter. 2: Transversal diameter (**A**) Transverse plane, dorsal recumbency,

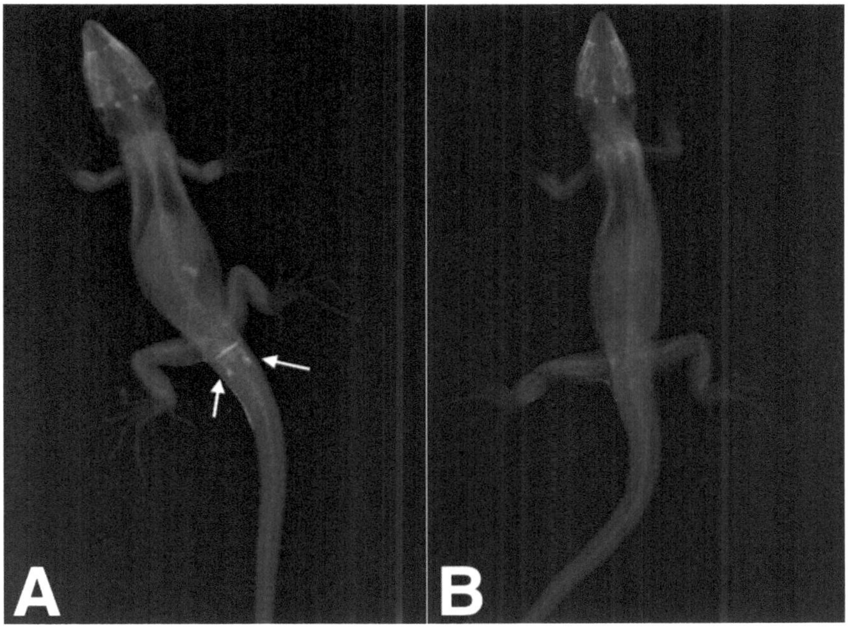

Figure 3. (**A**) Positive contrast radiographic study of a four-month old, 8 cm SVL Sierra Nevada lizard (*Timon nevadensis*); dorsoventral projection. White arrows indicate both the right and left hemipenes. (**B**) Negative contrast radiographic study of a four-month old Sierra Nevada lizard (*Timon nevadensis*); dorsoventral projection. There is no evidence of hemipenes in the postcloacal region.

4. Discussion

A variety of techniques have been used for sex determination in reptiles [1,12–14,19,21]. Laparoscopy and laparotomy are well described techniques [22–27]. While laparotomy for sex determination could be nowadays considered an outdated technique due to the invasiveness of the procedure, laparoscopy, due to the auxilium of a rigid endoscope directly inserted into the coelomic cavity, provides rapid and less invasive gonad visualization in medium size chelonians and lizards [24–26]. Probing was performed using a small, metallic, buttoned, straight, 1.0 × 60 mm lacrimal duct cannula. Even if the use of a rigid metallic cannula could be more traumatic than using an endovenous (EV) Teflon catether, we found the rigid cannula easier to introduce inside the cloacal rim without bending it or occluding its tip during contrast media infusion. Contrast radiographies identified at least one hemipene of all the male lizards classified as equivocal with probing. The hemipenes were visualized as a radiopaque elongated structure filled with contrast medium, inverted and directed cranio-caudally (Figure 1a). No adverse effects were seen after the injection of the contrast medium into the cloaca. Although sexual maturity of *Timon* lizards is reached at approximately two years of age [21], the ultrasound examination performed after 8 months (T2) detected and easily identified the gonads [23,24,28]. One animal died at 10 months and two animals at 11 months after the first measurements (T1) for unknown causes. Two lizards were dissected, and even if the cadaveric alterations were moderate to severe, the gonads were recognizable, and the results were consistent with the previously obtained data. One year after the first measurements at T1, the animals were visually rechecked (T3) and began to show signs of sexual dimorphism. In males, the presence of hemipenes was evident by the presence of hemipene bulges. Further research might be needed to confirm the results in a larger sample set. Nevertheless, the two-proportion z-test showed that the successful sex determination was 28.1% higher in contrast radiography compared to probing, which highlights the likely higher sensitivity of the former than the latter.

5. Conclusions

From these results, it seems that contrast radiography may have higher sensitivity in sex determination compared to probing. Even if we did not note any procedure-related traumatic complications, we recommend being very gentle during the contrast media injection, because of the thinness of the hemipenal pouches and the very small size of the lizards. Given the above, this technique could represent a valid and less invasive aid for sexing young and very small lizards.

Author Contributions: Conception and design: A.V., F.D.I., E.B. and M.R.; acquisition of data: A.V., M.A.; analysis and interpretation of data: M.A.; drafting the article: A.V. and F.D.I.; revising article for intellectual content: A.V., F.D.I., M.A., M.R. and E.B.; final approval of the completed article: A.V., F.D.I., M.A., M.R. and E.B.; agreed to be accountable for all aspects of the work, and to ensure that questions related to the accuracy or integrity of any part of the work are appropriately investigated and resolved: A.V., F.D.I., M.A., M.R. and E.B. All authors have read and agreed to the published version of the manuscript.

Funding: This research received no external funding.

Institutional Review Board Statement: Ethical approval for the study was given by the University of Parma (Organismo Preposto Al Benessere Degli Animali Comitato Etico Per La Sperimentazione Animale Prot. No. 20/CESA/2021).

Informed Consent Statement: Owners gave written consent for their pet's personal or clinical details along with any identifying images to be published in this study.

Data Availability Statement: Not applicable.

Conflicts of Interest: The authors declare that they have no competing interest.

Abbreviations

CT	computed tomography
EV	endovenous
ND	not determined
NT	near threatened
VD	ventro-dorsal

References

1. Ahmadzadeh, F.; Flecks, M.; Carretero, M.A.; Böhme, W.; Ihlow, F.; Kapli, P.; Rödder, D. Separate histories in both sides of the Mediterranean: Phylogeny and niche evolution of ocellated lizards. *J. Biogeogr.* **2016**, *43*, 1242–1253. [CrossRef]
2. Carretero, M.Á.; Salvador Milla, A. *Lagarto Bético–Timon Nevadensis (Buchholz, 1963)*; Museo Nacional de Ciencias Naturales: Madrid, Spain, 2017.
3. Braña, F. Sexual dimorphism in lacertid lizards: Male head increase vs female abdomen increase? *Oikos* **1996**, *75*, 511–523. [CrossRef]
4. Mateo, J.A.; Milla, A.S. *Lagarto Ocelado—Timon Lepidus (Daudin, 1802)*; Milla, S., Marco, A., Eds.; Museo Nacional de Ciencias Naturales: Madrid, Spain, 2017.
5. Pellett, S.; Cope, I. Visual identification of reptiles: Part 2—lizards. *Companion Anim.* **2013**, *18*, 399–405. [CrossRef]
6. Hailey, A.; Willemsen, R.E. Population density and adult sex ratio of the tortoise Testudo hermanni in Greece: Evidence for intrinsic population regulation. *J. Zool.* **2000**, *251*, 325–338. [CrossRef]
7. Golubović, A.; Arsovski, D.; Tomović, L.; Bonnet, X. Is sexual brutality maladaptive under high population density? *Biol. J. Linn. Soc.* **2018**, *124*, 394–402. [CrossRef]
8. Vetere, A.; Di Ianni, F.; Bertocchi, M.; Castiglioni, V.; Nardini, G. Unilateral ovarian torsion in a Moroccan eyed lizard (*Timon tangitanus*). *J. Exot. Pet Med.* **2022**, *41*, 46–47. [CrossRef]
9. Sykes, J.M.I.V. Updates and practical approaches to reproductive disorders in reptiles. *Vet. Clin. Exot. Anim. Pract.* **2010**, *13*, 349–373. [CrossRef]
10. Shea, G.M.; Reddacliff, G.L. Ossifications in the hemipenes of varanids. *J. Herpetol.* **1986**, *20*, 566–568. [CrossRef]

11. Arnold, E.N. The hemipenis of lacertid lizards (Reptilia: Lacertidae): Structure, variation and systematic implications. *J. Nat. Hist.* **1986**, *20*, 1221–1257. [CrossRef]
12. Di Ianni, F.; Volta, A.; Pelizzone, I.; Manfredi, S.; Gnudi, G.; Parmigiani, E. Diagnostic sensitivity of ultrasound, radiography and computed tomography for gender determination in four species of lizards. *Vet. Radiol. Ultrasound* **2015**, *56*, 40–45. [CrossRef]
13. Knotek, Z.; Cermakova, E.; Oliveri, M. Reproductive medicine in lizards. *Vet. Clin. Exot. Anim. Pract.* **2017**, *20*, 411–438. [CrossRef] [PubMed]
14. Denardo, D. Reproductive biology. In *Reptile Medicine and Surgery*; Mader, D., Ed.; Saunders-Elsiever: St. Louis, MO, USA, 2006; pp. 376–390.
15. Funk, R.S. Lizard reproductive medicine and surgery. *Vet. Clin. Exot. Anim. Pract.* **2002**, *5*, 579–613. [CrossRef]
16. Stahl, S.J. Veterinary management of snake reproduction. *Vet. Clin. Exot. Anim. Pract.* **2002**, *5*, 615–636. [CrossRef]
17. Morris, P.J.; Alberts, A.C. Determination of sex in white-throated monitors (*Varanus albigularis*), gila monsters (*Heloderma suspectum*), and beaded lizards (*H. horridum*) using two-dimensional ultrasound imaging. *J. Zoo Wildl. Med.* **1996**, *27*, 371–377.
18. Gartrell, B.D.; Girling, J.E.; Edwards, A.; Jones, S.M. Comparison of noninvasive methods for the evaluation of female reproductive condition in a large viviparous lizard, Tiliqua nigrolutea. *Zoo Biol.* **2002**, *21*, 253–268. [CrossRef]
19. Gnudi, G.; Volta, A.; Di Ianni, F.; Bonazzi, M.; Manfredi, S.; Bertoni, G. Use of ultrasonography and contrast radiography for snake gender determination. *Vet. Radiol. Ultrasound* **2009**, *50*, 309–311. [CrossRef]
20. Iannucci, A.; Altmanová, M.; Ciofi, C.; Ferguson-Smith, M.; Milan, M.; Pereira, J.C.; Pether, J.; Rehák, I.; Rovatsos, M.; Stanyon, R.; et al. Conserved sex chromosomes and karyotype evolution in monitor lizards (Varanidae). *Heredity* **2019**, *123*, 215–227. [CrossRef]
21. Manukyan, L.; Montandon, S.A.; Fofonjka, A.; Smirnov, S.; Milinkovitch, M.C. A living mesoscopic cellular automaton made of skin scales. *Nature* **2017**, *544*, 173–179. [CrossRef] [PubMed]
22. Urbanová, D.; Halán, M. The use of ultrasonography in diagnostic imaging of reptiles. *Folia Vet.* **2016**, *60*, 51–57. [CrossRef]
23. Berthelet, A.; Bulliot, C. Imaging of reptiles-a sexing technique for squamates by ultrasound. *Point Vét.* **2014**, *45*, 6–7.
24. Morris, P.J.; Jackintell, L.A.; Alberts, A.C. Predicting the gender of subadult Komodo dragons (*Varanus komodoensis*) using two-dimensional ultrasound imaging and plasma testosterone concentration. *Zoo Biol.* **1996**, *15*, 341–348. [CrossRef]
25. Wibbels, T.; Lutz, P.L.; Musick, J.A.; Wyneken, J. Critical approaches to sex determination in sea turtles. *Biol. Sea Turtles* **2003**, *2*, 103–134.
26. Cree, A.; Cockrem, J.F.; Brown, M.A.; Watson, P.R.; Guillette, L.J., Jr.; Newman, D.G.; Chambers, G.K. Laparoscopy, radiography, and blood analyses as techniques for identifying the reproductive condition of female tuatara. *Herpetologica* **1991**, *47*, 238–249.
27. Innis, C.J. Endoscopy and endosurgery of the chelonian reproductive tract. *Vet. Clin. Exot. Anim. Pract.* **2010**, *13*, 243–254. [CrossRef] [PubMed]
28. Mannion, P. (Ed.) *Diagnostic Ultrasound in Small Animal Practice*; John Wiley & Sons: Hoboken, NJ, USA, 2008.

animals

MDPI

Case Report

A Novel Application of 3D Printing Technology Facilitating Shell Wound Healing of Freshwater Turtle

Tsung-Fu Hung [1,*], Po-Jan Kuo [2,*], Fung-Shi Tsai [3], Pin-Huan Yu [4] and Yu-Shin Nai [5,*]

1. WeCare Animal Medical Center, 1F., No. 270, Daxing Rd., Taoyuan Dist., Taoyuan City 334, Taiwan
2. Tri-Service General Hospital, School of Dentistry, National Defense Medical Center, Taipei City 114, Taiwan
3. Momonga Exotic Animal Hospital, No. 20, Section 4, Chongxin Rd., Sanchong District, New Taipei City 241, Taiwan; junior10873@msn.com
4. Institute of Veterinary Clinical Science, National Taiwan University, Taipei City 106, Taiwan; pinhuan@ntu.edu.tw
5. Department of Entomology, National Chung Hsing University, Taichung City 402, Taiwan

* Correspondence: roachbug@gmail.com (T.-F.H.); kuopojan@gmail.com (P.-J.K.); ysnai@nchu.edu.tw (Y.-S.N.)

Simple Summary: This report describes how to apply the combination of 3D scanning, computer-aided design (CAD), and 3D printing to make a protective device for rescuing wild animals. In recent years, although 3D tools have become relatively low-cost and reachable, veterinary medical applications based on this technology are quite limited. The present article successfully extricates a wild freshwater turtle from an extensive shell defect within a short period. Integration of multiple sciences to 3D technology can provide a facile model for veterinary medical applications.

Abstract: Numerous cases and a shortage of resources usually limit wild animal rescue. New technology might save these severely injured wild animals from euthanasia by easing the requirement of intensive medication. Three-dimensional (3D) technologies provide precise and accurate results that improve the quality of medical applications. These 3D tools have become relatively low-cost and accessible in recent years. In the medical field of exotic animals, turtle shell defects are highly challenging because of inevitable water immersion. This report is the first attempt to apply the combination of 3D scanning, computer-aided design (CAD), and 3D printing to make a device that protects the wound from exposure to water or infection sources. The presented techniques successfully extricate a wild freshwater turtle from an extensive shell defect within a short period. Integration of multiple sciences to 3D technology can provide a facile model for veterinary medical applications.

Keywords: 3D printing technology; freshwater turtle; *Ocadia sinensis*; shell wound healing

Citation: Hung, T.-F.; Kuo, P.-J.; Tsai, F.-S.; Yu, P.-H.; Nai, Y.-S. A Novel Application of 3D Printing Technology Facilitating Shell Wound Healing of Freshwater Turtle. *Animals* **2022**, *12*, 966. https://doi.org/10.3390/ani12080966

Academic Editors: Lysimachos Papazoglou and Cinzia Benazzi

Received: 7 February 2022
Accepted: 5 April 2022
Published: 8 April 2022

1. Introduction

In exotic animal medicine, shell trauma is a frequent casualty for wildlife turtles and tortoises [1–3]. The shell fractures were usually restored by external fixation utilizing orthopaedic screws, orthopaedic pins, surgical wiring, epoxy, or super glue to bridge fracture segments [4–7]. These procedures provided instant protection and fixation of shell fracture [1,3]. However, a significant shell defect involving the shoulder, pelvic area, or penetrating punctures of the coelom could lead to a poor or even grave prognosis [3]. In the case of missing fragments, the shell usually cannot be stabilized and rehabilitated, which may be fatal without adequate medication and intensive nursing. Keeping the wound moist and clean is ideal for wound healing [1,3,7]. Although reptile wound healing is relatively slow compared with that of mammals, extensive shell loss becomes harsher and more challenging for freshwater turtles. The water turtle must maintain an everyday life for nutrition, urination, and defecation within the water environment [2]. Therefore, inevitable water immersion and faecal contamination may cause recurrent infections and delay wound healing. Thus, prolonged and intensive nursing becomes a medical burden [3].

Three-dimensional (3D) technology has been utilized in many medical applications thanks to its high precision and accuracy [8–10]. Recently, based on the development of 3D fabrication and medical imaging technology, increasingly affordable 3D printing technologies now make it possible to create highly customizable patient-tailored products [11].

Herein, we describe a feasible way to facilitate the recovery of a wild freshwater turtle from the delayed healing of severe and extensive shell defects, which reduces the intensity and frequency of medical care. The computer-aided design (CAD) and the fused deposition modelling (FDM) 3D printer could rapidly fabricate a customized device, which provided physical support of the defective shell structure. In addition, the device protected the delayed healing wound from water exposure. After the delivery of the device, the burden of intensive care was drastically decreased, and the wound infection eased rapidly, which favoured wound healing. We tracked the wound condition in the fifth month, and one year after the delivery of the device, the follow-up results showed a favourable prognosis and the subject was free from adoption without special care.

The purpose of this investigation was to develop novel attempts to apply 3D printing technology to wild freshwater turtles for wound healing; the technique provides an alternative option for exotic animal medicine.

2. Materials and Methods

2.1. Animal Background

A female freshwater turtle (*Ocadia sinensis*) was found and rescued by a local wildlife rescue association in Taiwan. This 30 cm long turtle had a severe shell defect on the left side (Figure 1A). The coelomic membrane was exposed on the fracture site, and debris, dirt, and fly larva were present around the affected area. After emergency wound management by a volunteer of the association, the turtle was transferred to the veterinary hospital. After two months of intensive care from a veterinarian, including systemic antibiotics, wet-to-dry docking [3], and supportive treatment with irrigation, flushing, and nutrition, the wound remained with minor infection, and healing was delayed.

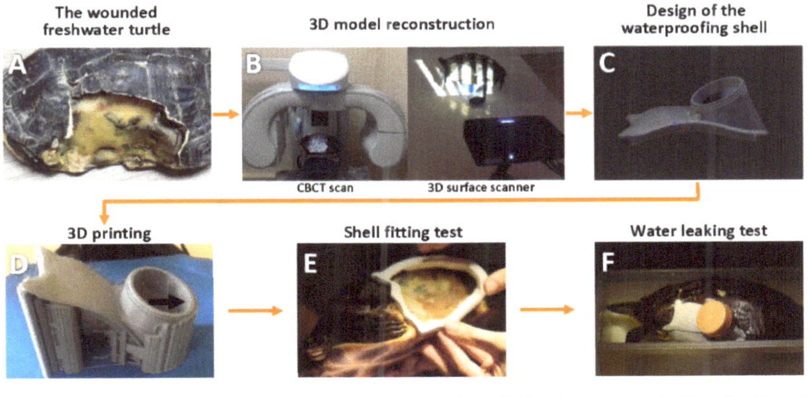

Figure 1. Flowchart of the treatment plan. (**A**) The female freshwater turtle (*Ocadia sinensis*) with a 13 cm long oval defect; (**B**) The wounded freshwater turtle is scanned by a CBCT and 3D surface scanner for reconstruction of the 3D model; (**C**) the reconstruction of a hard tissue 3D model and defect surface model are applied to design a waterproofing shell; (**D**) all 3D models of the waterproofing shell are used to fabricate the template by a fused deposition modelling (FDM) type Delta 3D print; (**E**) waterproofing shell fitting test is performed for secure fixation; (**F**) water leaking test.

2.2. Diagnosis and Treatment Plan

All procedures were reviewed and approved by the Bureau of Forestry, Council of Agriculture of Taiwan, the competent authority of wildlife utility (project 106 FD-08.3-C-28). In the veterinary hospital, the turtle presented with a 13 cm long oval shell lost on the left side involving the carapace, plastron, and most parts of the bridge (Figure 1A). The left anterior edge of the shell bridge was lost; thus, the support structure between the carapace and plastron was disconnected. A fistula wound with purulent discharge was found on the caudal side of the coelomic membrane, and several minor wounds spread around the membrane with purulent discharge. Systemic infection was found based on the increased quantity of white blood cells. Therefore, systemic antibiotic therapy was used (metronidazole 20 mg/kg b.i.d. with trimethoprim 30 mg/kg b.i.d. under subcutaneous injection [SC] for 14 days). The wound on the exposed choelomic membrane did not heal well, with recurrent infections even after 2 months of intensive care.

The daily care procedure included removing the dirty bandage, wound irrigation (sterile Ringer's solution 10 mL/kg SC, s.i.d.), and nutrition support (oral gavaging the commercial critical care diet at a dose of 3% of body weight), which may require a team of at least two people to complete. Moreover, the dressing was not waterproof to keep the defect dry. Therefore, the wound was difficult to keep clean and reduce possible bacterial infections at the wound site for this freshwater turtle, especially when there was a shortage of human medical resources. The novel device as the customized shell was designed and fabricated following the flow chart (Figure 1) to provide mechanical support and water isolation. All device design and test steps were performed and completed under the supervision of a veterinarian. In addition, the pain and stress behaviour of the turtle was carefully monitored during treatment.

2.3. The Bone Model and Surface Model of Defect Reconstruction

The turtle underwent cone-beam computed tomography (CBCT) (Newtom 5G, Imola, Italy) scanning (Figure 1B left). The CBCT scan was obtained with a 0.3 mm voxel size, power of 110 kVp, 5 mA, 10 s exposure, and all raw data were compacted in DICOM format for further radiographic diagnosis. The skeletal 3D model was reconstructed after adjusting the density until the tissue border was clearly identified in the software (OsiriX MD, Pixmeo, Geneva, Switzerland). The 3D reconstructed model was exported as a stereolithography (STL) file for further editing. The surface model of the defect was acquired by a structured light 3D surface scanner (Einscan-S, Shining 3D, Hangzhou, China) (Figure 1B right) and exported as an STL file. All 3D models were edited by Blender, a computer-aided design software (Blender, Blender Foundation, Amsterdam, The Netherlands). The bone and surface models were carefully aligned and merged to show the locale of the bone edge under the scutum to confirm any bone degeneration in the case of underlying infection.

The bone model was mainly used for diagnosis and skeletal structure analysis, which indicated the mechanical force support area. Moreover, the surface model provides the detailed architecture and morphology of the defect and surrounding shell structure for the device outline design. The edited models were processed by slicer software (Kisslicer, Jonathan Dummer, Orlando, FL, USA; Simplify3D, Simplify3D Inc., Cincinnati, OH, USA) and printed by a fused deposition modelling (FDM)-type Delta 3D printer (Atom 2, ALT, Taipei, Taiwan) with modified polylactic acid (PLA) filaments (Gypsum, Next Print, Tainan, Taiwan).

2.4. Design, Construction, and Fitting of the Protective Device

The shell defect formed several undercuts on the carapace, plastron, and caudal part of a bridge, which would facilitate fixation; however, the lost anterior edge of the skeletal bridge was a challenge for waterproofing owing to the unstable soft tissue during left forelimb movement. According to the reconstructed model, we designed a protective device (Figure 1C) and printed it (Figure 1D) as described previously. The device practicability was tested through fixation stability (Figure 1E), not interfering with animal activity, and waterproofing (Figure 1F). A rim was designed and fabricated with a snap-fit groove to tightly engage the overhanging shell's undercuts (Figure 2A). The junction between the shell edge/soft tissue and the rim was diminished, and the groove of the rim was covered over the edge of the defect for at least 5 mm to achieve secure fixation. The prototype of the rim showed promising engagement with the printed defect model (Figure 2B). After fixation, the rim was further remodelled to form a hull shape, with an opening kept above (Figure 2C). The rim with a hull structure was printed and tested again on the defect for stability (Figure 2D). To reduce the chance of leakage, the junction between the rim groove and shell defect edge/soft tissue was filled with polydimethylsiloxane (PDMS) as a sealant (Sylgard 184, Dow Corning, Midland, MI, USA). A layer of silicone sealant (Silicone sealant, Dow Corning, Midland, MI, USA) was applied to the outer surface between the shell and rim. Furthermore, a cover part was designed with a porthole to attach to the opening (Figure 2E), and it was printed and secured on the opening of the hull structure (Figure 2F). The edge of the cover, which contacts a layer of silicone sealant, coated the opening to close the waterproof opening. The porthole was designed with thread to fit a wide-mouth bottle cap to facilitate daily care [12].

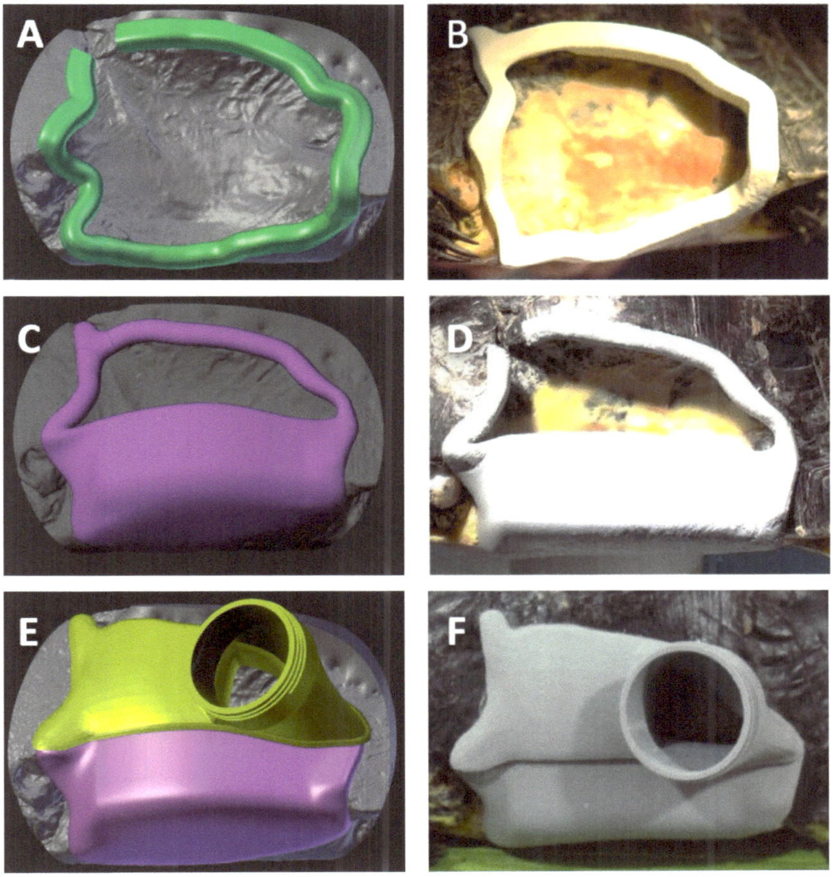

Figure 2. Design and fitting of the protective device on the shell defect. 3D reconstructed models of protective device: (**A**) the CAD model of snap-fit joints rim structure (green part); (**B**) printed and fitting tested on the defect; (**C**) a hull structure (purple part) was designed according to the previously designed rim structure; (**D**) printed and fitting tested again on the defect; (**E**) a model of the cover with a threaded porthole (yellow part) was designed; (**F**) printed and installed on the hull structure for further testing.

2.5. Daily Care and Wound Management

The hull and cover were fixed as designed and would not be removed during daily care. The porthole remained open during the dry docking and closed with the cap when the turtle was immersed in the water. The cap was removed approximately for 6~8 h daily to avoid high humidity that could induce fungal organism colonization. The caudal pocket wound of the defect was cleaned daily and debrided through the porthole. The device's margin fitness was assessed once every month, and the sealant was added if a crack or leakage was noted.

3. Results

3.1. Device Fitting and Leakage Test

The delivery of the rim structure attested to the feasibility of noninvasive fixation. The rim was adapted to the defect and allowed the turtle to stay in a room for 3 h. The fitness of the rim was well adapted, and the shape of the rim with hull structure was modified to slim (Figure 2D,E), while the original design of the rim was too thick that it might scratch and be pushed by the turtle's limbs. The hull structure was installed and rechecked as described above. The device attached to the defect was stable and was not reachable by animal activity. After delivery, animal behaviour presented no difference, including scratching, constant hiding, unwillingness to move, and activity change. Clean gauze was placed in the hull without the cover, and the patient was half-immersed and constrained in water. After 1 h, the gauze was dry, indicating no water leakage into the hull. Moreover, the total weight of the device was approximately 110 g, and it affected the physiological activities very little.

3.2. Patient Follow-Up

In the clinical observation, the affected area presented debris without applying the device to protect the defect. The exudate formation and chronic infection did not subside at this stage (Figure 3A). After fabricating and applying the protective device, the device could cover the defect and avoid wound contamination during the healing process. This device provided waterproofing and reduced the wound infection risk during immersion. The appetite and activity of the patient significantly increased one month after applying the device. In addition, under a veterinarian's assessment of pain and distress, the patient did not show any pain-like (nocifensive) behaviour. Keratinization was present on the coelomic membrane (Figure 3B), and the counts of white blood cells were in the normal range ($1.4 \times 10^4/\mu L$) at the one-month recall after device application.

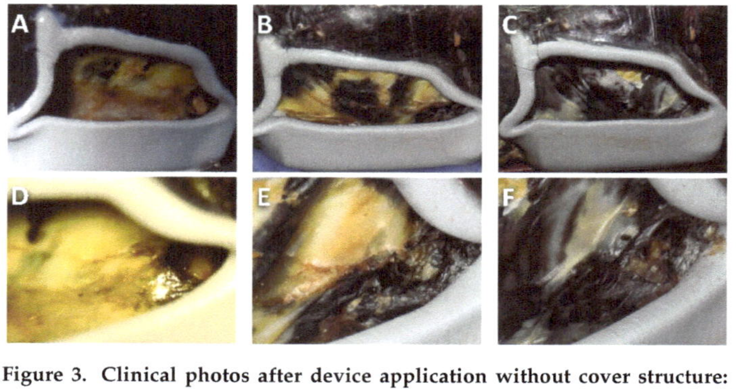

Figure 3. Clinical photos after device application without cover structure: (**A–C**) lateral view; (**D–F**) direct view of soft tissue damage at the defect posterior area; (**A**) fragile tissue at day 0; (**B**) small keratinized areas presentation at 3-month recall; (**C**) mature keratinized tissue formation at 5-month recall; (**D**) fragile tissue with exudate formation at day 0; (**E**) keratinized tissue formation at 3-month recall; (**F**) complete closure of fistula and mature keratinised tissue presentation at 5-month recall.

The matured keratinized tissue almost covered the defect surface at five months of reexamination, and the wound was solid during palpation (Figure 3C). The tissue damage on the posterior defect border healed without exudate discharge, and a leather-like granulation bed was in place (Figure 3D–F). Using the CBCT 3D model to assess hard tissue formation, bony bridge development was observed at the fracture site of the anterior area (Figure 4A,D). Ossification was present on the border and the central location of the defect. Moreover, the fenestration lesion on the posterior side was reduced (Figure 4C,F).

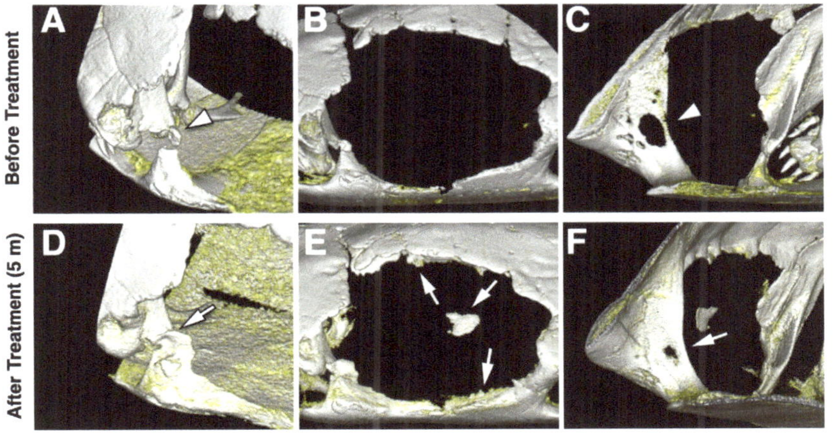

Figure 4. CBCT three-dimensional images of initial condition (A–C) and 5 months after device application (D–F): (**A**) hard tissue defect on the anterior border (arrowhead); (**B**) defect lateral view; (**C**) hard tissue fenestration on the posterior border (arrowhead); (**D**) fracture healing (arrow) on the anterior border; (**E**) new hard tissue formation (arrow) at defect border and centre area; (**F**) hard tissue healing (arrow) on the posterior border of the defect.

Through 12 months of follow-up, the device did not loosen and maintained stability. The turtle maintained normal physiologic functions, including swimming ability, buoyancy, and ambulation, during daily access to an underwater environment. No adverse effects inducing pain or stress were noted.

4. Discussion

One of the most common current issues encountered in wild freshwater turtles is shell injury. Most of these injuries are caused by road traffic accidents and predator attacks [3]. In the present report, we described the workflow of digital planning and 3D printed custom devices for a poor prognostic freshwater turtle with severe shell damage. At the 12-month follow-up, the favourable outcome suggested that this shell-type device provides protection and precisely fits the defect border with complex geometry. Several clinical reports have demonstrated the treatment of shell injury of freshwater turtles using different appliances. In contrast, there is insufficient information on healing outcomes. Compared with these studies, the present report characterized wound healing conditions by clinical and 3D radiographic findings.

Three-dimensional technology has become feasible and affordable in recent years. The powerful machines become facile and easy to maintain, while the consumable materials provide sustainable strength and easy storage at a reasonable price [13]. This technology has been utilized in many fields of medication with favourable outcomes [14]. This report applied 3D scanning, CAD software, and 3D printing to successfully and rapidly achieve this rescue at a low cost. The 1-year follow-up result further confirms the success of our treatment. These accessible 3D tools with high performance may provide an alternative way to help more animals with limited time and costs and replace human resources. With these 3D tools, the device's design can evolve quickly along with the need after tests and discussion. The power of CAD software and printer gives us the ability to respond to needs and validate ideas rapidly. When severe shell damage occurs in chelonians, the loss of shell specimens makes wound management more challenging for aquatic and semiaquatic chelonians. Adjunctive with the 3D technique, fabrication of a customized waterproof shield allows for normal husbandry and natural behaviours during shell wound healing.

The shell material and structure are integrated systems that provide load-bearing capacity and protect the internal organs from injury [15]. A previous study analysed the stress distribution of turtle shells using a finite element model [16]. The highest value was located on the inside surface of the four bridges of the turtle shell. The initial 3D imaging showed that the left anterior bridge was fractured (Figure 4A). Therefore, the device's material should provide strong resistance to the turtle shell. PLA was selected as a leading candidate material owing to its thermoplastic, relatively lightweight, and high-strength mechanical properties [16]. There were still several research limitations. Primarily, this is a single case report, and we did not measure the stress resistance distribution of this device vertically and horizontally. In addition, the longevity of this PLA device was not tested, nor was the application of bacterial cultures for the antibiotic. To make a firm attachment of the hull structure on the defect, we designed snap-in joints [17] in the groove to engage the overhanging shell. Compared with conventional screw fixation, this noninvasive approach avoids anaesthesia complications and the risk of infection. Under the initial condition, the physical condition of the chelonian with severe shell injury was not suitable for anaesthesia, and the bone defect was not stable for using pins or screws. Therefore, we considered some sealants to fill the device frame and the shell seam at the beginning of the device application. PDMS was chosen for this purpose. PDMS is a biocompatible material that has been used in many medical devices [18]. Thanks to its excellent conformability, PDMS is broadly used in the semiconductor industry to replicate surface features down to the subnano scale [19] and is also commercially available as a sealant. With the property of biocompatibility, no adverse effects to aquatic organisms, and resistance to biological fouling [20,21], which is essential when the device is constantly in water, PDMS becomes an adequate sealant to eliminate the risk of infection and chemical toxicity [22]. At a five-month examination, hard tissue healing was generally observed radiographically at the anterior bridge of the defect after application of the device (Figure 4D).

We framed the defect with a hull. The hull design enhanced the framework's strength to compensate for the weakened shell structure and acted as a watertight armor to protect either the vulnerable coelomic membrane or the clean environment from the contagious but inevitable water immersion. The hull is slightly above the waterline and left with a large opening for subsequent wound management (Figure 2C,D). The opening provided room to reach the wounds inside the defect. The opening with an irregular frame was enclosed by the corresponding cover with a porthole (Figure 2E,F). The cover was designed as a removable part when accessing the fundamental defect.

The porthole on the cover was designed and fabricated with thread that matched a wide bottle cap (Figure 1F) available at local convenience stores. This design followed the previous study, and it is the first case presented in the literature [12]. As a result, the porthole can be neatly opened and closed, and the purulent pocket wound can be easily accessed through the porthole for daily care. In addition, this design avoided frequent disengagement between the cover and hull, which further prevented the failure of the watertight structure. The daily care was manageable with one medical personnel, and the workforce was significantly reduced with this device.

The chelonian shell was still not fully recovered at the 1-year follow-up. However, the CBCT examination indicated that there was a significant improvement in wound healing of hard tissue (Figure 4). The physical condition of the turtle was stable, and it could walk smoothly without restricted movement with the device. The fixation of the device should be observed every month. Continuous maturation and keratinization of the coelomic membrane were observed, and the device was not removed during the 12-month follow-up.

In addition, the device could be removed when the shell fracture was fully closed with callus formation and bone healing.

5. Conclusions

This report ably demonstrated a rapid and practical process to rescue a severely injured wild freshwater turtle using advanced 3D technology with limited human resources and a relatively low-cost method. Thanks to the maturation of 3D technology, these 3D tools have become robust and affordable. These sophisticated tools gave us the chance to extricate an animal from the risk of euthanasia. This progression delivered an alternative point of view to apply 3D technology to wild animal rescue, and in all of veterinary medicine, which may help us approach new methods from conventional practice.

Author Contributions: T.-F.H., P.-J.K. and Y.-S.N. organized the experimental plan. T.-F.H. and P.-J.K. mainly carried out the experiments. F.-S.T. and P.-H.Y. provided the resources and were responsible for data curation. T.-F.H., P.-J.K. and Y.-S.N. drafted the manuscript. All authors have read and agreed to the published version of the manuscript.

Funding: This research received no external funding.

Institutional Review Board Statement: The animal study protocol was reviewed and approved by the Bureau of Forestry, Council of Agriculture of Taiwan, the competent authority of wildlife utility (project 106 FD-08.3-C-28).

Informed Consent Statement: In this case report, the wild animal was rescued under emergency situations in a veterinary hospital and was not subjected to an experimental animal study. This report denotes a concept of 3D technology application in wild animal rescue by describing an alternative procedure that indeed extricates a wild animal from the situation of euthanasia. From an auxiliary approach, all methods applied to the patient were noninvasive and designed to diminish any complications, while all medications were provided by a researcher who is a licenced veterinarian in Taiwan.

Data Availability Statement: Not applicable.

Conflicts of Interest: The authors declare no conflict of interest.

References

1. Vella, D. Management of aquatic turtle shell fractures. *Lab Anim.* **2009**, *38*, 52–53. [CrossRef] [PubMed]
2. Vella, D. Management of freshwater turtle shell injuries. *Lab Anim.* **2009**, *38*, 13–14. [CrossRef] [PubMed]
3. Fleming, G. Clinical Technique: Chelonian Shell Repair. *J. Exot. Pet Med.* **2008**, *17*, 246–258. [CrossRef]
4. Barten, S. Shell Damage. In *Reptile Medicine and Surgery*, 2nd ed.; Divers, S., Mader, D., Eds.; Elsevier Saunders: Philadelphia, PA, USA, 2005.
5. Kishimori, J.; Lewbart, G.; Marcellin-Little, D.; Roe, S.; Trogdon, M.; Henson, H.; Stoskopf, M.K. Chelonian Shell Fracture Repair Tecinques. *Exot. DVM* **2001**, *3*, 35–41.
6. Hernandez-Divers, J.S. Surgery: Principles and Techniques. In *BSAVA Manual of Reptiles*, 2nd ed.; Girling, S.J., Raiti, P., Eds.; Wiley: Hoboken, NJ, USA, 2004.
7. McArthur, S.; Hernandez-Divers, S. Surgery. In *Medicine and Surgery of Tortoises and Turtles*; McArthur, S., Wilkinson, R., Meyer, J., Eds.; Blackwell: Oxford, UK, 2008; pp. 403–463.
8. Hoang, D.; Perrault, D.; Stevanovic, M.; Ghiassi, A. Surgical applications of three-dimensional printing: A review of the current literature & how to get started. *Ann. Transl. Med.* **2016**, *4*, 456. [CrossRef] [PubMed]
9. Ma, L.; Zhou, Y.; Zhu, Y.; Lin, Z.; Wang, Y.; Zhang, Y.; Xia, H.; Mao, C. 3D-printed guiding templates for improved osteosarcoma resection. *Sci. Rep.* **2016**, *6*, 23335. [CrossRef] [PubMed]
10. Tack, P.; Victor, J.; Gemmel, P.; Annemans, L. 3D-printing techniques in a medical setting: A systematic literature review. *Biomed. Eng. Online* **2016**, *15*, 115. [CrossRef] [PubMed]
11. Bauermeister, A.J.; Zuriarrain, A.; Newman, M.I. Three-Dimensional Printing in Plastic and Reconstructive Surgery: A Systematic Review. *Ann. Plast. Surg.* **2016**, *77*, 569–576. [CrossRef] [PubMed]
12. Sypniewski, L.A.; Hahn, A.; Murray, J.K.; Chalasani, V.; Woods, L.; Piao, D.; Bartels, K.E. Novel Shell Wound Care in the Aquatic Turtle. *J. Exot. Pet Med.* **2016**, *25*, 110–114. [CrossRef]
13. He, Y.; Xue, G.H.; Fu, J.Z. Fabrication of low cost soft tissue prostheses with the desktop 3D printer. *Sci. Rep.* **2014**, *4*, 6973. [CrossRef] [PubMed]
14. Ventola, C.L. Medical Applications for 3D Printing: Current and Projected Uses. *Pharm. Ther.* **2014**, *39*, 704–711.

15. Cebra-Thomas, J.; Tan, F.; Sistla, S.; Estes, E.; Bender, G.; Kim, C.; Riccio, P.; Gilbert, S.F. How the turtle forms its shell: A paracrine hypothesis of carapace formation. *J. Exp. Zool. Part B Mol. Dev. Evol.* **2005**, *304*, 558–569. [CrossRef] [PubMed]
16. Zhang, W.; Wu, C.; Zhang, C.; Chen, Z. Numerical Study of the Mechanical Response of Turtle Shell. *J. Bionic Eng.* **2012**, *9*, 330–335. [CrossRef]
17. Klahn, C.; Singer, D.; Meboldt, M. Design Guidelines for Additive Manufactured Snap-Fit Joints. *Procedia CIRP* **2016**, *50*, 264–269. [CrossRef]
18. Tipnis, N.P.; Burgess, D.J. Sterilization of implantable polymer-based medical devices: A review. *Int. J. Pharm.* **2018**, *544*, 455–460. [CrossRef] [PubMed]
19. Odom, T.W.; Love, J.C.; Wolfe, D.B.; Paul, K.E.; Whitesides, G.M. Improved Pattern Transfer in Soft Lithography Using Composite Stamps. *Langmuir* **2002**, *18*, 5314–5320. [CrossRef]
20. Ng, J.M.; Gitlin, I.; Stroock, A.D.; Whitesides, G.M. Components for integrated poly(dimethylsiloxane) microfluidic systems. *Electrophoresis* **2002**, *23*, 3461–3473. [CrossRef]
21. Fendinger, N.J. Polydimethylsiloxane (PDMS): Environmental Fate and Effects. In *Organosilicon Chemistry Set: From Molecules to Materials*; VCH: Weinheim, Germany, 2005; pp. 626–638. [CrossRef]
22. Zhang, H.; Chiao, M. Anti-fouling Coatings of Poly(dimethylsiloxane) Devices for Biological and Biomedical Applications. *J. Med. Biol. Eng.* **2015**, *35*, 143–155. [CrossRef] [PubMed]

animals

Article

Radiation Dose Reduction in Different Digital Radiography Systems: Impact on Assessment of Defined Bony Structures in Bearded Dragons (*Pogona vitticeps*)

Natalie Steiner [1], Eberhard Ludewig [2], Wiebke Tebrün [3] and Michael Pees [1,*]

[1] Department of Small Mammal, Reptile and Avian Diseases, University of Veterinary Medicine, 30559 Hanover, Germany
[2] Division of Diagnostic Imaging, Department of Small Animals and Horses, University of Veterinary Medicine, 1210 Vienna, Austria
[3] Wimex Agrarprodukte Import and Export GmbH, 93128 Regenstauf, Germany
* Correspondence: michael.pees@tiho-hannover.de

Simple Summary: Digital radiography has long been established in veterinary clinics, leading to increased use of digital systems in reptile species as well. In this study, we used different digital radiography systems on cadavers of bearded dragons (*Pogona vitticeps*). We aimed to examine the impact of a radiation dose reduction on the image quality and the assessment of defined skeletal structures. We employed a blinded assessment and a defined scoring system to evaluate the techniques tested. Our results demonstrate that both a 50% and a 75% reduction in the radiation dose significantly decreased image assessments. These findings highlight the need for correct radiation dose protocols to produce high-quality radiographs in reptile species.

Abstract: Three different digital detector systems were used to study the effect of a defined radiation dose reduction on the image quality of digital radiographs in bearded dragons (*Pogona vitticeps*). A series of radiographs of seven bearded-dragon cadavers with a body mass ranging from 132 g to 499 g were taken in dorsoventral projection. The digital systems employed included two computed radiography systems (CR) (one system with a needle-based and one with a powdered-based scintillator) and one direct radiography system (DR). Three levels of the detector dose were selected: A standard dose (defined based on the recommended exposure value of the CR_P, D/100%), a half dose (D/50%), and a quarter dose (D/25%). Four image criteria and one overall assessment were defined for each of four anatomic skeletal regions (femur, rib, vertebra, and phalanx) and evaluated blinded by four veterinarians using a predefined scoring system. The results were assessed for differences between reviewers (interobserver variability), radiography systems, and dosage settings (intersystem variability). The comparison of the ratings was based on visual grading characteristic (VGC) analysis. Dose reduction led to significantly lower scores in all criteria by every reviewer, indicating a linear impairment of image quality in different skeletal structures in bearded dragons. Scores did not differ significantly between the different systems used, indicating no advantage in using a computed or direct radiography system to evaluate skeletal structures in bearded dragons. The correlation was significant ($p \leq 0.05$) for interobserver variability in 100% of the cases, with correlation coefficients between 0.50 and 0.59. While demonstrating the efficacy of the use of digital radiography in bearded dragons and the similar quality in using different computed or direct radiography systems, this study also highlights the importance of the appropriate level of detector dose and demonstrates the limits of post-processing algorithm to compensate for insufficient radiation doses in bearded dragons.

Keywords: reptiles; lizards; digital radiography; image quality; dose reduction

Citation: Steiner, N.; Ludewig, E.; Tebrün, W.; Pees, M. Radiation Dose Reduction in Different Digital Radiography Systems: Impact on Assessment of Defined Bony Structures in Bearded Dragons (*Pogona vitticeps*). *Animals* **2023**, *13*, 1613. https://doi.org/10.3390/ani13101613

Academic Editor: Tom Hellebuyck

Received: 30 March 2023
Revised: 9 May 2023
Accepted: 9 May 2023
Published: 11 May 2023

1. Introduction

Radiography in reptile medicine is an important diagnostic tool [1]. Various indications exist, such as suspected skeletal lesions, evaluation of the lungs, and changes in the gastrointestinal or reproductive tract [1–6]. Furthermore, radiography can also be helpful to assess the reproductive status or gender determination [7,8]. Due to their unique anatomy, reptile species often present some challenges for radiography such as small body sizes with miniature anatomical structures or the lack of internal fat tissue between organs and a lack of division between the thorax and abdomen, leading to limited soft tissue contrast. High-resolution screen film systems have been used to compensate to a certain degree for such limitations [1,2,9]. With the transition from screen film to digital radiography, reptile medicine was faced with the advantages and disadvantages of the new technology. Digital systems provide advantages such as a greater dynamic range, data transfer, a linear signal response, post-processing, and storage possibilities [9–14]. Even though screen-film systems have a superior spatial resolution, digital systems can differentiate very small attenuation differences better [12]. Various studies have already compared conventional to digital systems in reptiles, showing no less good image quality for diagnostic interpretation [9,12,15]. In contrast, digital images were superior for structures with high contrast such as bony structures and air–tissue boundaries as seen in the lungs [12]. Based on these results, the comparison of different digital systems amongst each other and the implementation of dose requirements for valuable diagnostic images is of great interest [11,13,16]. For a basic understanding, today, in veterinary medicine, different digital detector systems with different advantages are used: First, digital detector systems can further be divided into computed radiography (CR) and direct radiography (DR) [10,14,17,18]. DR includes flat panel systems, where a scintillator converts incoming X-rays directly into light [10,18]. CR uses storage phosphor image plates with a separate read-out process [14,18]. DR systems were long described as superior to phosphor storage panel images [19]. However, within the last 20 years, the basic principle of CR was modified to achieve both a higher detective quantum efficiency (DQE, the efficiency of a detector in converting X-ray energy into image signal) and spatial resolution. In particular, the introduction of needle phosphor plates (NIP) combined with a novel line-to-line CR stimulation and light collection technology improved the DQE of storage phosphor systems greatly [20–22]. This superiority can be used either for a higher signal-to-noise ratio (SNR) or dose reduction [13,21–23]. Wirth, S. et al. [19] even described a possible dose reduction in radiography of human skeletal structures of up to 75% with a NIP and a line-to-line scanner in comparison to flat panel systems and power-based phosphor detectors without inferior image quality. Concerning the importance of radiation exposure reduction for the patient and medical staff, the potential for dose reduction in digital systems seems therefore promising. In addition to the DQE in NIP systems, features such as post-image processing and the higher dynamic range in CR systems also reveal higher dose reduction potential than in DR systems. Unfortunately, even with the knowledge of dose reduction without the loss of image quality, taking this principle into clinical practice does show various difficulties. In digital radiography, there is no inverse correlation between dose and image contrast, therefore "film-blackening" as an indicator of overexposure no longer exists as seen in conventional radiography systems before. What generally can be seen when assessing digital radiography is a reciprocal relationship between the dose and the signal-to-noise ratio. However, the subjective and individual sensitivity to recognize an increase in noise is relatively low. Digital systems do therefore have a risk of substantially increasing the patient's dose without even being aware of it. Furthermore, inadequate image processing or suboptimal image display also offers the potential for decreasing diagnostic information in digital radiography systems, leading to an increase in dose without the need for clinical evaluation [24]. Regarding the lack of subjective visual control ability of the optimal diagnostically relevant radiation dose, the implementation of dose indicators and dose monitoring is mandatory for digital radiography.

This study aimed to investigate the effect of a defined dose reduction for diagnostic imaging in skeletal structures of bearded dragons (*Pogona vitticeps*), regarding a possible loss in image quality, for three different digital detector technologies. The results could contribute to standardization in radiographic settings and dose control for reptile species. Regarding the importance of a reduction of radiation exposure for medical staff and the patient, the radiation dose should always be kept as low as possible, while achieving a diagnostically valuable image.

2. Materials and Methods

2.1. Radiographic Settings

For the radiographs, three different digital detector systems were used (Table 1).

Table 1. Technical equipment and exposure settings used in the experimental setup. Al: Aluminum equivalent.

System	X-ray System	Detector System	Exposure Setting
FP—100% FP—50% FP—25%	PHILIPS Bucky Diagnost TH Grid: no Focus-Detector Distance 110 cm Focus size: 0.6 × 0.6 mm^2 Filtration: 2.5 mm Al	TRIXELL Digital flat-panel detector 4343 RC-E (Detector size: 43 × 43 cm^2)	42 kVp, 5 mAs 42 kVp, 2.5 mAs 42 kVp, 1.25 mAs
CR_P—100% CR_P—50% CR_P—25%		FUJI HR/PHILIPS AC 500 (Screen size: 18 × 24 cm^2)	42 kVp, 5 mAs 42 kVp, 2.5 mAs 42 kVp, 1.25 mAs
CR_N—100% CR_N—50% CR_N—25%		AGFA DX-S (Screen size 18 × 24 cm^2)	42 kVp, 5 mAs 42 kVp, 2.5 mAs 42 kVp, 1.25 mAs

Radiographs were taken using a bucky table unit (Philips Bucky Diagnost TH, Philips Healthcare, Hamburg, Germany). The animals were directly placed on the detector. An anti-scatter grid was not used. Exposure settings were adjusted on the base of the dose requirements of the CR_P system and evaluated for a reference animal (animal 1 in Table 2). Exposure settings of 42 kVp and 5 mAs were identified to generate images with a dose indicator value (lgM of 1.94) equivalent to a detector dose level of 2.5 μGy (D/100%).

Table 2. Weight and snout-to-vent length of animals used in the study. Animals are listed in order of radiographic examination.

Number of the Animal	Weight [g]	Snout to Vent Length [mm]
1	499	237
2	397	208
3	334	214
4	318	190
5	132	147
6	198	160
7	164	155

The detector dose was subsequently reduced to half the dosage by halving the mAs value to 2.5 (D/50%) and to a quarter, quartering the mAs value to 1.25 (D/25%). Tube voltage was kept constant at 42 kVp for each image, leading to three different settings for each radiographic system (D/100% 42 kVp, 5 mAs, D/50% 42 kVp 2.5 mAs, D/25% 42 kVp 1.25 mAs). Dose-Area Product (DAP) measurements were performed for all systems to monitor the uniformity of exposure. The size of the exposed field of 20 × 30 cm^2 and the focus-detector distance of 110 cm were kept constant for all animals and images. System-specific processing algorithms were used. In pre-studies, the parameters of these processing algorithms were evaluated regarding detail visibility.

2.2. Procedure

In this study, seven bodies of Inland Bearded Dragons (*Pogona vitticeps*) with a mean body mass of 292 g and a mean snout-to-vent length of 164 mm (Table 2) were used. Animal bodies were selected from patient animals that had to be euthanized due to various reasons not visibly affecting the skeletal system, with the owner's agreement. Animals were euthanized and stored at −18 °C for different time periods between one and six months. All radiographs were taken only in the dorsoventral position. Evaluation of the radiographs was fully blinded, using a DICOM Anonymizer (https://sourceforge.net/projects/dicomanonymizer/, accessed on 10 July 2021) with randomly chosen unconnected four-digit numbers, not in any relation to the specific animal or exposure setting. Reviewers were four veterinarians with a minimum of two years of experience in a clinic specializing in reptile medicine and therefore regular training in the assessment of reptile radiographs. Assessments were conducted independently. The workstation was equipped with two medical grey-scale monitors (EIZO MX240W, matrix: 1920 × 1200 pixel, dot pitch: 0.27 mm; luminance: 320 cd/m^2, contrast ratio: 850:1; Avnet Technology, Nettetal, Germany). Commercial medical image analysis software was used (GOP-View XR2-T, Contextvision, Stockholm, Sweden). To become familiar with the assessment system, prior to the study, the reviewers went through a training period evaluating 25 randomly chosen radiographs that were not used for the study results. Evaluation time per image was unlimited. The ambient light and other conditions of the viewing environment fulfilled the requirements for medical image interpretation.

2.3. Scoring System

For this study, exclusively skeletal structures were examined. Four different structures (femur, ribs, vertebra, and phalanx) were chosen based on different bone architecture or features such as the differentiation of bone to soft tissue and surroundings (Table 3). The femur was essentially chosen to evaluate the differentiation between corticalis and spongiosa, joint structures were evaluated based on the left front phalanx, and the last left rib was used to further evaluate the details and structure of the corticalis. For each anatomical structure, four different characteristics (Table 3) were evaluated using a four-scaled scoring system, ranging from 1 (optimal evaluation) to 4 (insufficient evaluation). Scoring systems were used according to the modification of Körner, M. et al. [25] (Table 4).

Table 3. Definition of criteria for radiographic assessment.

Criterion	Description
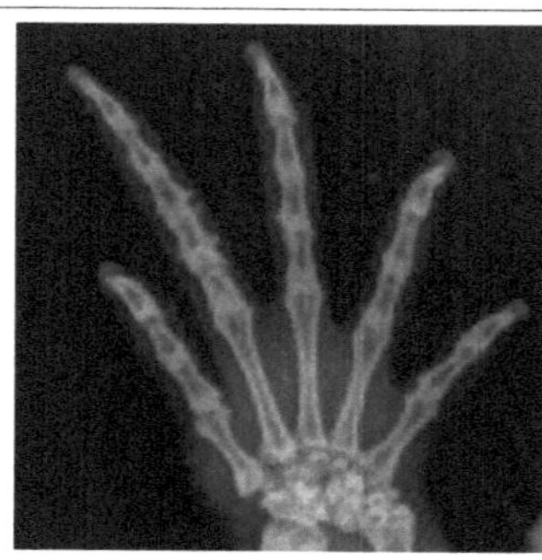	Left phalanx of the forelimb: identifiability of the joint contours of the interphalangeal joints: visualization and demarcation of the bone contour to the joint space, demarcation to the surrounding tissue.

Table 3. *Cont.*

Criterion	Description
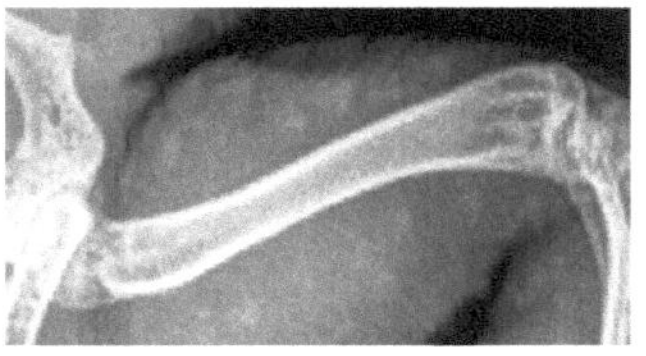	Right femur: assessability of the trabeculae and cortical bone: etail of the trabeculae and cortical bone, delineability of the structures from the surrounding area.
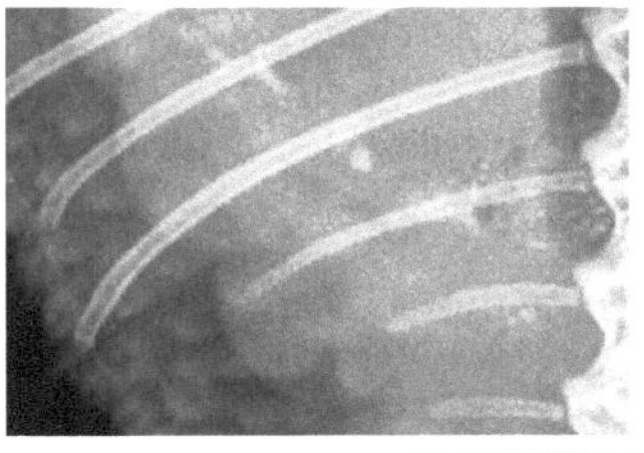	Left last rib: assessability of cancellous bone and cortical bone along the bone:demarcability of cancellous bone to cortical bone, demarcability of structures to surrounding area.
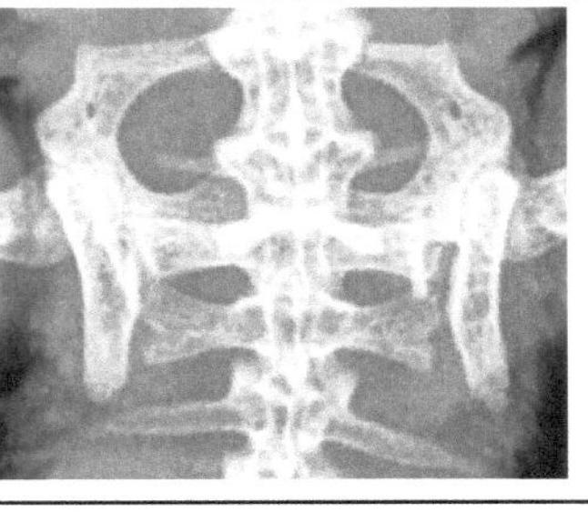	Cruciate vertebrae, 1st–3rd caudal vertebrae: identifiability of the individual vertebral bodies:assessability of the architecture and external contour of the vertebral bodies, delineability of the vertebrae from each other and from the surrounding area.

Table 4. Definition of the scores for radiographic assessment.

Scoring	Assessment
1	Optimal impression, structure completely evaluable, no limitation for clinical interpretation.
2	Good impression, structure evaluable, minor limitation for clinical interpretation.
3	Acceptable impression, detail representation limited, clinical interpretation restricted.
4	Insufficient impression, no interpretation possible.

2.4. Statistical Analysis

In this study, four criteria were evaluated in seven animals using three different radiographic systems comparing three different dose levels, leading to 252 criteria evaluated by each of the four reviewers and resulting in 1008 examinations in total. Interobserver variability was addressed using Spearman's rank correlation coefficient to test for possible differences between the reviewers themselves. Additionally, interobserver agreement was assessed using the Wilcoxon signed-rank test. Calculations were made using SPSS (IBM SPSS Statistics 20, IBM, Armonk, NY, USA). Mean values, scoring frequencies, and 95% confidence intervals were calculated for each criterion, and an overall assessment was conducted to facilitate comparison between systems and reviewers.

A visual grading characteristics (VGC) analysis was applied for intersystem comparisons. The obtained VGC curve graphically demonstrates the comparison of the two systems. In case of an equal rating, the curve would be diagonal resulting in an area under the curve (AUC) of 0.5. The more one system is rated superior, the more the curve moves to the respective axis, therefore changing the area under the curve value towards 0.0 or 1.0 [26]. Commercial software was used for the calculation of AUC and comparisons (Sigma Plot 11, Systat Software Inc., San José, CA, USA). For all calculations, the correlation was considered to be significant with $p \leq 0.05$ and highly significant with $p \leq 0.001$. The strength of the coefficient was assessed in accordance with recommended standards [27], with correlations between 0.1 and 0.3 considered low, correlations between 0.3 and 0.5 considered moderate, and over 0.5 considered high.

3. Results

3.1. Dose Effects

3.1.1. Scoring of the Different Criteria

For all criteria in all systems, scores given by the reviewers were lower after the dose reduction, regardless of halving or quartering the full dose (D50% or D25%).

Scores regarding all criteria together were significant to highly significant, ranging from $p = 0.008$ to ≤ 0.001.

In all four criteria, comparisons of D25% to D100% reached significantly lower scores than the D50% to D100%. The only exception was the phalanx using the CR_P, but this was still nearly significant. The ribs showed the least significant difference between scores after the dose reduction. In both CR systems, a significant difference in scoring was only seen after the dose reduction from D100% to D25%, and in the FP system, a reduction from D100% to D50% already scored significantly different. The vertebra always scored significantly less with decreasing dosages independently of the radiography system used. For the phalanx, a dose reduction led to significantly lower scores in all comparisons for FP and CR_N, but using the CR_P, no significant difference was seen comparing the different dosages used. The femur showed different results with each system used. The CR_N comparison of D50% to D25% showed no significant difference, while D100% to D50% and D100% to D25% did, with the latter even being highly significant. Using the CR_P, only the comparison of D100% to D25% was significant, whereas while using the FP, every comparison of dosages showed a significant to highly significant difference in scoring (Table 5).

Table 5. Summary of the statistical analyses stating significant occurrence in interobserver variability through statistically calculated AUC (area under the curve) values. CR_N = needle-based detector system, CR_P = powder-based detector system, FP = flat panel detector system. For interpretation of the AUC values see Section 2.4.

System	Criterion	Dose Comparison	AUC	95% Confidence Interval	p-Value
CR_N	Femur	100–50	0.727	0.595–0.859	0.004
		100–25	0.811	0.700–0.923	<0.001
		50–25	0.630	0.484–0.776	0.095
	Phalanx	100–50	0.678	0.536–0.820	0.022
		100–25	0.781	0.657–0.906	<0.001
		50–25	0.653	0.509–0.797	0.049
	Rib	100–50	0.645	0.500–0.791	0.062
		100–25	0.773	0.647–0.899	<0.001
		50–25	0.635	0.489–0.781	0.082
	Vertebra	100–50	0.695	0.555–0.834	0.012
		100–25	0.833	0.724–0.942	<0.001
		50–25	0.662	0.520–0.805	0.037
	All	100–50	0.683	0.613–0.753	<0.001
		100–25	0.796	0.737–0.855	<0.001
		50–25	0.645	0.573–0.717	<0.001

Table 5. *Cont.*

System	Criterion	Dose Comparison	AUC	95% Confidence Interval	*p*-Value
CR_P	Femur	100–50	0.591	0.441–0.741	0.241
		100–25	0.730	0.594–0.866	0.003
		50–25	0.651	0.506–0.796	0.052
	Phalanx	100–50	0.617	0.470–0.764	0.132
		100–25	0.651	0.507–0.794	0.053
		50–25	0.528	0.376–0.681	0.719
	Rib	100–50	0.557	0.405–0.708	0.466
		100–25	0.679	0.539–0.819	0.022
		50–25	0.617	0.470–0.765	0.132
	Vertebra	100–50	0.667	0.523–0.810	0.032
		100–25	0.771	0.645–0.897	<0.001
		50–25	0.656	0.511–0.800	0.046
	All	100–50	0.605	0.531–0.678	0.007
		100–25	0.701	0.633–0.770	<0.001
		50–25	0.603	0.529–0.677	0.008
FP	Femur	100–50	0.746	0.618–0.874	0.002
		100–25	0.851	0.748–0.954	<0.001
		50–25	0.659	0.516–0.803	0.041
	Phalanx	100–50	0.673	0.531–0.815	0.026
		100–25	0.823	0.714–0.931	<0.001
		50–25	0.709	0.573–0.845	0.007
	Rib	100–50	0.682	0.541–0.824	0.019
		100–25	0.771	0.643–0.899	<0.001
		50–25	0.621	0.474–0.769	0.120
	Vertebra	100–50	0.702	0.562–0.843	0.009
		100–25	0.790	0.670–0.911	<0.001
		50–25	0.680	0.536–0.824	0.021
	All	100–50	0.699	0.630–0.768	<0.001
		100–25	0.803	0.745–0.861	<0.001
		50–25	0.661	0.590–0.732	<0.001

3.1.2. Scoring of the Different Systems

Comparing the different radiography systems used, the FP system showed in all criteria, except one, significantly lower scores when reducing the dosage. Only for the criterion for the rib was the reduction from D50% to D25% not significant, but it was significant in the comparison of D100% to D50% and D100% to D25%. Regarding the CR_N system, only the criterion for vertebra showed significantly lower scores with every dose reduction. For the femur and the rib, only a reduction to a quarter of the dosage (D100% to D25%) showed a significantly lower score. For the criterion for the phalanx using the CR, none of the dosages compared showed any significant difference. The CR_N system showed significant differences in all reductions for the phalanx and vertebra but not for the femur, with no significant difference between D50% to D25%, and the ribs only showed significance in using a quarter of the dosage (D100% to D25%). Further details can be seen in Table 5.

3.1.3. Comparison of the Different Systems

Comparisons of the different systems with each other were performed using the scoring of the overall assessment of the different criteria, while also assessing the three different dosages used (D100%, D50%, D25%). There was no significant difference between any of the systems with none of the dosages. AUC ranged between 0.50 and 0.59, with a mean of 0.55. Details can be seen in Table 6.

Table 6. Summary of the statistical analyses stating significant occurrence in intersystem variability through statistically calculated AUC values. CR_N = needle-based detector system, CR_P = powder-based detector system, FP = flat panel detector system. For interpretation of the AUC values see Section 2.4.

	Dose Comparison	AUC	95% Confidence Interval	*p*-Value
CR_N vs. CR_P	100%	0.556	0.403–0.708	0.476
	50%	0.540	0.388–0.693	0.606
	25%	0.541	0.388–0.693	0.600
CR_N vs. FP	100%	0.547	0.394–0.700	0.550
	50%	0.581	0.430–0.732	0.298
	25%	0.589	0.438–0.741	0.251
CR_P vs. FP	100%	0.496	0.343–0.649	1.039
	50%	0.537	0.385–0.690	0.635
	25%	0.549	0.397–0.700	0.534

3.2. *Interobserver Variability*

The average scores given by the reviewers can be seen in Table 7 for each criterion and in Table 8 for each radiography system. Interobserver correlations were calculated for all criteria and overall assessments, as well as scoring between the different systems, adding up to 48 rank correlation values. Correlations were significant in 100% (48/48) of the cases and highly significant in 95.8% (46/48) of the cases. Reviewer 3 showed lower scores with a Spearman's value of 0.33 and 0.38 in comparison to reviewers 1 and 2. Reviewer scores for the phalanx were again highly significantly correlated in all cases, with a Spearman's value of 0.60 to 0.79, with a mean of 0.70. Scores for the ribs showed a Spearman's value of 0.45 to 0.68, with a mean of 0.57, being highly significant in all the cases, and the vertebrae showed a Spearman's value for reviewer's scores of 0.45 to 0.68, with a mean of 0.62. The interobserver agreement for all criteria was highly significant in all cases. The agreements ranged from moderate (0.44) to high (0.66). Interobserver variability was also tested for the different systems, again showing highly significant correlations for every reviewer with each system. Spearman's ranks ranged from 0.45 to 0.71 (average of 0.57) for CR_P, from 0.38 to 0.68 (average of 0.54) for FP, and from 0.80 to 0.69 (average of 0.60) for CR_N. Details on Spearman's rank correlation can be found in Table 9 for each criterion and radiography system. Reviewers' ratings were highly significantly different, except reviewer 4 compared to reviewer 1.

Table 7. Average scoring from all reviewers for the different criteria and radiography systems. CR_N = needle-based detector system, CR_P = powder-based detector system, FP = flat panel detector system.

System	Dose	Femur	Phalanx	Rib	Vertebra
CR_P	100	2.04	2.18	2.39	2.29
	50	2.25	2.61	2.61	2.82
	25	2.64	2.71	3.07	3.25

Table 7. *Cont.*

System	Dose	Femur	Phalanx	Rib	Vertebra
CR_N	100	1.89	2.18	2.03	2.11
	50	2.54	2.71	2.82	2.71
	25	2.86	3.14	3.29	3.21
FP	100	1.82	2.11	2.18	2.29
	50	2.50	2.75	2.86	2.96
	25	2.89	3.36	3.25	3.39

Table 8. Average scoring of each reviewer for the different radiography systems used. Mean as average for all reviewer scores. CR_N = needle-based detector system, CR_P = powder-based detector system, FP = flat panel detector system.

Radiography System	Reviewer 1	Reviewer 2	Reviewer 3	Reviewer 4	Mean
CR_P	2.50	2.37	2.87	2.55	2.57
CR_N	2.85	2.29	2.76	2.70	2.65
FP	2.68	2.35	3.10	2.67	2.70

Table 9. Spearman's rank correlation coefficient comparing each reviewer with each other regarding scoring of the different criteria and radiography systems used. * showing significant correlation; ** showing highly significant correlation ($p \leq 0.001$). CR_N = needle-based detector system, CR_P = powder-based detector system, FP = flat panel detector system.

	Comparison between Reviewers					
	1–2	1–3	1–4	2–3	2–4	3–4
Overall	0.645 **	0.490 **	0.663 **	0.438 **	0.511 **	0.615 **
Femur	0.629 **	0.330 *	0.624 **	0.375 *	0.526 **	0.428 **
Phalanx	0.707 **	0.595 **	0.753 **	0.655 **	0.793 **	0.667 **
Rib	0.680 **	0.533 **	0.622 **	0.454 **	0.473 **	0.679 **
Vertebra	0.741 **	0.571 **	0.754 **	0.495 **	0.608 **	0.543 **
CR_P	0.631 **	0.487 **	0.711 **	0.453 **	0.526 **	0.622 **
CR_N	0.684 **	0.586 **	0.640 **	0.479 **	0.501 **	0.690 **
FP	0.679 **	0.479 **	0.650 **	0.379 **	0.518 **	0.557 **

4. Discussion

4.1. General Aspects

Digital radiography systems have long been established in veterinary medicine and are therefore also used in exotic pet medicine. The awareness of the need for standardized protocols and dosage optimization has slowly begun to arise in the world of exotic pets. Even though many authors describe the use of radiography in reptiles [2,4,6,8,28], no to little information exists on technical settings to achieve images with optimal diagnostic value. The main conclusion of the present study is that a dose reduction in digital radiography systems may limit the image quality of skeletal structures in bearded dragons, and it is conceivable that this information can also cautiously be transferred to other small to medium lizard species. Assuming that the data in this study reflect real conditions such as in a clinical setting using living patients, a general dose reduction leads to decreased reviewer scores regarding image quality, although image processing algorithms might still produce a reasonable overall image impression.

4.2. Evaluation Methods

In the present study, different skeletal structures of varying thickness and structure were used to have a broader spectrum of differences such as the transition from spongiosa to corticalis (femur, criterion 1) or demarcation to the joint (phalanx, criterion 2). Visceral structures were not assessed due to the use of dead bodies and the general challenge of assessing visceral structures with digital radiography in reptiles [2,9]. Visual grading characterizing (VGC) analysis was used and is recommended to evaluate the performance of different radiography systems. By using visual grading, anatomical criteria can be evaluated objectively with a link to clinical interpretation. Still, bias can occur, in this case primarily regarding the reviewers rating the systems. The reviewers were chosen regarding their experience with radiography in reptile species with a minimum of two years of experience in regularly evaluating radiographs in a specialized clinic. A training session was conducted beforehand to reduce divergence in scoring and achieve standardization. However, scores did differ significantly from each other (except for reviewer 1 compared to 4), showing the importance of subjectiveness and individual experience by different human beings even when using a scoring system. The disadvantages and advantages of VGC analysis have already been described in other studies such as Tebrün, W. et al. [13], Båth, M. et al. [29], and Månsson, L. G. [30]. Despite a lack of agreement in scores, the tendency with which the reviewers did score the criteria was the same. Therefore, assessments were significantly correlated for all criteria and all reviewers. These correlations prove the validity of the method, as the interpretation of the study results exclusively relies on the comparison of the scores between the system and dosages.

4.3. Effect of Dose Reduction

Data showed that in every system, a dose reduction led to significantly worse scores for most of the criteria, especially in the "double" reduction from 100% to 25%. Regarding the criteria, the vertebra seems to be the most sensitive structure with significantly worse scores after every reduction with each system used. In contrast, the ribs showed the least influence of the dose reduction, only receiving decreased scores after a reduction of 25%. The effect of dose reduction for the criterion femur varied. In the ribs and femur, the reviewers had to evaluate the differentiation between the bone, whilst in the criteria for vertebra and phalanx, the demarcation to the surrounding tissue and joint space was addressed. The vertebra and phalanx are much more delicate structures than the femur and rib, indicating a possible greater impact of dose reduction on smaller structures. In particular, the joint space with a fine surface and superimpositions of other structures could therefore be more affected than more solid structures such as the femur. The ribs are much more delicate than the femur, impeding the evaluation of this criterion and possibly leading to a greater impact in dose reduction, again indicating a greater influence on smaller structures. In contrast, the femur showed the best scores, indicating that the thicker and bigger the structure, the higher the chance of being able to evaluate variances.

Reduced image quality with a decreased dosage results from a decreased signal-to-noise ratio. Noise is unavoidable in images produced by medical imaging, as no force distributes photons from the X-ray beam uniformly over the surface [31]. In digital radiography, the production of noise primarily depends on the elements of the radiography system, such as the detector, X-ray source, controller circuits, and others. Different techniques have been developed to reduce noise such as collimators or different types of filters [31]. Despite all these techniques, noise still exists and cannot be fully erased. Decreasing the dosage increases the effect of noise due to the lower number of photons emitted, leading to a higher signal-to-noise ratio. In this study, the dosage was decreased to half or a quarter of the reference dose. Studies such as Uffmann, M. et al. [24] described a low sensitivity of the human eye in knowingly detecting image noise, with complaints only after a 50% reduction in dosage. Therefore, the question arises as to whether diagnostic information is lost even before these reduction steps and if the human eye is a good evaluation tool for evaluating image quality, even though a dose reduction in digital radiography systems could be possible due to the higher detective quantum efficiency (DQE). Digital radiography systems, especially needle-based phosphorous systems, show a higher DQE than conventional screen film systems [21], allowing one to reduce the dosage while maintaining image quality.

4.4. Comparison of Detector Systems

In the present study, only digital radiography systems were used, as most veterinary practices today have changed from conventional to digital radiography. Three different systems were compared, namely one flat panel system (FP), a powder-based storage phosphor system (CR_P), and a needle-based storage phosphor system (CR_N), and as described above, they have individual pros and cons. Wirth, S. et al. [19] describes a better evaluation of bone structures in human limbs due to a clearer distinction of cortical structure, articular surface, and cortical delineation of the phalanx in CR_N systems compared to CR_P systems and FP systems with the possibility of a dose reduction of approximately 75% without a loss in image quality. This superiority of CR_N systems could not be reproduced in our study when assessing the skeletal structures in bearded dragons. Our results showed no significant difference in scores between the systems. There was a tendency for better performances in the CR_N systems compared to the others, but with no significance. The flat panel system seemed to be the most sensitive system with worsening scores after every dose reduction, again with no significant difference between the other systems. The reason for this discrepancy can possibly be found in the use of a small reptile species, with structures even smaller and finer than in the phalanx of the human skeletal system. On such small levels, the benefits of CR_N systems over the others could be nullified, leading to similar image quality in all systems used.

In general, it was not the aim of the study to validate the different technical systems, but rather only to assess the impact of dose reduction in each system. Therefore, with this study design and focused on the skeletal system in bearded dragons, we only conclude that all three systems seem to produce diagnostically reliable images with a possible beneficial use for CR_N systems.

4.5. Limitations of the Study

The study was conducted using dead animals, therefore only skeletal structures were chosen for assessment. No conclusion can be drawn regarding soft tissue representation with decreased radiation dosages. Using dead animals also leads to a lack of movement due to respiration or heart action, which must be kept in mind as it could have a possible impact on image quality in living animals. Soft tissue structures are surely more prone to respiration artifacts than skeletal structures, but overall influence could still be possible. Even though only animal bodies without detectable abnormalities in the bony structures were used, the animals were not undergoing specific tests regarding the underlying disease. The underlying disease could individually affect bone density or shape and can therefore not be fully ruled out. Regarding the different animals, it has to be mentioned that the size ranged from 134 to 499 g. We see this variation as minor as all the results point in the same direction and this body mass range also presents the typical sizes of bearded dragons presented in practice. Due to the limited number of individuals, we did not calculate statistical evaluation depending on the size, but this could be the focus of further studies. The duration of freezing after euthanasia varied from one to six months and could theoretically lead to different stages of decomposition of the animal body. As we only selected skeletal structures, we expect only a small impact on bone structure due to freezing, making this a minor limitation, which still should be kept in mind, regarding future studies. Reviewer numbers were rather low, allowing a higher risk for variation between the observations. Only one species and a limited number of animals were used, although the overall number of assessments allowed for detailed statistical comparisons. This study was conducted using only bearded dragons. Translation of the results to other reptile species should be made cautiously as reptile species vary greatly in size, shape, and even anatomical structures. Nevertheless, bearded dragons were chosen due to the high popularity of this species in European countries and therefore the higher clinical impact than rarer species. Further studies should go into detail comparing delicate structures in different settings to provide more in-depth information, as well as extend to the use of different species. Finally, smaller steps in dose reduction could perhaps allow for more detailed results. Additionally, the scales in reptiles could influence image quality and further aggravate the possibility of dose reduction while maintaining image quality [9]

5. Conclusions

The study results demonstrate that with digital radiography systems, an optimal dosage according to the needs of the digital system is essential. A general dose reduction is not recommended in reptile species as it will cause a loss of clinically relevant information even if the image quality subjectively appears sufficient. Further studies with smaller steps in dosage reduction should be conducted, as well as methods such as using artificial intelligence to replace the high subjectiveness of the reviewers and insensitivity of the human eye. This study shows the highly difficult aspect of defining a minimal dose to reliably answer specific diagnostic questions, regarding the vast variety of influences on image quality such as patient-related, system-related, and observer-related factors.

Author Contributions: Conceptualization, E.L. and M.P.; methodology, E.L. and M.P.; software, E.L. and M.P.; validation, E.L. and M.P.; formal analysis, E.L., M.P., N.S. and W.T.; investigation, E.L., M.P., N.S. and W.T.; resources, E.L. and M.P.; data curation, W.T. and N.S.; writing—original draft preparation, N.S. and W.T.; writing—review and editing, E.L., M.P., N.S. and W.T.; visualization, M.P. and N.S.; supervision, E.L. and M.P.; project administration, M.P. All authors have read and agreed to the published version of the manuscript.

Funding: This research received no external funding.

Institutional Review Board Statement: Ethical review and approval were waived for this study as no living animals were used. Carcasses used in this study had to be euthanized for other reasons and were used with the owners' agreement.

Informed Consent Statement: Not applicable.

Data Availability Statement: The data are provided as an online supplement to this article, https://doi.org/10.17605/OSF.IO/8H956, last accessed on 15 April 2023.

Acknowledgments: We thank the colleagues who served as independent reviewers for contributing to our study. The authors wish to thank Martin Enderlein (†), who contributed significantly to the conception of the study and data acquisition but passed away unexpectedly before the study could be finished and published.

Conflicts of Interest: The authors declare no potential conflict of interest. Wimex Agrarprodukte Import and Export GmbH had no role in the design of the study; in the collection, analyses, or interpretation of data; in the writing of the manuscript, or in the decision to publish the results.

References

1. Gumpenberger, M. Diagnostic imaging of reproductive tract disorders in reptiles. *Vet. Clin. Exot. Anim. Pract.* **2017**, *20*, 327–343. [CrossRef] [PubMed]
2. Schumacher, J.; Toal, R.L. Advanced radiography and ultrasonography in reptiles. *Semin. Avian Exot. Pet Med.* **2001**, *10*, 162–168. [CrossRef]
3. Grosset, C.; Daniaux, L.; Guzman, D.S.-M.; Weber, E.S., III; Zwingenberger, A.; Paul-Murphy, J. Radiographic anatomy and barium sulfate contrast transit time of the gastrointestinal tract of bearded dragons (*Pogona vitticeps*). *Vet. Radiol. Ultrasound* **2014**, *55*, 241–250. [CrossRef]
4. Ahranjani, B.A.; Shojaei, B.; Tootian, Z.; Masoudifard, M.; Rostami, A. Anatomical, radiographical and computed tomographic study of the limbs skeleton of the Euphrates soft shell turtle (*Rafetus euphraticus*). *Vet. Res. Forum* **2016**, *7*, 117–124. [CrossRef]
5. Banzato, T.; Russo, E.; Finotti, L.; Zotti, A. Development of a technique for contrast radiographic examination of the gastrointestinal tract in ball pythons (*Python regius*). *Am. J. Vet. Res.* **2012**, *73*, 996–1001. [CrossRef]
6. Mans, C. Clinical update on diagnosis and management of disorders of the digestive system of reptiles. *J. Exot. Pet. Med.* **2013**, *22*, 141–162. [CrossRef]

7. Chen, T.-Y.; Lee, Y.-T.; Chi, C.-H. Observation of reproductive cycle of female yellow-margined box turtle (*Cuora flavomarginata*) using radiography and ultrasonography. *Zoo Biol.* **2011**, *30*, 689–698. [CrossRef]
8. Di Ianni, F.; Volta, A.; Pelizzone, I.; Manfredi, S.; Gnudi, G.; Parmigiani, E. Diagnostic sensitivity of ultrasound, radiography and computed tomography for gender determination in four species of lizards. *Vet. Radiol. Ultrasound* **2015**, *56*, 40–45. [CrossRef]
9. Bochmann, M.; Ludewig, E.; Pees, M. Vergleich der Bildqualität konventioneller und digitaler Radiographie bei Echsen. *Tierärztliche Prax.* **2011**, *39*, 259–267. [CrossRef]
10. Körner, M.; Weber, C.H.; Wirth, S.; Pfeifer, K.-J.; Reiser, M.F.; Treitl, M. Advances in digital radiography: Physical principles and system overview. *Radiographics* **2007**, *27*, 675–686. [CrossRef]
11. McEntee, M.; Frawley, H.; Brennan, P.C. A comparison of low contrast performance for amorphous silicon/caesium iodide direct radiography with a computed radiography: A contrast detail phantom study. *Radiography* **2007**, *13*, 89–94. [CrossRef]
12. Bochmann, M.; Ludewig, E.; Krautwald-Junghanns, M.-E.; Pees, M. Comparison of the image quality of a high-resolution screen-film system and a digital flat panel detector system in avian radiology. *Vet. Radiol. Ultrasound* **2011**, *52*, 256–261. [CrossRef] [PubMed]
13. Tebrün, W.; Ludewig, E.; Köhler, C.; Böhme, J.; Pees, M. Needle-based storage-phosphor detector radiography is superior to a conventional powder-based storage phosphor detector and a high-resolution screen-film system in small patients (budgerigars and mice). *Sci. Rep.* **2019**, *9*, 10057. [CrossRef] [PubMed]
14. Ludewig, E.; Pees, M.; Morgan, J.P. Clinical technique: Digital radiography in exotic pets—Important practical differences compared with traditional radiography. *J. Exot. Pet. Med.* **2012**, *21*, 71–79. [CrossRef]
15. Pees, M.; Bochmann, M.; Krautwald-Junghanns, M.-E.; Schmidt, V.; Ludewig, E. Vergleichende röntgenologische Darstellung des Respirationstraktes von Schlangen mittels konventioneller hochauflösender Film-Folien-Kombination und einem digitalen Detektorsystem. *Berl. Munch. Tierarztl. Wochenschr.* **2010**, *123*, 177–185.
16. Uffmann, M.; Prokop, M.; Eisenhuber, E.; Fuchsjäger, M.; Weber, M.; Schaefer-Prokop, C. Computed radiography and direct radiography: Influence of acquisition dose on the detection of simulated lung lesions. *Invest. Radiol.* **2005**, *40*, 249–256. [CrossRef]
17. Jimenez, D.A.; Armbrust, L.J.; O'Brien, R.T.; Biller, D.S. Artifacts in digital radiography. *Vet. Radiol. Ultrasound* **2008**, *49*, 321–332. [CrossRef]
18. Carter, C.; Vealé, B. *Digital Radiography and PACS*, 4th ed.; Elsevier Health Sciences: St. Louis, MO, USA, 2022.
19. Wirth, S.; Treitl, M.; Reiser, M.F.; Körner, M. Imaging performance with different doses in skeletal radiography: Comparison of a needle-structured and a conventional storage phosphor system with a flat-panel detector. *Radiology* **2009**, *250*, 152–160. [CrossRef]
20. Marshall, N.W.; Smet, M.; Hofmans, M.; Pauwels, H.; De Clercq, T.; Bosmans, H. Technical characterization of five x-ray detectors for paediatric radiography applications. *Phys. Med. Biol.* **2017**, *62*, N573–N586. [CrossRef] [PubMed]
21. Körner, M.; Treitl, M.; Schaetzing, R.; Pfeifer, K.-J.; Reiser, M.; Wirth, S. Depiction of low-contrast detail in digital radiography: Comparison of powder-and needle-structured storage phosphor systems. *Invest. Radiol.* **2006**, *41*, 593–599. [CrossRef]
22. Marshall, N.W.; Lemmens, K.; Bosmans, H. Physical evaluation of a needle photostimulable phosphor based CR mammography system. *Med. Phys.* **2012**, *39*, 811–824. [CrossRef] [PubMed]
23. Schaetzing, F. *Advances in Digital Radiography: RSNA Categorical Course in Diagnostic Radiology Physics*; Radiology Society of North America (RSNA): Oak Brook, IL, USA, 2003; pp. 7–22. Available online: https://websites.umich.edu/~ners580/ners-bioe_481/lectures/pdfs/RSNA2003_CR_Schaetzing.pdf (accessed on 5 February 2023).
24. Uffmann, M.; Schaefer-Prokop, C. Digital radiography: The balance between image quality and required radiation dose. *Eur. J. Radiol.* **2009**, *72*, 202–208. [CrossRef] [PubMed]
25. Körner, M.; Wirth, S.; Treitl, M.; Reiser, M.; Pfeifer, K.-J. Initial clinical results with a new needle screen storage phosphor system in chest radiograms. *RoFo Geb. Rontgenstrahlen Bildgeb. Verfahr.* **2005**, *177*, 1491–1496. [CrossRef]
26. Ludewig, E.; Richter, A.; Frame, M. Diagnostic imaging—Evaluating image quality using visual grading characteristic (VGC) analysis. *Vet. Res. Commun.* **2010**, *34*, 473–479. [CrossRef] [PubMed]
27. Mukaka, M.M. Statistics corner: A guide to appropriate use of correlation coefficient in medical research. *Malawi Med. J.* **2012**, *24*, 69–71.
28. Banzato, T.; Hellebuyck, T.; Van Caelenberg, A.; Saunders, J.H.; Zotti, A. A review of diagnostic imaging of snakes and lizards. *Vet. Rec.* **2013**, *173*, 43–49. [CrossRef] [PubMed]
29. Båth, M.; Månsson, L.G. Visual grading characteristics (VGC) analysis: A non-parametric rank-invariant statistical method for image quality evaluation. *Brit. J. Radiol.* **2007**, *80*, 169–176. [CrossRef]
30. Månsson, L.G. Methods for the evaluation of image quality: A review. *Radiat. Prot. Dosim.* **2000**, *90*, 89–99. [CrossRef]
31. Manson, E.N.; Ampoh, V.A.; Fiagbedzi, E.; Amuasi, J.H.; Flether, J.J.; Schandorf, C. Image noise in radiography and tomography: Causes, effects and reduction techniques. *Curr. Trends Clin. Med. Imaging* **2019**, *2*, 555620.

animals

MDPI

Article

Establishment of a Real-Time PCR Assay for the Detection of *Devriesea agamarum* in Lizards

Maria Brockmann [1,*], Christoph Leineweber [1], Tom Hellebuyck [2], An Martel [2], Frank Pasmans [2], Michaela Gentil [1], Elisabeth Müller [1] and Rachel E. Marschang [1]

1 Laboklin GmbH & Co. KG, Steubenstr. 4, 97688 Bad Kissingen, Germany
2 Department of Pathobiology, Pharmacology and Zoological Medicine, Ghent University, Salisburylaan 133, B-9820 Merelbeke, Belgium
* Correspondence: brockmann@laboklin.com

Simple Summary: Bacterial infections can play an important role in dermatitis in lizards. The bacterial species *Devriesea* (*D.*) *agamarum* is a known cause of dermatitis, cheilitis and even fatal disease in lizards. Disease has most often been reported in *Uromastyx* species, but other lizards may also be affected. However, some are asymptomatic carriers, increasing the risk of spreading *D. agamarum*. Usually, *D. agamarum* is detected with culture-based methods. It was the aim of this study to establish a real-time PCR assay to expand diagnostic options in routine diagnostics. The presented assay is able to detect *D. agamarum* in clinical samples, decreasing laboratory turn-around time in comparison to conventional culture-based detection methods. This enables a fast therapeutic approach for affected animals and decreases the risk of spread.

Abstract: (1) Background: *Devriesea* (*D.*) *agamarum* is a potential cause of dermatitis and cheilitis in lizards. The aim of this study was to establish a real-time PCR assay for the detection of *D. agamarum*. (2) Methods: Primers and probe were selected targeting the 16S rRNA gene, using sequences of 16S rRNA genes of *D. agamarum* as well as of other bacterial species derived from GenBank. The PCR assay was tested with 14 positive controls of different *D. agamarum* cultures as well as with 34 negative controls of various non-*D. agamarum* bacterial cultures. Additionally, samples of 38 lizards, mostly *Uromastyx* spp. and *Pogona* spp., submitted to a commercial veterinary laboratory were tested for the presence of *D. agamarum* using the established protocol. (3) Results: Concentrations of as low as 2×10^4 colonies per mL were detectable using dilutions of bacterial cell culture (corresponding to approximately 200 CFU per PCR). The assay resulted in an intraassay percent of coefficient of variation (CV) of 1.31% and an interassay CV of 1.80%. (4) Conclusions: The presented assay is able to detect *D. agamarum* in clinical samples, decreasing laboratory turn-around time in comparison to conventional culture-based detection methods.

Keywords: *Devriesea agamarum*; polymerase chain reaction; *Uromastyx* sp.; *Pogona vitticeps*; bearded dragon; lizard; reptile; cheilitis; dermatitis

Citation: Brockmann, M.; Leineweber, C.; Hellebuyck, T.; Martel, A.; Pasmans, F.; Gentil, M.; Müller, E.; Marschang, R.E. Establishment of a Real-Time PCR Assay for the Detection of *Devriesea agamarum* in Lizards. *Animals* **2023**, *13*, 881. https://doi.org/10.3390/ani13050881

Academic Editor: Simone Taddei

Received: 18 January 2023
Revised: 20 February 2023
Accepted: 24 February 2023
Published: 28 February 2023

1. Introduction

Devriesea (*D.*) *agamarum* is a bacterial species known to cause dermatitis and cheilitis in lizards. Disease has most often been described in *Uromastyx* spp. [1–5]. However, *D. agamarum* can also infect other lizards [6,7]. It has been reported in captive [8,9] as well as in free-ranging lizards [7]. Clinical signs of disease generally include dermatitis or cheilitis, often described with a yellow crusty appearance [10]. Disease outbreaks with extensive mortality have also been reported, especially if lizards developed septicaemia. [7,11]. Bearded dragons have been described to asymptomatically carry *D. agamarum* in their oral cavities [2,3]. Treatment of affected animals usually includes debridement of dermal lesions and systemic use of antibiotics—especially cephalosporines are considered effective [4,12]—and may also require disinfection of the enclosure [13]. Autovaccines have also been discussed as a treatment method [14].

Therefore, a fast and reliable diagnostic approach is an important consideration, both in clinical disease with suspected *D. agamarum* infection and in entry controls. *D. agamarum* is relatively easily cultured at 37 °C but also grows at temperatures of 25–42 °C on Columbia agar with 5% sheep blood [11]. Diagnosis can, however, be complicated in laboratories with limited experience with this pathogen. Matrix-Assisted Laser Desorption/Ionisation Time-of-Flight Mass Spectrometry (MALDI-TOF MS), one of the most frequently used standard techniques for the identification of bacteria in routine laboratory diagnostics, may not (yet) be able to identify *D. agamarum* when working with standardized databases [15]. However, this issue is likely to be overcome as databases expand. Currently, laboratories can improve *D. agamarum* identification by implementing and expanding their own MALDI-TOF MS databases or using 16S rRNA gene sequencing to identify cultured but unidentified isolates. Another option would be a specific PCR assay for *D. agamarum*, which might prove especially valuable if other infectious agents, such as viral or fungal pathogens, are also suspected, allowing concurrent testing from the same sample. The aim of this study was, therefore, to develop a PCR assay for the detection of *D. agamarum*.

2. Materials and Methods

2.1. Bacterial Isolates Used to Establish the qPCR

In total, 14 *D. agamarum* isolates were used in this study as positive controls.

Three *D. agamarum* isolates (GenBank accession numbers: MT664091-93) were obtained from routine diagnostic submissions at Laboklin GmbH & Co. KG (Bad Kissingen Germany) in 2019 [15], while 11 were isolated between 2005 and 2009 at the Faculty of Veterinary Medicine, Ghent University (Table 1). Non-*D. agamarum* isolates (n = 34) were obtained from the German Collection of Microorganisms and Cell Cultures (DSMZ, Braunschweig, Germany). Some were included in order to determine the ability of the assay to exclude a broad spectrum of different bacterial species. Others, like *Brachybacterium* sp. or *Dermabacter* sp., were included as their sequences were described to be highly similar to *D. agamarum* [11] (Table 2). These 34 isolates served to determine the specificity of the PCR.

Table 1. *Devriesea agamarum* isolates used to establish the PCR.

No.	Isolate	Host	Sample Material	Laboratory
1	MT664091.1/0219 Bf	*Brachylophus fasciatus*	Swab (skin)	Laboklin GmbH & Co. KG, Germany
2	MT664092.1/0919 Ur	*Uromastyx* sp.	Swab (skin)	Laboklin GmbH & Co. KG, Germany
3	MT664093.1/0319 Ur	*Uromastyx* sp.	Swab (skin)	Laboklin GmbH & Co. KG, Germany
4	IMP2 vacc	*Agama impalearis*	Swab (dermatitis), liver	Faculty of Veterinary Medicine, Ghent University
5	30.7	*Uromastyx dispar*	Swab (dermatitis, cheilitis)	Faculty of Veterinary Medicine, Ghent University
6	34.1	*Uromastyx acanthinura*	Swab (dermal abscess)	Faculty of Veterinary Medicine, Ghent University
7	23	*Pogona vitticeps*	Swab (oral cavity)	Faculty of Veterinary Medicine, Ghent University
8	24	*Pogona vitticeps*	Swab (oral cavity)	Faculty of Veterinary Medicine, Ghent University
9	25	*Pogona vitticeps*	Swab (oral cavity)	Faculty of Veterinary Medicine, Ghent University
10	26	*Crotaphytus collaris*	Swab (dermatitis)	Faculty of Veterinary Medicine, Ghent University

Table 1. *Cont.*

No.	Isolate	Host	Sample Material	Laboratory
11	28	*Eublepharis macularius*	Swab (oral cavity)	Faculty of Veterinary Medicine, Ghent University
12	30b	*Ctenonotus gingivus*	Swab (cloaca)	Faculty of Veterinary Medicine, Ghent University
13	4d	*Iguana delicatissima*	Swab (dermal abscess)	Faculty of Veterinary Medicine, Ghent University
14	L26	*Pogona vitticeps*	Swab (oral cavity)	Faculty of Veterinary Medicine, Ghent University

Table 2. Bacterial species used as negative controls to determine the specificity of the PCR.

No.	Bacterial Species	DSMZ Number	Original Depositor (Acc. to DSMZ)
1	*Acinetobacter baumannii*	DSM 30007	J.V. Cook
2	*Bacillus atrophaeus*	DSM 675	E. McCoy
3	*Cytobacillus firmus*	DSM 359	G. Bredemann
4	*Brachybacterium faecium*	DSM 4810	H.E. Schefferle
5	*Dermabacter hominis*	DSM 30958	C. Moissl-Eichinger
6	*Enterococcus faecalis*	DSM 2570	Kaiser-Permanente
7	*Enterococcus faecium*	DSM 20477	A. Grumbach
8	*Enterococcus faecium*	DSM 2146	J.M. Skerman (*Streptococcus faecalis*)
9	*Escherichia coli*	DSM 1103	F. Schoenknecht
10	*Escherichia coli*	DSM 1576	G.C. Crooks
11	*Flavobacterium psychrophilum*	DSM 21280	J.-F. Bernardet
12	*Klebsiella oxytoca*	DSM 5175	R. Hugh
13	*Klebsiella pneumoniae*	DSM 26371	H. Dalton
14	*Klebsiella pneumoniae*	DSM 30104	CDC, Atlanta; 298-53
15	*Listeria monocytogenes*	DSM 19094	H. Seeliger
16	*Mycobacterium phlei*	DSM 750	IPH
17	*Nocardia nova*	DSM 44481	N. F. Conant
18	*Proteus hauseri*	DSM 30118	K.B. Lehmann
19	*Proteus mirabilis*	DSM 4479	CDC PR 14
20	*Pseudomonas aeruginosa*	DSM 1128	C.P. Hegarty
21	*Pseudomonas aeruginosa*	DSM 1117	A. Madeiros
22	*Salmonella enterica*	DSM 19587	CDC
23	*Salmonella enterica*	DSM 17420	CDC
24	*Staphylococcus aureus*	DSM 2569	E.H. Gerlach
25	*Staphylococcus aureus*	DSM 1104	F. Schoenknecht
26	*Staphylococcus aureus*	DSM 46320	W. Witte
27	*Staphylococcus aureus*	DSM 799	AMC
28	*Streptococcus dysgalactiae*	DSM 20662	T.M. Higgs
29	*Staphylococcus epidermidis*	DSM 1798	FDA
30	*Streptococcus equi* ssp. *equi*	DSM 20561	R.E.O. Williams
31	*Staphylococcus felis*	DSM 7377	S. Igimi
32	*Staphylococcus intermedius*	DSM 20373	V. Hajek
33	*Streptococcus pyogenes*	DSM 11728	E. Mortimer
34	*Yersinia enterocolitica*	DSM 4780	J.M. Coffey

2.2. DNA Preparation

Pure cultures of each strain were incubated in 750 µL lysis buffer (MagNA Pure DNA Tissue Lysis Buffer, Roche, Mannheim, Germany) and 75 µL proteinase K (proteinase K, lyophilisiert, ≥30 U/mg, Carl Roth GmbH & Co KG, Karlsruhe, Germany) for one hour at 65 °C. From this, 200 µL were utilized for automated nucleic acid (NA) extraction using the MagNA Pure 96 DNA and Viral NA Small Volume Kit (Roche, Mannheim, Germany) according to the manufacturer's instructions. The resulting NAs were eluted in a volume of 100 µL. The isolated NAs were kept at −18 °C until the PCR tests were performed. The DNA used for the dilution series was extracted manually using the QIAamp® DNA MicroKit (50) (Qiagen, Hilden, Germany), and the DNA concentration was measured with a spectrophotometer (NanoDrop 2000, Thermo Fisher Scientific, Inc., Wilmington, NC, USA).

2.3. Design of Primers and Oligonucleotide Probe, PCR Protocol and Optimisation of Annealing Temperature

The 16S rRNA gene was selected as the target region based on the availability of sequence data from a variety of isolates. Sequences of *D. agamarum* (NZ_LN849456.1, LN849456.1, NR_044368.1, EU009865.1, KF647330.1) were retrieved from GenBank, and multiple sequence alignment was performed with other sequences of different bacterial species (e.g., *Agromyces* species, *Arthobacter* species, *Brachybacterium* species, *Dermabacter* species, *Pseudomonas* species) using MUSCLE (https://www.ebi.ac.uk/Tools/msa/muscle/ last accessed on 20 February 2023). The primers and probe were designed using primer3 (https://primer3.ut.ee/ last accessed on 20 February 2023).

Reactions included 1.0 µL of each primer (10 µM), 1.0 µL of the probe (2 µM), 4.0 µL DNA Process Control Detection Kit qPCR Reaction Mix and 5.0 µL template DNA in a total volume of 20.0 µL. Amplification was performed with a LightCycler 96 (Roche, Mannheim, Germany) in a 96-well format.

The following protocol was used: Preincubation at 95 °C for 30 s followed by 40 cycles of two-step-amplification (95 °C for 5 s and 60 °C for 30 s). PCR-grade water (Roche, Mannheim, Germany) served as a negative control. While all other bacterial samples were negative, *Dermabacter hominis* produced positive PCR results at an annealing temperature of 60 °C (Figure 1). Therefore, DNA of *Dermabacter hominis* (DSM 30958) from the DSMZ as well as DNA of *D. agamarum* (GenBank Accession number: MT664092.1/0919Ur) [15], was tested in duplicate with different annealing temperatures (protocol 1: 65.0 °C, protocol 2: 66.0 °C, protocol 3: 67.0 °C) using a LightCycler 96 (Roche, Mannheim, Germany).

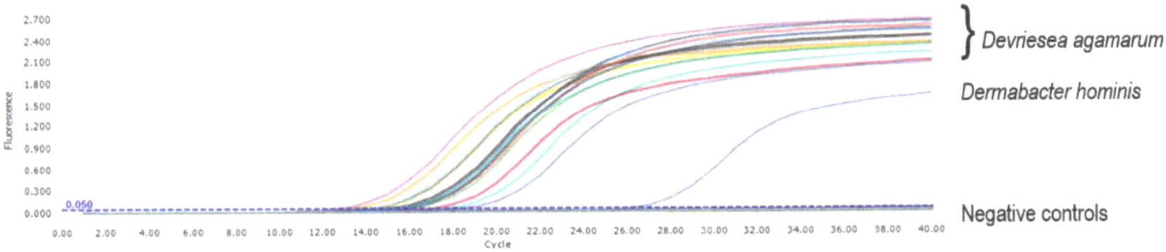

Figure 1. Two-step amplification carried out at 60 °C: All *Devriesea agamarum* isolates tested resulted in positive signals with CT values ranging from 12.61 to 18.54. *Dermabacter hominis* (blue curve) also gave a positive signal with a CT value of 26.29.

2.4. Determination of Specificity at 66 °C and of Repeatability and Sensitivity of the Assay

Various non-*D. agamarum* bacterial species (Table 2) were tested in duplicate for positive reactions at 66 °C. The following protocol was used: Preincubation at 95 °C for 30 s followed by 40 cycles of two-step-amplification (95 °C for 5 s and 66 °C for 30 s). PCR-grade water (Roche, Mannheim, Germany) served as a negative control. DNA of 14 *D. agamarum* isolates (Table 1) was also tested.

To assess the intraassay repeatability of the PCR, standard deviations were calculated for 10-fold serial dilutions in triplicate on a single plate. For the interassay reproducibility, standard deviations were calculated for a 10-fold serial dilution series which was amplified three times daily for two days. The standard deviations of the CT values were used to calculate the percent of coefficient of variation (CV%).

Detection limit for bacterial cell dilutions: In order to determine the sensitivity of the assay, a culture of *D. agamarum* (isolate MT664091.1/0219Bf) was used, starting with a dilution (D0) of 0.5 McFarland (1.5×10^8 per mL). This suspension (D0) was 10-fold serially diluted (D1–D10) in duplicate, and 1 mL of each dilution was inoculated onto Columbia agar with defibrinated sheep blood (Becton Dickinson GmbH, Heidelberg, Germany/Oxoid GmbH, Wesel, Germany), incubated at 36 °C and checked for growth after 30 h. Colony forming units (CFU) were counted if the result was expected to be between 0 and 300 CFU. These results were then used to determine the limit of detection of the assay. DNA was extracted from 200 µL of each dilution (D1–D10) as described above, and PCR was carried out in duplicate. The detection of *Dermabacter hominis* (DSM 30958) was quantified in the same way.

Detection limit for DNA from pure culture: To evaluate the assay's sensitivity, PCR was carried out in triplicate using serial 10-fold dilutions of DNA prepared from colonies of *D. agamarum* (isolate MT664091.1/0219Bf).

Spiked-in matrix: The assay was also evaluated using a spiked-in matrix (*D. agamarum*-negative-tested lizard skin with a known concentration of target DNA). For these assays, 10 µL of the above-described dilutions D0 to D6 were inoculated onto *D. agamarum*-negative-tested lizard skin. The skin was incubated in 500 µL lysis buffer, and 50 µL proteinase K and NA were extracted from 200 µL of this suspension as described above and eluted in a total volume of 100 µL NA. The PCR was carried out in duplicate as described above.

2.5. Testing of Samples Submitted to a Commercial Veterinary Laboratory

Clinical samples from lizards for which appropriate material (skin, crusts, dry swab) was submitted to a commercial veterinary laboratory between March 2022 and December 2022 and for which the submitting veterinarian indicated an interest in *D. agamarum* diagnostics were tested using the established protocol to evaluate the PCR for use with clinical samples. Some of the samples were derived from animals showing clinical signs, others from asymptomatic animals tested in the context of a health check—often when *D. agamarum* had been isolated from animals in the group previously. If suitable material was available (skin or swab in a transport medium), bacteriology was performed (as previously described [15]). Identification of isolates was based on growth characteristics on agar plates (Columbia Agar with defibrinated sheep blood and Endo Agar, Becton Dickinson GmbH, Heidelberg, Germany), biochemical parameters and MALDI-TOF MS. In doubt, colonies were also re-checked via PCR. If the results of the bacteriological culture were available, the results of the PCR and culture would be compared. Samples were considered PCR positive if the cycle threshold (CT) was <35.0 and equivocal if the CT was ≥35.0, but a signal was obtained. If swabs of different origins regarding the localisation (e.g., dermal and oral) of the same animal were received, all samples were tested separately. Amplicons from positive samples were sequenced (ABI PRISM 3130 XL Genetic Analyser, Applied Biosystems, Foster City, CA, USA) and sequences were analysed by BLAST (https://blast.ncbi.nlm.nih.gov/Blast.cgi, last accessed on 20 February 2023).

3. Results

3.1. Development of the PCR

The selected primer and probe sequences (Eurofins MWG Operon, Ebersberg, Germany) are shown in Table 3. The product size was expected to be 246 base pairs.

Table 3. Primers and probe used in the PCR for detection of *Devriesea agamarum*.

Forward Primer Devag16_For	GATGACTGCAGAGATGTGGTG
Reverse Primer Devag16_Rev	TTTGTACCGGCCATTGTAGCAT
Oligonucleotide probe FAM BHQ1	CATGTTGCCAGCACTTCGG

Optimisation of annealing temperature was performed using three protocols with different annealing temperatures. In protocol 1 (65 °C), *D. agamarum* DNA was detected with CT values of 14.58 and 16.49 and *Dermabacter hominis* with CT values of 29.43 and 29.62. In protocol 2 (66 °C), *D. agamarum* DNA was detected with CT values of 16.22 and 16.14, and *Dermabacter hominis* showed values of 34.28 and 35.19. In protocol 3 (67 °C), *D. agamarum* was detected with CT values of 15.57 and 16.57, while *Dermabacter hominis* was not detected. Therefore, an annealing temperature of 66 °C was chosen for all further analyses as it was considered sufficient to discriminate between pure cultures of *D. agamarum* and *Dermabacter hominis* DNA without losing sensitivity for the detection of *D. agamarum*.

3.2. Specificity, Repeatability and Sensitivity of the Assay

Specificity of the assay using the protocol with 66 °C as an annealing temperature was determined using 14 *D. agamarum* isolates (Table 1) and 34 non-*D. agamarum* bacterial isolates (Table 2) in duplicate.

No signal was obtained from any bacterial DNA from isolates other than *D. agamarum* except for *Dermabacter hominis*. The CT value obtained using the DNA from pure cultures of *Dermabacter hominis* were high (34.55 and 34.48) in comparison to those reached using DNA from *D. agamarum* (12.08–18.14) but still below the threshold set for clinical samples. As these are the results for DNA extracted from pure culture, 66 °C was considered sufficient for further testing of samples without prior cultivation as samples without prior cultivation are expected to yield a lower pathogen level. The intraassay CV was calculated to be 1.31%, while the interassay CV was 1.80%.

Detection limit for bacterial cell dilutions: A positive PCR signal was detected for *D. agamarum* dilutions D0–D4. An equivocal signal was detected for D5 and D6. In culture, D4 corresponded to 2×10^4 colonies per mL. Therefore, the assay sensitivity was 2×10^4 colonies per mL with dilutions of bacterial cell culture serving as a template. This corresponds to approximately 200 CFU per PCR. For the serially diluted culture of *Dermabacter hominis*, no positive PCR signals were observed. Two dilutions (D0 and D1) resulted in equivocal CT values, with D0 being set to 0.5 McFarland (1.5×10^8 per mL).

Detection limit for DNA from pure culture: The DNA concentration was determined to be 42.5 ng/µL (A260/A280: 1.94). *D. agamarum* DNA was detectable in dilutions up to 1:10^5. Therefore, DNA concentrations of as low as 425 fg/µL were detectable in the PCR.

Spiked-in matrix: Spiked-in matrixes produced clearly positive results up to skin spiked with 10 µL of D2 with D2 corresponding to 2×10^6 colonies per mL (approximately 360 CFU per PCR considering dilution during sample preparation).

3.3. Testing of Clinical Samples

In order to test the use of the developed PCR for clinical samples, a total of 48 samples from 38 lizards were tested for the presence of *D. agamarum* (Table 4). The samples were derived from several species, mostly agamids (*Pogona* spp. and *Uromastyx* spp.), and were of different origins (zoological collection, animal rescue centre, private owner) from Germany and the Netherlands. Some of these animals were asymptomatic and were tested in the context of a health check. Others showed clinical signs such as skin lesions, hyperkeratosis or stomatitis (Table 4). Samples were mostly derived from the oral cavity or skin/crusts. Of the 38 animals tested, *D. agamarum* was detected by PCR in 16 animals (42.10%). A further five animals (13.16%) were considered to have equivocal results, and 17 animals (44.74%) were negative for *D. agamarum*. One animal (animal 3) that tested negative proved to be infected with a fungus of the family Onygenaceae. A bacteriological examination was performed for 33 of the 38 animals, but *D. agamarum* was not cultured. However, in most of these cases, various other bacterial species (Table 4) were cultured when bacteriology was performed.

Table 4. Results of bacteriological culture and *Devriesea agamarum* PCR of clinical samples from 38 lizards.

Animal No.	Species	Clinical Signs and History and Additional Information	Country of Origin	Time from Sampling to Sample Preparation:	Sample Material	Bacterial Culture	PCR Result
1	*Uromastyx* sp.	Suspicious skin lesion Same wildlife park as animal 2	NL	unknown	Skin	N.D.	Positive (CT 23.71)
2	*Uromastyx* sp.	Suspected lesions/ dermatitis Same wildlife park as animal 1	NL	unknown	Swabs (skin)	N.D.	Positive (CT 30.5[illegible])
3	*Pogona* sp.	Black discoloration of the scales after shedding, especially on the tail.	NL	5d	Skin	+ *Glutamibacter creatinolyticus* + *Micrococcus* sp. + aerobic spore-forming bacteria	Negative
					Swab (skin)		Negative
4	*Uromastyx* sp.	Partner to animal 5	NL	2d	Swab (oral mucosa)	++ *Pantoea agglomerans* ++ *Pseudomonas aeruginosa* ++ aerobic spore-forming bacteria	Positive (CT 28.4[illegible])
5	*Uromastyx* sp.	Partner to animal 4	NL	2d	Swab (oral mucosa)	++ *Exiguobacterium mexicanum* ++ *Kluyvera intermedia* ++ *Pseudomonas chlororaphis* ++ aerobic spore-forming bacteria	Negative
6	*Uromastyx* sp.	Reported to have been treated with antibiotics	NL	2d	Swab (oral mucosa)	++ *Acinetobacter variabilis* +++ *Arthrobacter globiformis* + *Bacillus cereus*	Positive (CT 28.4[illegible])
7	*Pogona vitticeps*	No clinical signs	DE	1d	Swab (oral mucosa)	++ *Proteus mirabilis*	Positive (CT 33.1[illegible])
8	*Pogona vitticeps*	No clinical signs	DE	1d	Swab (oral mucosa)	+++ *Enterobacter cloacae* + *Pseudomonas aeruginosa*	Equivocal (CT 35.71)
9	*Pogona vitticeps*	No clinical signs Partner to animal 10 Confiscated	DE	1d	Swab (oral mucosa)	++ *Bordetella hinzii*	Negative
10	*Pogona vitticeps*	No clinical signs Partner to animal 9 Confiscated	DE	1d	Swab (oral mucosa)	+ *Bordetella hinzii* + *Peribacillus muralis* ++ *Enterococcus faecalis* + *Pantoea agglomerans*	Negative
11	*Pogona vitticeps*	No clinical signs Abandoned	DE	1d	Swab (oral mucosa)	++ *Morganella morganii* ++ *Klebsiella oxytoca*	Positive (CT 27.56)
12	*Pogona vitticeps*	No clinical signs Abandoned	DE	1d	Swab (oral mucosa)	++ *Klebsiella oxytoca* + *Proteus mirabilis*	Negative
13	*Pogona henrylawsoni*	No clinical signs Abandoned juvenile same enclosure as animals 14 and 15	DE	unknown	Swab (oral mucosa)	+ *Aeromonas hydrophila* (+) aerobic spore-forming bacteria	Negative
14	*Pogona henrylawsoni*	No clinical signs Abandoned juvenile same enclosure as animals 13 and 15	DE	unknown	Swab (oral mucosa)	++ *Proteus mirabilis* (+) aerobic spore-forming bacteria	Negative
15	*Pogona henrylawsoni*	No clinical signs Abandoned juvenile same enclosure as animals 13 and 14	DE	unknown	Swab (oral mucosa)	+ *Aeromonas hydrophila* (+) alpha-hemolytic streptococci	Negative

Table 4. *Cont.*

Animal No.	Species	Clinical Signs and History and Additional Information	Country of Origin	Time from Sampling to Sample Preparation:	Sample Material	Bacterial Culture	PCR Result
16	*Pogona vitticeps*	No clinical signs	DE	unknown	Swab (oral mucosa)	+ *Aeromonas hydrophila* + *Pseudomonas aeruginosa* (+) aerobic spore-forming bacteria	Positive (CT 27.20)
17	*Pogona vitticeps*	No clinical signs	DE	unknown	Swab (oral mucosa)	+++ *Klebsiella oxytoca*	Equivocal (CT 35.26)
18	*Pogona vitticeps*	No clinical signs	DE	unknown	Swab (oral mucosa)	++ *Klebsiella oxytoca*	Equivocal (CT 35.07)
19	Scincidae	Unknown	DE	2d	Skin	N.D.	Negative
20	Iguanidae	Hyperkeratosis (dorsal)	DE	2d	Skin	N.D.	Negative
21	*Uromastyx* sp.	Skin lesion	DE	2d	Skin	N.D.	Positive (CT 21.34)
22	*Uromastyx* sp.	Skin lesion Animals 22, 23 and 24 kept in the same enclosure	DE	unknown	Swab (outer skin of the mouth)	+ *Pseudomonas aeruginosa* + *Serratia marcescens*	Positive (CT 28.45)
					Swab (oral mucosa)	+ *Pseudomonas aeruginosa* + *Serratia marcescens*	Positive (CT 34.89)
					Skin	N.D.	Positive (CT 31.56)
					Swab (skin)	N.D.	Positive (CT 30.88)
23	*Uromastyx* sp.	Skin lesion Animals 22, 23 and 24 kept in the same enclosure	DE	unknown	Swab (outer skin of the mouth)	+ *Enterobacter cloacae*	Positive (CT 32.20)
					Swab (oral mucosa)	+ *Enterobacter cloacae*	Positive (CT 32.06)
					Skin	N.D.	Positive (CT 31.97)
					Swab (Skin)	N.D.	Positive (CT 34.05)
24	*Uromastyx* sp.	Skin lesion Animals 22, 23 and 24 kept in the same enclosure	DE	unknown	Swab (outer skin of the mouth)	+ *Pseudomonas synxantha*	Equivocal (CT 35.24)
					Swab (oral mucosa)	+ *Pantoea agglomerans*	Negative
					Skin	N.D.	Positive (CT 33.84)
					Swab (Skin)	N.D.	Equivocal (CT 35.75)
25	*Corucia zebrata*	Multiple animals with minimal to moderate stomatitis	NL	2d	Swab (skin)	N.D.	Negative
					Skin	+++ *Pseudomonas aeruginosa*	Negative
26	*Chlamydosaurus kingii*	No clinical signs Kept with animal 27	DE	5d	Swab (oral mucosa)	+ *Proteus mirabilis* ++ *Serratia marcescens*	Negative
27	*Chlamydosaurus kingii*	No clinical signs Kept with animal 26	DE	5d	Swab (oral mucosa)	+ *Klebsiella oxytoca* + *Stenotrophomonas maltophilia* + aerobic spore-forming bacteria	Negative
28	*Laemanctus serratus*	No clinical signs Kept with animals 29 and 30	DE	5d	Swab (oral mucosa)	+ *Klebsiella oxytoca* + *Morganella morganii*	Positive (CT 31.20)
29	*Laemanctus serratus*	No clinical signs Kept with animals 28 and 30	DE	5d	Swab (oral mucosa)	+ *Deinococcus proteolyticus* + *Morganella morganii* + *Serratia marcescens*	Positive (CT 34.35)
30	*Laemanctus serratus*	No clinical signs Kept with animals 28 and 29	DE	5d	Swab (oral mucosa)	+ *Morganella morganii* + *Proteus mirabilis*	Positive (CT 29.10)

Table 4. *Cont.*

Animal No.	Species	Clinical Signs and History and Additional Information	Country of Origin	Time from Sampling to Sample Preparation:	Sample Material	Bacterial Culture	PCR Result
31	*Laudakia stellio picea*	No clinical signs Kept with animal 32	DE	5d	Swab (oral mucosa)	(+) aerobic spore-forming bacteria	Equivocal (CT 35.80)
32	*Laudakia stellio picea*	No clinical signs Kept with animal 31	DE	5d	Swab (oral mucosa)	+ *Pseudomonas* sp. + *Serratia marcescens* + aerobic spore-forming bacteria	Equivocal (CT 36.87)
33	*Pogona vitticeps*	No clinical signs Kept with animal 34	DE	5d	Swab (oral mucosa)	(+) *Staphylococcus epidermidis*	Positive (CT 34.07)
34	*Pogona vitticeps*	No clinical signs Kept with animal 33	DE	5d	Swab (oral mucosa)	+ *Staphylococcus aureus* (+) aerobic spore-forming bacteria	Positive (CT 32.61)
35	*Pogona vitticeps*	Juvenile No clinical signs Kept with animals 36, 37 and 38	DE	5d	Swab (oral mucosa)	(+) aerobic spore-forming bacteria	Negative
36	*Pogona vitticeps*	Juvenile No clinical signs Kept with animals 35, 37 and 38	DE	5d	Swab (oral mucosa)	+ *Achromobacter xylosoxidans*	Negative
37	*Pogona vitticeps*	Juvenile No clinical signs Kept with animals 35, 36 and 38	DE	5d	Swab (oral mucosa)	+ *Serratia marcescens* (+) aerobic spore-forming bacteria	Negative
38	*Pogona vitticeps*	Juvenile No clinical signs Kept with animals 35, 36 and 37	DE	5d	Swab (oral mucosa)	+ *Proteus mirabilis*	Negative

Legend: N.D. = not done, d = days, DE = Germany, NL = the Netherlands.

4. Discussion

D. agamarum is an important pathogen causing skin lesions and, in some cases, systemic disease in lizards. Depending on the species, some animals can be inapparent carriers, while others may develop severe diseases. Diagnosis of the causative agent is therefore important in order to facilitate treatment as well as to prevent the spread of disease. Since animals may suffer when untreated and the risks of spreading increase with time, a fast diagnostic approach is important. The detection of *D. agamarum* is commonly achieved via culture, followed in some cases by 16S rRNA gene sequencing [11,16].

The PCR developed in this study provides a time-saving tool compared to culture and bacterial identification. Detection of *D. agamarum* and concurrent bacteriological examination was performed in 33 of the 38 animals, resulting in 13 of 33 clearly PCR-positive animals but no culture-positives. *D. agamarum* is expected to be abundantly present in symptomatic animals. Culturing of *D. agamarum* is not considered difficult and has been successfully performed in this laboratory before [15]. However, a successful culture depends on the quality of the submitted samples. Appropriate samples include affected tissue below hyperkeratotic crusts or inside of the crusts as well as organs in septicaemic lizards and subcutaneous granulomas. In asymptomatic animals as well as in symptomatic animals, isolation from the oral cavity, gastrointestinal tract or healthy skin may be challenging. *D. agamarum* was cultured in the laboratory in which the study was performed during the study period, but these samples were excluded from the study as no suitable material for concurrent PCR testing (e.g., dry swab, skin) was available. In this study, six animals (1, 2, 21 and 22–24) were known to have been symptomatic and had positive PCR results. Bacteriological culture was performed in three of these animals (22–24). In animal 24, a positive PCR result was only obtained from the skin sample, which was not tested by bacteriological culture. Animal 6 might have been symptomatic (no information was received, but due to the reported previous treatment, it seemed likely). It was treated with antibiotics prior to sampling, which might have influenced the bacteriology results. Possible reasons for the failure of culturing *D. agamarum* out of positive clinical samples in this study include previous antibiotic treatment, incorrect sampling techniques, contamination with (oral) microbiota, increased transport time, inappropriate transport conditions or overgrowth by other bacteria. The latter is especially important as in the presented cases, no selective media for gram-positive bacteria were used, and various different bacterial species are expected to be present on the skin [17,18].

PCR analysis is useful if the performance of bacterial culture is difficult, e.g., due to previous treatment with antibiotics, inadequate preanalytical conditions (such as increased or decreased temperature, increased transport time, inadequate transport medium), or in cases in which overgrowth by other bacteria make detection challenging or impossible. The detection of *D. agamarum* via PCR can also simplify concurrent PCR testing for other known pathogens, e.g., viral or fungal pathogens known to cause dermatitis [19,20], since the same extracted nucleic acids can be used. In general, PCR is advantageous when culturable samples are unavailable, for example, when older samples are tested or stored DNA is examined. However, bacterial DNA can persist in the environment [21], and *D. agamarum* has been shown to survive for several months in the environment, depending on the conditions [13]. A PCR could therefore detect bacteria even in cases in which these were not responsible for clinical signs or in which no replication-competent bacteria were present.

The PCR developed here was not 100% specific. Pure cultures of *Dermabacter hominis* did result in a weak positive signal. However, if diluted, only equivocal results were observed. *Dermabacter hominis* is genetically closely related to *D. agamarum* [2,11,22]. *Dermabacter hominis* is associated with the human microbiome [23]. It is occasionally described in human clinical samples such as abscesses or blood cultures [24–26] but is usually found to be of minor clinical significance [27]. So far, its clinical importance in reptiles is very unclear. Contamination during sampling or sample preparation should be considered a possible option leading to false equivocal results. However, clinical samples are expected to yield less bacterial DNA, making false equivocal results less likely. The 16S rRNA gene is known to be highly conserved between bacterial species, which makes it a useful target if the aim is to identify different bacterial species. It is a commonly used target for bacterial detection, and therefore a large amount of sequence data is available for a wide range of bacterial species. However, it may not be ideal for differentiating closely related bacteria. Currently, the availability of sequence data for *D. agamarum* other than the 16S rRNA gene is limited, but in the future, other targets may prove to be better options. In the meantime, especially equivocal CT values should be evaluated with caution in the face of clinical signs, sampling and sample preparation, and ideally, retesting is recommended. Possibly, skin samples might prove more useful than swabs as they yielded lower CT values in two of the three animals for which both sample types were available, but this might be highly dependent on the sampling method.

The PCR protocol developed in this study proved helpful for the detection of *D. agamarum* in clinical samples. *D. agamarum* was detected in oral swabs from clinically healthy *Pogona* species and serrated casquehead iguana (*Laemanctus serratus*), while the results in which equivocal results were obtained were also from clinically healthy *Pogona* species as well as from clinically healthy black hardun (*Laudakia stellio picea*). *Pogona* species have previously been shown to be possible inapparent carriers of *D. agamarum* and a possible source of infection for more sensitive species [2,3].

Therefore, this PCR protocol may not only be useful for clinical cases but also as a screening tool. However, the number of tested samples is still small, and testing of larger sample numbers is necessary in order to confirm the usefulness of this method for clinical practice.

5. Conclusions

A real-time PCR was developed that is able to detect *D. agamarum* in clinical samples. The assay provides a fast method for the detection of this important pathogen of lizards but should be evaluated with further samples in the clinical context.

Author Contributions: Conceptualization, M.B., E.M. and R.E.M.; Data curation, M.B., C.L. and R.E.M.; Formal analysis, M.B.; Methodology, M.B., M.G. and R.E.M.; Resources, M.B., C.L., T.H., A.M. and F.P.; Supervision, R.E.M.; Validation, M.B., M.G. and R.E.M.; Visualization, M.B.; Writing—original draft, M.B.; Writing—review & editing, C.L., T.H., A.M., F.P., M.G. and R.E.M. All authors have read and agreed to the published version of the manuscript.

Funding: The research received support in the form of salaries for authors [M.B., C.L., M.G., R.E.M.] from Laboklin GmbH & CO KG, a veterinary laboratory offering diagnostic services, including bacteriological and molecular biological examinations. The author E.M. is the head of the funding company.

Institutional Review Board Statement: Not applicable.

Informed Consent Statement: Not applicable.

Data Availability Statement: Data is contained within the article.

Conflicts of Interest: Laboklin GmbH & Co. KG is a veterinary laboratory offering diagnostic services, including bacteriological and molecular biological examinations. This does not alter our adherence to sharing data and materials.

References

1. Haesendonck, R.; van Nieuwerburgh, F.; Haesebrouck, F.; Deforce, D.; Pasmans, F.; Martel, A. Genome Sequence of *Devriesea agamarum*, isolated from agamid lizards with dermatitis. *Genome Announc.* **2015**, *3*, e00949-15. [CrossRef] [PubMed]
2. Hellebuyck, T.; Martel, A.; Chiers, K.; Haesebrouck, F.; Pasmans, F. *Devriesea agamarum* causes dermatitis in bearded dragons (*Pogona vitticeps*). *Vet. Microbiol.* **2009**, *134*, 267–271. [CrossRef] [PubMed]
3. Devloo, R.; Martel, A.; Hellebuyck, T.; Vranckx, K.; Haesebrouck, F.; Pasmans, F. Bearded dragons (*Pogona vitticeps*) asymptomatically infected with *Devriesea agamarum* are a source of persistent clinical infection in captive colonies of dab lizards (*Uromastyx* sp.). *Vet. Microbiol.* **2011**, *150*, 297–301. [CrossRef] [PubMed]
4. Lukac, M.; Horvatek-Tomic, D.; Prukner-Radovcic, E. Findings of *Devriesea agamarum* associated infections in spiny-tailed lizards (*Uromastyx* sp.) in Croatia. *J. Zoo Wildl. Med.* **2013**, *44*, 430–434. [CrossRef]
5. Schmidt-Ukaj, S.; Hochleithner, M.; Richter, B.; Hochleithner, C.; Brandstetter, D.; Knotek, Z. A survey of diseases in captive bearded dragons: A retrospective study of 529 patients. *Vet. Med.* **2017**, *62*, 508–515. [CrossRef]
6. Rossier, C.; Hoby, S.; Wenker, C.; Brawand, S.G.; Thomann, A.; Brodard, I.; Jermann, T.; Posthaus, H. Devrieseasis in a plumed basilisk (*Basiliscus plumfrons*) and chinese water dragons (*Physignathus cocincinus*) in a zoologic collection. *J. Zoo Wildl. Med.* **2016**, *47*, 280–285. [CrossRef]
7. Hellebuyck, T.; Questel, K.; Pasmans, F.; van Brantegem, L.; Philip, P.; Martel, A. A virulent clone of *Devriesea agamarum* affects endangered Lesser Antillean iguanas (*Iguana delicatissima*). *Sci. Rep.* **2017**, *7*, 12491. [CrossRef]
8. Schmidt-Ukaj, S.; Loncaric, I.; Klang, A.; Spergser, J.; Häbich, A.-C.; Knotek, Z. Infection with *Devriesea agamarum* and *Chrysosporium guarroi* in an inland bearded dragon (*Pogona vitticeps*). *Vet. Dermatol.* **2014**, *25*, 555–558, e97. [CrossRef]
9. Gallego, M.; Juan-Sallés, C.; Hellebuyck, T. *Devriesea agamarum* associated cheilitis in a North African spiny-tailed lizard (*Uromastyx acanthinura*) in Spain. *Open Vet. J.* **2018**, *8*, 224–228. [CrossRef]
10. Hedley, J.; Whitehead, M.L.; Munns, C.; Pellett, S.; Abou-Zahr, T.; Calvo Carrasco, D.; Wissink-Argilaga, N. Antibiotic stewardship for reptiles. *J. Small Anim. Pract.* **2021**, *62*, 829–839. [CrossRef]
11. Martel, A.; Pasmans, F.; Hellebuyck, T.; Haesebrouck, F.; Vandamme, P. *Devriesea agamarum* gen. nov., sp. nov., a novel actinobacterium associated with dermatitis and septicaemia in agamid lizards. *Int. J. Syst. Evol. Microbiol.* **2008**, *58*, 2206–2209. [CrossRef] [PubMed]
12. Hellebuyck, T.; Pasmans, F.; Haesebrouck, F.; Martel, A. Designing a successful antimicrobial treatment against *Devriesea agamarum* infections in lizards. *Vet. Microbiol.* **2009**, *139*, 189–192. [CrossRef] [PubMed]
13. Hellebuyck, T.; Pasmans, F.; Blooi, M.; Haesebrouck, F.; Martel, A. Prolonged environmental persistence requires efficient disinfection procedures to control *Devriesea agamarum*-associated disease in lizards. *Lett. Appl. Microbiol.* **2011**, *52*, 2[illegible]–32. [CrossRef] [PubMed]
14. Hellebuyck, T.; van Steendam, K.; Deforce, D.; Blooi, M.; van Nieuwerburgh, F.; Bullaert, E.; Ducatelle, R.; Haesebrouck, F.; Pasmans, F.; Martel, A. Autovaccination confers protection against *Devriesea agamarum* associated septicemia but not dermatitis in bearded dragons (*Pogona vitticeps*). *PLoS ONE* **2014**, *9*, e113084. [CrossRef] [PubMed]
15. Brockmann, M.; Aupperle-Lellbach, H.; Gentil, M.; Heusinger, A.; Müller, E.; Marschang, R.E.; Pees, M. Challenges in microbiological identification of aerobic bacteria isolated from the skin of reptiles. *PLoS ONE* **2020**, *15*, e0240085. [CrossRef] [PubMed]
16. Bauwens, L.; Vercammen, F.; Hendrickx, F.; Pasmans, F.; Martel, A. Prevalence of *Devriesea agamarum* in the lizard collection of The Royal Zoological Society of Antwerp. *J. Zoo Aquar. Res.* **2014**, *2*, 88.

17. Brockmann, M.; Aupperle-Lellbach, H.; Müller, E.; Heusinger, A.; Pees, M.; Marschang, R.E. Aerobes Keimspektrum und Resistenzlage bei Hautläsionen von Reptilien. *Tierarztl. Prax. Ausg. K Kleintiere. Heimtiere.* **2020**, *48*, 78–88. [CrossRef]
18. Pasmans, F.; Martel, A.; Jacobson, E.R. Bacterial Diseases of Reptiles. In *Infectious Diseases and Pathology of Reptiles: Color Atlas and Text*; Jacobson, E.R., Ed.; CRC Press: Boca Raton, FL, USA, 2021; pp. 705–794.
19. Marschang, R.E.; Origgi, F.C.; Stenglein, M.D.; Hyndman, T.H.; Wellehan, J.F.; Jacobson, E.R. Viruses and Viral Diseases of Reptiles. In *Infectious Diseases and Pathology of Reptiles: Color Atlas and Text*; Jacobson, E.R., Ed.; CRC Press: Boca Raton, FL, USA, 2021; pp. 575–703.
20. Paré, J.A.; Conley, K.J. Mycotic Diseases of Reptiles. In *Infectious Diseases and Pathology of Reptiles: Color Atlas and Text*; Jacobson, E.R., Ed.; CRC Press: Boca Raton, FL, USA, 2021; pp. 795–857.
21. Young, G.; Turner, S.; Davies, J.K.; Sundqvist, G.; Figdor, D. Bacterial DNA persists for extended periods after cell death. *J. Endod.* **2007**, *33*, 1417–1420. [CrossRef]
22. Bayram, L.C.; Abay, S.; Saticioglu, İ.B.; Güvenc, T.; Ekebas, G.; Aydi, F. Panophthalmitis in a Gentoo Penguin (*Pygoscelis Papua*) from the Antarctic Peninsula: Evaluation of Microbiological and Histopathological Analysis Outcomes. *Res. Sq.* **2021**, Preprint. [CrossRef]
23. Swaney, M.H.; Nelsen, A.; Sandstrom, S.; Kalan, L.R. Sweat and sebum preferences of the human skin microbiota. *Microbiol. Spectr.* **2023**, *11*, e04180-22. [CrossRef]
24. Gómez-Garcés, J.L.; Oteo, J.; García, G.; Aracil, B.; Alós, J.I.; Funke, G. Bacteremia by *Dermabacter hominis*, a rare pathogen. *J. Clin. Microbiol.* **2001**, *39*, 2356–2357. [CrossRef] [PubMed]
25. Lee, H.-J.; Cho, C.-H.; Kwon, M.-J.; Nam, M.-H.; Lee, K.-N.; Lee, C.-K. A Patient with Fatal Septicemia Caused by a Rare Pathogen *Dermabacter Hominis*. *Infect Chemother* **2011**, *43*, 86. [CrossRef]
26. Fernández-Natal, I.; Sáez-Nieto, J.A.; Medina-Pascual, M.J.; Albersmeier, A.; Valdezate, S.; Guerra-Laso, J.M.; Rodríguez, H.; Marrodán, T.; Parras, T.; Tauch, A.; et al. *Dermabacter hominis*: A usually daptomycin-resistant gram-positive organism infrequently isolated from human clinical samples. *New Microbes New Infect.* **2013**, *1*, 35–40. [CrossRef] [PubMed]
27. Schaub, C.; Dräger, S.; Hinic, V.; Bassetti, S.; Frei, R.; Osthoff, M. Relevance of *Dermabacter hominis* isolated from clinical samples, 2012–2016: A retrospective case series. *Diagn. Microbiol. Infect. Dis.* **2020**, *98*, 115118. [CrossRef]

animals

MDPI

Article

Antibodies against Two Testudinid Herpesviruses in Pet Tortoises in Europe

Christoph Leineweber and Rachel E. Marschang *

Laboklin GmbH & Co. KG, Steubenstr. 4, 97688 Bad Kissingen, Germany
* Correspondence: rachel.marschang@gmail.com; Tel.: +49-97172020

Simple Summary: Herpesviruses are important pathogens in tortoises and cause latent infections. Serological testing is therefore an important tool for the detection of herpesvirus-infected tortoises. This retrospective study describes the detection of antibodies against two herpesviruses in pet tortoises in Europe. Of the 1728 samples tested, antibodies against one or both of the viruses used were detected in 122 (7.06%) of the tortoises. Detection rates differed depending on virus type, host species, and year of sampling. For individual viruses, detection rates also differed depending on season and country of origin. A better understanding of both the herpesviruses' prevalences and the immune response to infection will help protect these animals in future.

Abstract: Herpesviruses are important pathogens of tortoises, and several serologically and genetically distinct virus types have been described in these animals. Virus neutralization testing is commonly used in Europe to determine previous infection with the two types most often found in pet European tortoises, testudinid herpesvirus (TeHV) 1 and 3. In this retrospective study, the results of serological testing for antibodies against each of these viruses in serum or plasma samples from 1728 tortoises were evaluated, and antibody detection rates were compared based on virus type, host species, year, season, and country of origin. Antibodies (titer 2 or higher) against at least one of the two viruses used were detected in a total of 122 (7.06%; 95% CI 5.95–8.37%) of the animals tested. The antibody detection rates differed significantly depending on the tortoise species ($p < 0.0001$) and the year of sampling (TeHV1 $p = 0.0402$; TeHV3 $p = 0.0482$) for both virus types. For TeHV1, antibody detection rates differed significantly ($p = 0.0384$) by season. The highest detection rate was in summer (5.59%; 95% CI 4.10–7.58%), and the lowest was in fall (1.25%; CI 0.53–2.87%). TeHV1 antibody detection rates did not differ significantly ($p = 0.7805$) by country, whereas TeHV3 antibody detection rates did ($p = 0.0090$). The highest detection rate, 12.94% (95% CI 7.38–21.70%), was found in samples from Italy. These results support previous hypotheses on the species' susceptibility to TeHV1 and 3 and the use of serology as a diagnostic test for the detection of herpesvirus-infected tortoises.

Keywords: tortoise; *Testudo*; testudinid herpesvirus; virus neutralization; serology

Citation: Leineweber, C.; Marschang, R.E. Antibodies against Two Testudinid Herpesviruses in Pet Tortoises in Europe. *Animals* **2022**, *12*, 2298. https://doi.org/10.3390/ani12172298

Academic Editor: Mark Mitchell

Received: 27 July 2022
Accepted: 1 September 2022
Published: 5 September 2022

1. Introduction

Herpesviruses have often been described in tortoises and can be associated with severe disease, most often of the upper digestive and upper respiratory tract [1]. These viruses are considered a significant threat to the health of pet tortoises and can cause significant disease outbreaks with high morbidity and mortality rates [1]. So far, four different herpesviruses have been described in tortoises. Testudinid herpesvirus 1 (TeHV1) was first described in association with a disease outbreak among Russian (*Testudo* [*Agrionemys*] *horsfieldii*) and pancake tortoises (*Malacochersus tornieri*) in Japan [2]. It has also been described in pet tortoises in Europe repeatedly [3–5]. Testudinid herpesvirus 2 (TeHV2) has been described in Agassiz's desert tortoises (*Gopherus agassizii*) in the USA [6] and has also been reported in a single case in a Texas tortoise (*Gopherus berlandieri*) kept in Spain [4]. Testudinid herpesvirus 3 (TeHV3) is in the species *Testudinid alphaherpesvirus 3*, the only tortoise herpesvirus species classified by the international committee on taxonomy of viruses (ICTV). TeHV3 has been found in a wide range of species in Europe and in other parts of the world [1,4]. Testudinid herpesvirus 4 (TeHV4) has so far only been detected in African species in the USA and in Europe [4,7,8]. All of these viruses have been shown to cluster within the subfamily *Alphaherpesvirinae* together with other chelonian herpesviruses in the genus *Scutavirus* [7,9].

Since herpesviruses cause latent infections, detection of clinically inapparent infected animals that may not be shedding virus is an important diagnostic tool. Serological testing has been described for the detection of antibodies against TeHV1, 2, and 3. The virus for which the most data on serological testing have been published is TeHV3. This virus has been used in neutralization tests and in enzyme-linked immunosorbent assays (ELISAs) for the detection of antibodies. TeHV3 isolated from a Hermann's tortoise (*Testudo hermanni*) in terrapene heart cells (TH-1) has been used in a number of studies for the detection of antibodies in several tortoise species [10–12]. An ELISA has also been described for the detection of antibodies against TeHV3 in Mediterranean tortoises [13]. A study comparing herpesviruses isolated from tortoises showed that the two types included in the study (later shown to be TeHV3 and a single TeHV1 isolate) did not cross react serologically [14], demonstrating the necessity of using the virus type of interest for serological testing. In contrast, a study using an ELISA with a TeHV3-specific antigen for the detection of antibodies in wild Agassiz's desert tortoises revealed a high (30.9%) antibody prevalence [15]. The authors hypothesized that this could be due to cross-reactivity between TeHV3 and TeHV2, since only TeHV2 was found in tortoises from the same population in virus detection studies. So far, only TeHV1 and TeHV3 have been isolated in cell cultures and are available for serological testing. These are also the two virus types most often found in pet tortoises in Europe [3,4].

The aim of this retrospective study was to provide an overview of serological testing for antibodies against TeHV1 and TeHV3 in pet tortoises in Europe. It was hypothesized that antibody detection rates would differ significantly depending on the host species and virus type and that these rates would correlate to polymerase chain reaction (PCR)-based virus detection rates previously reported in pet tortoises in Europe according to host species, virus type, country of origin, and season.

2. Materials and Methods

Plasma and serum samples submitted to a commercial veterinary laboratory (Laboklin GmbH & Co. KG, Bad Kissingen, Germany) between January 2016 and December 2020 were evaluated in this retrospective study. Samples were submitted for diagnostic testing by veterinarians and owners. Reasons for testing were not provided. All samples were heat treated at 56 °C for 30 min for compliment inactivation and stored at 4 °C for up to 5 days before testing. Neutralizing antibodies against a TeHV1 isolate from a Russian tortoise (1301/B99R/97, GenBank DQ343883.1) and a TeHV3 isolate from a Hermann's tortoise (4295, [16]) in TH-1 were detected as described previously [10,11], except that plates for both antibodies were read after 7 days of incubation at 28 °C in order to lower turn-around time. Plasma from Hermann's tortoises that had previously tested negative for both TeHV1 and TeHV3 antibodies were used as negative controls; plasma from Russian tortoises that had previously tested positive for antibodies against TeHV1 but not TeHV3 was diluted to a titer of 8 and used as a positive control for TeHV1 antibody testing; and plasma from spur-thighed tortoises (*Testudo graeca*) that had previously tested positive for a antibodies against TeHV3 but not TeHV1 was diluted to a titer of 8 and used as a positive control for TeHV3 antibody testing. Titers between 2 and 4 were considered suspect positive, titers of 8 low positive and titers of 16 and greater were considered positive. Plasma or serum dilutions associated with cytotoxicity were not evaluated, and only higher dilutions in which cytotoxicity was no longer observed were included. If no neutralization was detected in these higher dilutions, the sample was considered negative. Tests in which virus titration controls or positive or negative plasma controls did not provide the expected results were not evaluated, and samples were retested.

The statistical analyses were carried out using the statistical analysis software (SAS) (SAS Institute, Cary, NC, USA) for the calculation of the positivity rates. The 95% binomial confidence intervals were calculated based on the Wilson procedure [17]. Binary logistic regression was used with cut-off for significance of $p \leq 0.05$ for comparisons of the positivity rates between the factors species, year, season, and country of sample origin. The seasons were divided as follows: March to May were classified as spring, June to August as summer, September to November as fall, and December to February as winter samples.

3. Results

A total of 1728 samples from tortoises were evaluated. Antibodies (titer 2 or higher) against TeHV1 only were detected in 41 samples (2.37%; 95% CI 1.75–3.20%), and against TeHV3 only in 62 samples (3.59%; 95% CI 2.81–4.58%). Antibodies against both viruses were detected in samples from 19 animals (1.1%; 95% CI 0.71–1.71%), including one Hermann's tortoise, five spur-thighed tortoises, three marginated tortoises (*Testudo marginata*), four Russian tortoises, two radiated tortoises (*Astrochelys radiata*), one leopard tortoise (*Stigmochelys pardalis*), and three tortoises for which the species was not provided. A total of 122 (7.06%; 95% CI 5.95–8.37%) of the animals tested had detectable antibodies against at least one of the herpesviruses used. The antibody detection rates differed significantly ($p < 0.0001$) for TeHV1 and TeHV3 (Table 1) depending on the tortoise species (Figure 1). There were significant differences (TeHV1 $p = 0.0402$; TeHV3 $p = 0.0482$) between years of sampling (Table 2). For TeHV1, antibody detection rates differed significantly ($p = 0.0384$) by season. The highest detection rate was in summer (5.59%; 95% CI 4.10-7.58%), and the lowest was in fall (1.25%; CI 0.53–2.87%) (Table 3). The detection rates for antibodies against TeHV3 did not differ significantly ($p = 0.2617$) by season (Table 3). TeHV1 antibody detection rates did not differ significantly ($p = 0.7805$) (Figure 3, Supplementary Table S1) by country. In contrast, TeHV3 antibody detection rates did differ significantly by country ($p = 0.0090$): no antibodies were detected in samples from Spain, Austria, Belgium, Luxemburg, Czech Republic, Poland, or Norway; and the highest detection rate, 12.94% (95% CI 7.38–21.70%), was found in samples from Italy (Figure 2, Supplementary Table S1).

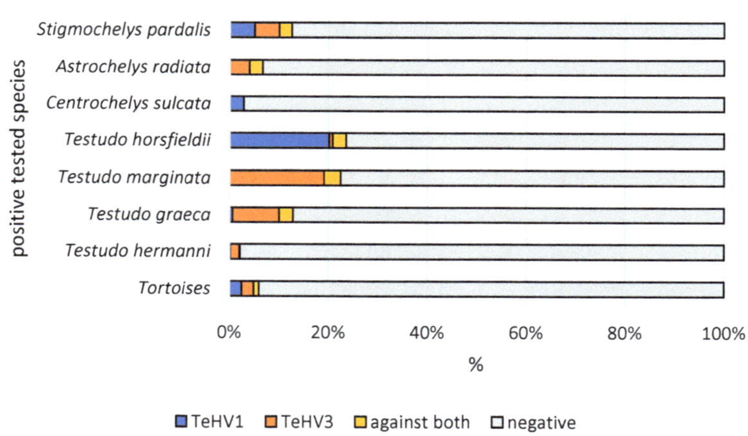

Figure 1. Herpesvirus (TeHV1 and TeHV3) antibody detection rates (titer 2 or higher) of the positive tested species (Hermann's tortoises, *Testudo hermanni*; spur-thighed tortoises, *Testudo graeca*; marginated tortoises, *Testudo marginata*; Russian tortoises, *Testudo horsfieldii*; African spurred tortoise, *Centrochelys sulcata*; radiated tortoise, *Astrochelys radiata*; and leopard tortoise, *Stigmochelys pardalis*).

Table 1. Detection rates of antibodies against testudinid herpesvirus 1 (TeHV1) and testudinid herpesvirus 3 (TeHV3) depending on species tested.

Species	Total Tested		TeHV1 Antibodies				TeHV3 Antibodies			
			Titer < 2	Titer 2 to 4	Titer 8	Titer ≥ 16	Titer < 2	Titer 2 to 4	Titer 8	Titer ≥ 16
Tortoise, species unknown	285	n	275	3	3	4	275	1	5	4
		%	96.50	1.05	1.05	1.40	96.50	0.35	1.75	1.40
		CI	93.66–98.08	0.36–3.04	0.36–3.04	0.55–3.55	93.66–98.08	0.06–1.96	0.75–4.03	0.55–3.55
Testudo hermanni	797	n	795	2	0	0	781	10	2	4
		%	99.75	0.25	0	0	98.00	1.25	0.25	0.50
		CI	99.09–99.93	0.07–0.91	0–0.48	0–0.48	96.76–98.76	0.68–2.29	0.07–0.91	0.19–1.28
Testudo graeca	180	n	174	3	1	2	158	9	3	10
		%	96.66	1.67	0.56	1.11	87.77	5.00	1.67	5.56
		CI	92.92–98.47	0.57–4.79	0.10–3.09	0.30–3.96	82.19–91.79	2.65–9.23	0.57–4.79	3.05–9.93
Testudo marginata	89	n	86	3	0	0	69	11	1	8
		%	96.63	3.37	0	0	77.53	12.36	1.12	8.99
		CI	90.55–98.85	1.15–9.45	0–4.14	0–4.14	67.82–84.96	7.04–20.79	0.02–6.09	4.63–16.75
Testudo horsfieldii	144	n	111	11	6	16	139	3	1	1
		%	77.08	7.64	4.17	11.11	96.54	2.08	0.69	0.69
		CI	69.56–8.19	4.32–13.16	1.93–8.80	6.96–17.29	92.13–98.51	0.71–5.94	0.12–3.82	0.12–3.82
Testudo kleinmanni	5	n	5	0	0	0	5	0	0	0
		%	100	0	0	0	100	0	0	0
		CI	56.55–100	0–43.45	0–43.45	0–43.45	56.55–100	0–43.45	0–43.45	0–43.45
Centrochelys sulcata	35	n	34	1	0	0	35	0	0	0
		%	97.14	2.86	0	0	100	0	0	0
		CI	85.46–99.49	0.51–14.54	0–9.89	0–9.89	90.11–100	0–9.89	0–9.89	0–9.89
Astrochelys radiata	75	n	73	1	0	1	70	3	1	1
		%	97.34	1.33	0	1.33	93.34	4.00	1.33	1.33
		CI	90.78–99.26	0.23–7.17	0–4.87	0.23–7.17	85.32–97.12	1.37–11.11	0.23–7.17	0.23–7.17
Stigmochelys pardalis	40	n	37	0	1	2	37	0	1	2
		%	92.50	0	2.50	5.00	92.50	0	2.50	5.00
		CI	80.14–97.42	0–8.76	0.44–12.88	1.38–16.50	80.14–97.42	0–8.76	0.44–12.88	1.38–16.50
Aldabrachelys gigantea	12	n	12	0	0	0	12	0	0	0
		%	100	0	0	0	100	0	0	0
		CI	75.75–100	0–24.25	0–24.25	0–24.25	75.75–100	0–24.25	0–24.25	0–24.25
Chelonoidis nigra	2	n	2	0	0	0	2	0	0	0
		%	100	0	0	0	100	0	0	0
		CI	34.24–100	0–65.76	0–65.76	0–65.76	34.24–100	0–65.76	0–65.76	0–65.76
Chelonoidis carbonarius	14	n	14	0	0	0	14	0	0	0
		%	100	0	0	0	100	0	0	0
		CI	78.47–100	0–21.53	0–21.53	0–21.53	78.47–100	0–21.53	0–21.53	0–21.53
Chelonoidis denticulatus	3	n	3	0	0	0	3	0	0	0
		%	100	0	0	0	100	0	0	0
		CI	43.85–100	0–56.15	0–56.15	0–56.15	43.85–100	0–56.15	0–56.15	0–56.15
Geochelone elegans	9	n	9	0	0	0	9	0	0	0
		%	100	0	0	0	100	0	0	0
		CI	70.09–100	0–29.91	0–29.91	0–29.91	70.09–100	0–29.91	0–29.91	0–29.91

Table 1. *Cont.*

Species	Total Tested		TeHV1 Antibodies				TeHV3 Antibodies			
			Titer < 2	Titer 2 to 4	Titer 8	Titer ≥ 16	Titer < 2	Titer 2 to 4	Titer 8	Titer ≥ 16
Geochelone platynota	7	n	7	0	0	0	7	0	0	0
		%	100	0	0	0	100	0	0	0
		CI	64.57–100	0–35.43	0–35.43	0–35.43	64.57–100	0–35.43	0–35.43	0–35.43
Astrochelys yniphora	5	n	5	0	0	0	5	0	0	0
		%	100	0	0	0	100	0	0	0
		CI	56.55–100	0–43.45	0–43.45	0–43.45	56.55–100	0–43.45	0–43.45	0–43.45
Kinixys sp.	3	n	3	0	0	0	3	0	0	0
		%	100	0	0	0	100	0	0	0
		CI	43.85–100	0–56.15	0–56.15	0–56.15	43.85–100	0–56.15	0–56.15	0–56.15
Indotestudo elongata	4	n	4	0	0	0	4	0	0	0
		%	100	0	0	0	100	0	0	0
		CI	51.01–100	0–48.99	0–48.99	0–48.99	51.01–100	0–48.99	0–48.99	0–48.99
Gopherus berlandieri	1	n	1	0	0	0	1	0	0	0
		%	100	0	0	0	100	0	0	0
		CI	20.65–100	0–79.35	0–79.35	0–79.35	20.65–100	0–79.35	0–79.35	0–79.35
Malacochersus tornieri	6	n	6	0	0	0	6	0	0	0
		%	100	0	0	0	100	0	0	0
		CI	60.97–100	0–39.03	0–39.03	0–39.03	60.97–100	0–39.03	0–39.03	0–39.03
Homopus sp.	1	n	1	0	0	0	1	0	0	0
		%	100	0	0	0	100	0	0	0
		CI	20.65–100	0–79.35	0–79.35	0–79.35	20.65–100	0–79.35	0–79.35	0–79.35
Manouria sp.	1	n	1	0	0	0	1	0	0	0
		%	100	0	0	0	100	0	0	0
		CI	20.65–100	0–79.35	0–79.35	0–79.35	20.65–100	0–79.35	0–79.35	0–79.35
Pyxis sp.	6	n	6	0	0	0	6	0	0	0
		%	100	0	0	0	100	0	0	0
		CI	60.97–100	0–39.03	0–39.03	0–39.03	60.97–100	0–39.03	0–39.03	0–39.03
Chersina angulata	2	n	2	0	0	0	2	0	0	0
		%	100	0	0	0	100	0	0	0
		CI	34.24–100	0–65.76	0–65.76	0–65.76	34.24–100	0–65.76	0–65.76	0–65.76
Psammobates sp.	2	n	2	0	0	0	2	0	0	0
		%	100	0	0	0	100	0	0	0
		CI	34.24–100	0–65.76	0–65.76	0–65.76	34.24–100	0–65.76	0–65.76	0–65.76
Total	1728	n	1668	24	11	25	1647	37	14	30
		%	96.53	1.39	0.64	1.45	95.31	2.14	0.81	1.74
		CI	95.56–97.29	0.94–2.06	0.36–1.14	0.98–2.13	94.21–96.21	1.56–2.94	0.48–1.36	1.22–2.47

n: number in each category; CI: 95% confidence interval.

Evaluation of the results from samples from the five *Testudo* species (n = 1215) also showed some differences. There was a significant difference in antibody detection rates between the individual *Testudo* species for both virus types ($p < 0.0001$) (Table 1; Figure 1). No significant difference was found between the years of sampling for TeHV1 ($p = 0.5328$), but the differences were significant for TeHV3 ($p = 0.0377$) (Figure 4). Detection rates were found to differ significantly between seasons for TeHV1 ($p = 0.0425$) but not for TeHV3 ($p = 0.6922$) (Figure 5). For this subset of animals, as for all of the tortoises combined, country of sample origin did not significantly impact antibody detection rate for TeHV1 ($p = 0.4751$), whereas significant ($p < 0.0001$) differences were detected in detection rates for antibodies against TeHV3 based on country. The highest positivity rate (21.15%; 95% CI 12.24–34.03%) was found in Italy (Figure 6).

Table 2. Detection rates of antibodies against testudinid herpesvirus 1 (TeHV1) and testudinid herpesvirus 3 (TeHV3) depending on year of sampling. Results shown as: number (n), percent (%), 95% confidence interval (CI).

Year	Total Tested		TeHV1 Antibodies				TeHV3 Antibodies			
			Titer < 2	Titer 2 to 4	Titer 8	Titer ≥ 16	Titer < 2	Titer 2 to 4	Titer 8	Titer ≥ 16
2016	305	n	299	2	1	3	285	5	5	10
		%	98.03	0.66	0.33	0.98	93.44	1.64	1.64	3.28
		CI	95.77–99.09	0.18–2.37	0.06–1.84	0.33–2.85	90.09–95.71	0.70–3.78	0.70–3.78	1.79–5.93
2017	303	n	292	4	3	4	290	6	2	5
		%	96.37	1.32	0.99	1.32	95.71	1.98	0.66	1.65
		CI	93.62–97.96	0.51–3.34	0.34–2.87	0.51–3.34	92.80–97.48	0.91–4.25	0.18–2.70	0.71–3.80
2018	381	n	377	1	1	2	369	6	0	6
		%	98.96	0.26	0.26	0.52	96.86	1.57	0.0	1.57
		CI	97.33–99.59	0.05–1.47	0.05–1.47	0.14–1.89	94.58–98.19	0.72–3.39	0–1.00	0.72–3.39
2019	350	n	331	9	4	6	326	13	5	6
		%	94.58	2.57	1.14	1.71	93.15	3.71	1.43	1.71
		CI	91.68–96.50	1.36–4.81	0.44–2.90	0.79–3.68	90.00–95.35	2.18–6.24	0.61–3.30	0.79–3.68
2020	389	n	369	8	2	10	377	7	2	3
		%	94.86	2.06	0.51	2.57	96.92	1.80	0.51	0.77
		CI	92.19–96.65	1.05–4.01	0.14–1.85	1.40–4.67	94.69–98.23	0.87–3.67	0.14–1.85	0.26–2.24
Total	1728		1668	24	11	25	1647	37	14	30

Table 3. Detection rates of antibodies against testudinid herpesvirus 1 (TeHV1) and testudinid herpesvirus 3 (TeHV3) depending on season of sampling. Results shown as: number, percent, 95% confidence interval (CI).

Season	Total Tested		TeHV1 Antibodies				TeHV3 Antibodies			
			Titer < 2	Titer 2 to 4	Titer 8	Titer ≥ 16	Titer < 2	Titer 2 to 4	Titer 8	Titer ≥ 16
Spring	529	n	515	5	5	4	505	9	6	9
		%	97.34	0.95	0.95	0.76	95.47	1.70	1.13	1.70
		CI	95.60–98.41	0.41–2.20	0.41–2.20	0.30–1.93	93.33–96.93	0.90–3.20	0.52–2.45	0.90–3.20
Summer	680	n	642	17	6	15	642	17	5	16
		%	94.41	2.50	0.88	2.21	94.41	2.50	0.74	2.35
		CI	92.42–95.90	1.57–3.97	0.40–1.91	1.34–3.61	92.42–95.90	1.57–3.97	0.32–1.72	1.45–3.78
Fall	402	n	397	2	0	3	384	11	3	4
		%	98.75	0.50	0	0.75	95.51	2.74	0.75	1.00
		CI	97.13–99.47	0.14–1.80	0–0.95	0.26–2.18	93.03–97.15	1.54–4.84	0.26–2.18	0.39–2.54
Winter	117	n	114	0	0	3	116	0	0	1
		%	97.44	0	0	2.56	99.15	0	0	0.85
		CI	92.74–99.13	0–3.18	0–3.18	0.87–7.26	95.32–99.85	0–3.18	0–3.18	0.15–4.68
Total	1728		1668	24	11	25	1647	37	14	30

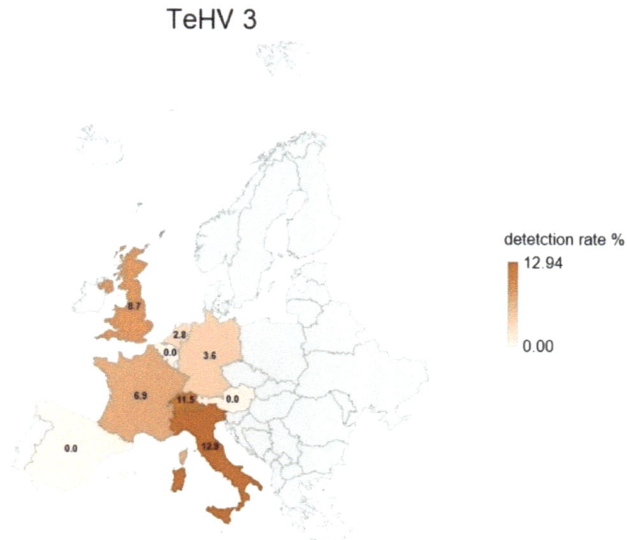

Figure 2. Herpesvirus (TeHV3) antibody detection rates (titer 2 or higher) divided into the countries of origin with more than 20 tested samples. Countries with lower numbers of samples submitted are shown in grey.

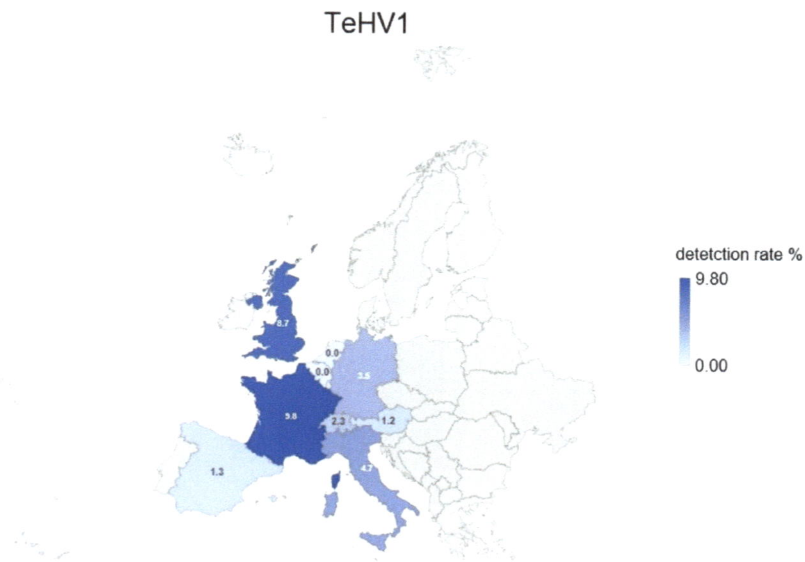

Figure 3. Herpesvirus (TeHV1) antibody detection rates (titer 2 or higher) divided into the countries of origin with more than 20 tested samples. Countries with lower numbers of samples submitted are shown in grey.

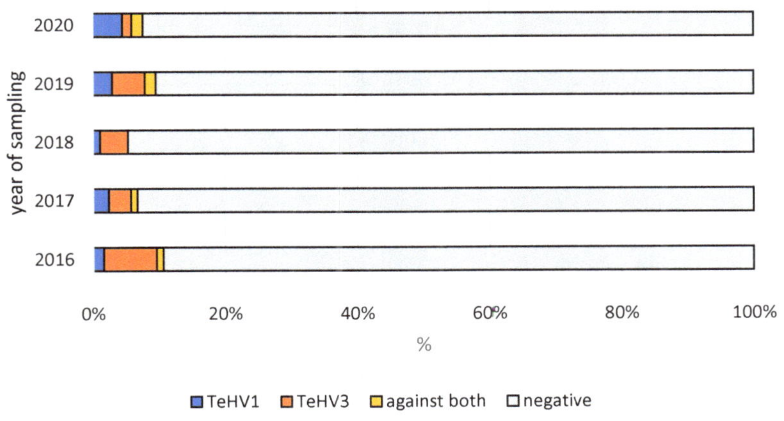

Figure 4. Herpesvirus (TeHV1 and TeHV3) antibody detection rates (titer 2 or higher) in five *Testudo* species (Hermann's tortoises, *Testudo hermanni*; spur-thighed tortoises, *Testudo graeca*; marginated tortoises, *Testudo marginata*; Russian tortoises, *Testudo horsfieldii*; and Egyptian tortoises, *Testudo kleinmanni*) for the different years of sampling.

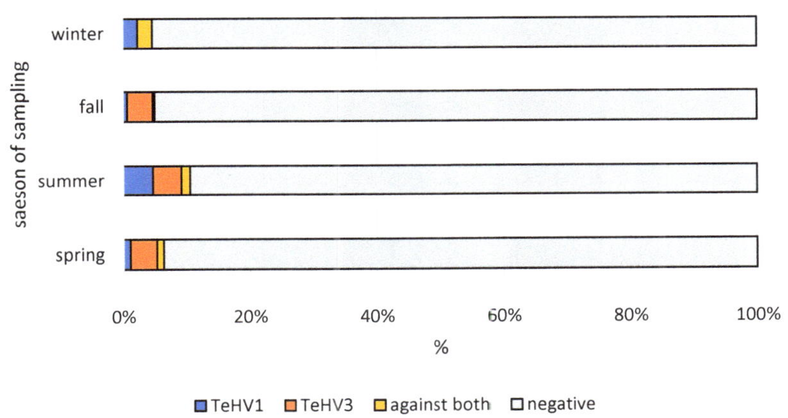

Figure 5. Herpesvirus (TeHV1 and TeHV3) antibody detection rates (titer 2 or higher) in five *Testudo* species (Hermann's tortoises, *Testudo hermanni*; spur-thighed tortoises, *Testudo graeca*; marginated tortoises, *Testudo marginata*; Russian tortoises, *Testudo horsfieldii*; and Egyptian tortoises, *Testudo kleinmanni*) for the different seasons of sampling.

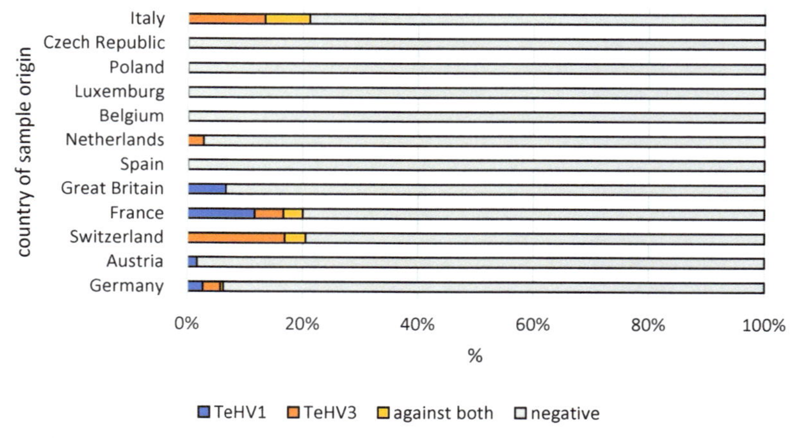

Figure 6. Herpesvirus (TeHV1 and TeHV3) antibody detection rates (titer 2 or higher) in five *Testudo* species (Hermann's tortoises, *Testudo hermanni*; spur-thighed tortoises, *Testudo graeca*; marginated tortoises, *Testudo marginata*; Russian tortoises, *Testudo horsfieldii*; and Egyptian tortoises, *Testudo kleinmanni*) for the different countries of origin.

4. Discussion

Serology is an important tool in the diagnosis of herpesvirus infections in tortoises. It is also an important part of quarantine examinations in these animals. Since infected animals remain carriers, recognition of previous infection is a key to keeping infection out of a naïve group. A number of instances of herpesvirus introductions into previously negative collections followed by severe disease outbreaks have been documented [5,10,18] and both virus detection and serology have been used to detect herpesvirus infection in clinically healthy tortoises [5,18,19].

The antibody detection rate found here is similar to that found in a study reporting on PCR-based herpesvirus detection in chelonians in Europe over the same period [4]. In that study, an overall positivity rate of 6.22% was reported for chelonians in the family Testudinidae [4]. The vast majority of the herpesviruses detected in tortoises in that study were either TeHV1 (45.37%) or TeHV3 (53.24%). Studies comparing both virus detection and serology in the same animals have often shown higher virus detection rates than antibody detection rates [10,18,19]. In those cases in which samples were collected during an outbreak [10], it is possible that some animals had not yet developed a measurable immune response following infection. Both virus shedding and antibody detection have been reported in clinically healthy tortoises with no history of a recent disease outbreak [19]. In a study in which tortoises were repeatedly tested for the presence of antibodies over time, it was shown that animals that tested positive could also sporadically test negative [12]. In the present study, only serological testing, not virus detection, was carried out, so that direct comparison of the two could not be performed.

Host species has been shown to be a significant factor in infection rate for both TeHV1 and 3. TeHV1 has most often been detected in Russian tortoises, and this species is hypothesized to be the original host of this virus, although numerous other tortoise species have also been shown to be susceptible to infection [2–4,20], and infection has been associated with disease in several of these [2,20]. TeHV3 has been detected in a wide range of tortoise species [4]. It has been hypothesized that this virus may have co-evolved in spur-thighed tortoises [21], and this species and the closely related marginated tortoises have been reported to be more resistant to disease than some other species, e.g., Hermann's tortoises, in outbreaks in mixed collections [10,18]. These host specificities are reflected in the antibody detection rates reported here. The highest rate of antibodies against TeHV1 was found in Russian tortoises. Distribution of the titers also showed a higher percentage of animals with titers > 16 in this species (Table 1, Figure 1). For TeHV3, the highest percentage of antibody positive animals were found for spur-thighed and marginated tortoises (Table 1, Figure 1). Previously published PCR-based virus detection rates in different animals in Europe over the same time period [4] revealed positivity rates that are comparable to the antibody detection rates found in this study. However, there were significant differences in PCR-based virus detection and antibody detection rates in Hermann's tortoises between the two studies. A total of 4.66% of the samples from Hermann's tortoises tested by PCR for herpesviruses (50 of 1072) were virus positive, the vast majority of which were TeHV3 [4]. In contrast, in the present study, antibodies against one of the viruses used were found in a significantly lower proportion of animals (2.25%, 18 of 797) (p = 0.0059), with extremely low titers of 2 or 4 in the majority of cases (12 of 18) (Table 1). While this is based on a different set of animals, so that direct comparison is not possible, it does support previous observations indicating that Hermann's tortoises may be less able to mount a strong antibody response to infection with TeHV3 than, e.g., spur-thighed and marginated tortoises. This may help explain the apparent increased pathogenicity of these viruses in this species, although the immune response to herpesvirus infection in tortoises is not yet well understood and requires further study.

In several of the animals included in this study, antibodies were detected against both of the viruses used. Dual infection with TeHV1 and TeHV3 has been reported previously [3]. A study using serology for the detection of antibodies against a range of viruses in wild-caught spur-thighed tortoises in Turkey indicated that TeHV1 and 3 were co-circulating in that group [11]. Although TeHV1 and 3 have not been found to cross react serologically, it is possible that some of the antibodies detected were specific for other, similar antigens. It is important to note that virus neutralization testing with TeHV1 and 3 has not been validated for all of the species included in the present study and that some of the animals included could have been previously exposed to other herpesviruses not included in this study, with unknown cross reactivity to TeHV1 and 3. However, the use of a virus neutralization test indicates that the antibodies detected would have a functional use in the tortoises. The actual effect of neutralizing antibodies on disease development in herpesvirus infected tortoises has not, however, been sufficiently studied. In the present study, no information was provided on the clinical status of the animals tested

Comparison of other factors that might influence virus and antibody detection rates between this study and the previous study on PCR-based virus detection [4] showed several other differences as well. Both PCR-based virus detection and antibody detection rates varied significantly depending on the year, but the years in which detection was highest for the two viruses differed between the two studies. Antibody detection rates may reflect a longer-term prevalence of these viruses in pet tortoises in Europe, and the results of the present study indicate that the immune response is subject to the same variations as virus detection by PCR. However, it is important to note that the samples used in both of these studies were submitted by veterinary practices from animals with unknown clinical status and histories, so that neither can fully reflect the true prevalence of herpesvirus infection in these animals.

Seasonality was also shown to influence PCR-based virus detection rates. The highest detection rates were found in the spring for both viruses [4]. In contrast, antibodies against both virus types were most often found in samples submitted in the summer. Since the immune system and antibody production in tortoises are dependent on environmental factors, this may reflect the reactivity of the immune system in summer rather than virus activity.

Both PCR-based virus detection and anti-TeHV3 antibody detection were significantly influenced by the country from which the samples were submitted. In both cases, the highest percentage of positive samples was found in samples from Italy, with most of these being TeHV3 and anti-TeHV3 antibodies [4]. Reasons for the high TeHV3 detection rate in Italy are not known. In the case of antibodies, it is possible that environmental factors could play a role in support of immunological reaction to infection, as the majority of the tortoises tested (Hermann's, spur-thighed, and marginated tortoises) can be found in the wild there.

The reason for testing was not known for the animals included in this study, limiting the interpretation of results. It is likely that the animals sampled were biased toward those with a history of disease, since most were submitted by veterinarians, although submission for quarantine examinations and health checks likely also occurred in many cases.

Future studies exploring true prevalence and dynamics of herpesvirus infection in pet tortoises in Europe would be useful. This is, however, likely to be very difficult to organize, making data from diagnostic samples, as presented here, worth reporting. In addition, it would be helpful to continue studies of the herpesviruses that can infect tortoises. It is likely that not all herpesviruses capable of infecting this group of animals have been discovered yet. Finally, the differences in antibody detection rates between individual species and the possible implications for interactions between host species and specific viruses deserve further study. This should include studies elucidating the immune response to infection in various tortoise species and viral factors that are capable of influencing these reactions.

5. Conclusions

Antibody detection is an important tool in preventing the spread of herpesvirus infections in tortoises. Immune response to infection may, however, depend on several factors including virus type, host species, and environmental factors.

Supplementary Materials: The following supporting information can be downloaded at: https://www.mdpi.com/article/10.3390/ani12172298/s1. Table S1: Detection rates of antibodies against testudinid herpesvirus 1 (TeHV1) and testudinid herpesvirus 3 (TeHV3) depending on country of origin. Results shown as: number, percent, 95% confidence interval (CI).

Author Contributions: Conceptualization, C.L. and R.E.M.; methodology, C.L. and R.E.M.; validation, C.L. and R.E.M.; formal analysis, C.L.; investigation, C.L. and R.E.M.; resources, C.L. and R.E.M.; data curation, C.L.; writing—original draft preparation, R.E.M.; writing—review and editing, C.L. and R.E.M.; visualization, C.L.; supervision, R.E.M.; project administration, C.L. All authors have read and agreed to the published version of the manuscript.

Funding: This research received no external funding.

Institutional Review Board Statement: Not applicable.

Data Availability Statement: The data that support the findings of this study are available from the corresponding author upon reasonable request.

Conflicts of Interest: Both authors are employed by a commercial laboratory. This employment had no role in the design of the study, in the analyses or interpretation of data; in the writing of the manuscript; or in the decision to publish the results.

References

1. Marschang, R.E.; Origgi, F.C.; Stenglein, M.D.; Hyndman, T.H.; Wellehan, J.F.X.; Jacobson, E.R. Viruses and viral diseases of reptiles. In *Infectious Diseases and Pathology of Reptiles, Color Atlas and Text*, 2nd ed.; Jacobson, E.R., Garner, M.M., Eds.; CRC Press: Boca Raton, LA, USA, 2021; Volume 1, pp. 575–703.
2. Une, Y.; Uemura, K.; Nakano, Y.; Kamiie, J.; Ishibashi, T.; Nomura, Y. Herpesvirus infection in tortoises (*Malacochersus tornieri* and *Testudo horsfieldii*). *Vet. Pathol.* **1999**, *36*, 624–627. [CrossRef] [PubMed]
3. Kolesnik, E.; Obiegala, A.; Marschang, R.E. Detection of *Mycoplasma* spp., herpesviruses, topiviruses, and ferlaviruses in samples from chelonians in Europe. *J. Vet. Diagn. Investig.* **2017**, *29*, 820–832. [CrossRef] [PubMed]
4. Leineweber, C.; Müller, E.; Marschang, R.E. Herpesviruses in captive chelonians in Europe between 2016 and 2020. *Front. Vet. Sci.* **2021**, *8*, 733299. [CrossRef] [PubMed]
5. Schüler, L.; Picquet, P.; Leineweber, C.; Dietz, J.; Müller, E.; Marschang, R.E. A testudinid herpesvirus 1 (TeHV1)-associated disease outbreak in a group of Horsfield's tortoises (*Testudo horsfieldii*). *Tierarztl. Prax. Ausg. K Kleintiere Heimtiere* **2021**, *49*, 462–467. [CrossRef] [PubMed]
6. Johnson, A.J.; Pessier, A.P.; Wellehan, J.F.X.; Brown, R.; Jacobson, E.R. Identification of a novel herpesvirus from a California desert tortoise (*Gopherus agassizii*). *J. Vet. Microbiol.* **2005**, *111*, 107–116. [CrossRef] [PubMed]

7. Bicknese, E.J.; Childress, A.L.; Wellehan, J.F., Jr. A novel herpesvirus of the proposed genus Chelonivirus from an asymptomatic bowsprit tortoise (*Chersina angulata*). *J. Zoo Wildl. Med.* **2010**, *41*, 353–358. [CrossRef] [PubMed]
8. Kolesnik, E.; Mittenzwei, F.; Marschang, R.E. Detection of testudinid herpesvirus type 4 in a leopard tortoise (*Stigmochelys pardalis*). *Tierärztl. Prax. Kleintiere* **2016**, *44*, 283–286. [CrossRef] [PubMed]
9. Gatherer, D.; ICTV Herpesvirales Study Group. 18 New Species in the Family Herpesvirdae. 2018. Available online: https://talk.ictvonline.org//taxonomy/p/taxonomyhistory?taxnode_id=201856398 (accessed on 12 September 2019).
10. Marschang, R.E.; Gravendyck, M.; Kaleta, E.F. Herpesviruses in tortoises: Investigations into virus isolation and the treatment of viral stomatitis in *Testudo hermanni* and *T. graeca*. *J. Vet. Med. B* **1997**, *44*, 385–394. [CrossRef] [PubMed]
11. Marschang, R.E.; Schneider, R.M. Antibodies against viruses in wild-caught spur-thighed tortoises (*Testudo graeca*) in Turkey. *Vet. Rec.* **2007**, *161*, 102–103. [CrossRef] [PubMed]
12. Marschang, R.E.; Milde, K.; Bellavista, M. Virus isolation and vaccination of Mediterranean tortoises against a chelonid herpesvirus in a chronically infected population in Italy. *Dtsch. Tierarztl. Wochenschr.* **2001**, *108*, 376–379. [PubMed]
13. Origgi, F.C.; Klein, P.A.; Mathes, K.; Blahak, S.; Marschang, R.E.; Tucker, S.J.; Jacobson, E.R. An enzyme linked immunosorbent assay (ELISA) for detecting herpesvirus exposure in *Mediterranean tortoises*. *J. Clin. Microbiol.* **2001**, *39*, 3156–3163. [CrossRef] [PubMed]
14. Marschang, R.E.; Frost, J.W.; Gravendyck, M.; Kaleta, E.F. Comparison of 16 chelonid herpesviruses by virus neutralization tests and restriction endonuclease digestion of viral DNA. *J. Vet. Med. B* **2001**, *48*, 393–399. [CrossRef] [PubMed]
15. Jacobson, E.R.; Berry, K.H.; Wellehan, J.F.X.; Origgi, F.; Childress, A.L.; Braun, J.; Schrenzel, M.; Yee, J.; Rideout, B. Serologic and molecular evidence for *Testudinid herpesvirus* 2 infection in wild Agassiz's desert tortoises, *Gopherus agassizii*. *J. Wildl. Dis.* **2012**, *48*, 747–757. [CrossRef] [PubMed]
16. Gandar, F.; Wilkie, G.S.; Gatherer, D.; Kerr, K.; Marlier, D.; Diez, M.; Marschang, R.E.; Mast, J.; Dewals, B.G.; Davison, A.J.; et al. The genome of a tortoise herpesvirus (*Testudinid herpesvirus* 3) has a novel structure and contains a large region that is not required for replication in vitro or virulence in vivo. *J. Virol.* **2015**, *89*, 11438–11456. [CrossRef] [PubMed]
17. Wilson, E.B. Probable inference, the law of succession, and statistical inference. *J. Am. Stat. Assoc.* **1927**, *22*, 209–212. [CrossRef]
18. Marenzoni, M.L.; Santoni, L.; Felici, A.; Maresca, C.; Stefanetti, V.; Sforna, M.; Franciosini, M.P.; Proietti, P.C.; Origgi, F.C. Clinical, virological and epidemiological characterization of an outbreak of *Testudinid herpesvirus* 3 in a chelonian captive breeding facility: Lessons learned and first evidence of TeHV3 vertical transmission. *PLoS ONE* **2018**, *13*, e0197169.
19. Martel, A.; Blahak, S.; Vissenaekens, H.; Pasmans, F. Reintroduction of clinically healthy tortoises: The herpesvirus *Torjan horse*. *J. Wildl. Dis.* **2009**, *45*, 218–220. [CrossRef] [PubMed]
20. Stöhr, A.C.; Marschang, R.E. Detection of a tortoise herpesvirus type 1 in a Hermann's tortoise (*Testudo hermanni boettgeri*) in Germany. *J. Herpetol. Med. Surg.* **2010**, *20*, 61–63. [CrossRef]
21. Marschang, R.E. Viruses infecting reptiles. *Viruses* **2011**, *3*, 2087–2126. [CrossRef] [PubMed]

animals

MDPI

Case Report

Concurrent Detection of a Papillomatous Lesion and Sequence Reads Corresponding to a Member of the Family *Adintoviridae* in a Bell's Hinge-Back Tortoise (*Kinixys belliana*)

Johannes Hetterich [1,*], Monica Mirolo [2], Franziska Kaiser [2], Martin Ludlow [2], Wencke Reineking [3], Isabel Zdora [3], Marion Hewicker-Trautwein [3], Albert D. M. E. Osterhaus [2] and Michael Pees [1]

[1] Department of Small Mammal, Reptile and Avian Medicine and Surgery, University of Veterinary Medicine Hannover Foundation, Bünteweg 9, 30559 Hannover, Germany; michael.pees@tiho-hannover.de

[2] Research Center for Emerging Infections and Zoonoses (RIZ), University of Veterinary Medicine Hannover Foundation, Bünteweg 17, 30559 Hannover, Germany; monica.mirolo@tiho-hannover.de (M.M.); martin.ludlow@tiho-hannover.de (M.L.); albert.osterhaus@tiho-hannover.de (A.D.M.E.O.)

[3] Department of Pathology, University of Veterinary Medicine Hannover Foundation, Bünteweg 17, 30559 Hannover, Germany; isabel.zdora@tiho-hannover.de (I.Z.); marion.hewicker-trautwein.ir@tiho-hannover.de (M.H.-T.)

* Correspondence: johannes.hetterich@tiho-hannover.de

Simple Summary: This article describes the diagnostic evaluation and therapy of a tortoise with a bulging oral lesion of unknown origin. Initially, the animal owner indicated that a similar oral mass at the same location had been surgically removed four years ago. At that time, the abnormal oral structure had not been evaluated further. This time, the mass was removed surgically and further examined by both veterinary pathologists (veterinarians specialised in the examination of animal diseases and body structures) and virologists. No causal agents, which have been described before to trigger comparable lesions in tortoises, were found. Parts of the removed lesion were examined in further virological studies to scan the tissue material for potential new infectious agents, which might be connected to the abnormal oral lesion. Indeed, the authors were able to detect virus material within the tissue mass belonging to a comparably new virus family. The exact influence of these agents on the origin of the lesion, though, remains unclear.

Abstract: An adult male Bell's hinge-back tortoise (*Kinixys belliana*) was admitted to a veterinary clinic due to a swelling in the oral cavity. Physical examination revealed an approximately 2.5 × 1.5 cm sized, irregularly shaped tissue mass with villiform projections extending from its surface located in the oropharyngeal cavity. An initial biopsy was performed, and the lesion was diagnosed as squamous papilloma. Swabs taken for virological examination tested negative with specific PCRs for papillomavirus and herpesvirus. Further analysis of the oropharyngeal mass via metagenomic sequencing revealed sequence reads corresponding to a member of the family *Adintoviridae*. The tissue mass was removed one week after the initial examination. The oral cavity remained unsuspicious in follow-up examinations performed after one, five and twenty weeks. However, a regrowth of the tissue was determined 23 months after the initial presentation. The resampled biopsy tested negative for sequence reads of *Adintoviridae*. Conclusively, this report presents the diagnostic testing and therapy of an oral cavity lesion of unknown origin. The significance of concurrent metagenomic determination of adintovirus sequence reads within the tissue lesion is discussed.

Keywords: reptile medicine; emerging diseases; neoplasia; next-generation-sequencing; metagenomic examination; surgery; adintovirus

Citation: Hetterich, J.; Mirolo, M.; Kaiser, F.; Ludlow, M.; Reineking, W.; Zdora, I.; Hewicker-Trautwein, M.; Osterhaus, A.D.M.E.; Pees, M. Concurrent Detection of a Papillomatous Lesion and Sequence Reads Corresponding to a Member of the Family *Adintoviridae* in a Bell's Hinge-Back Tortoise (*Kinixys belliana*). *Animals* **2024**, *14*, 247. https://doi.org/10.3390/ani14020247

Academic Editor: Tom Hellebuyck

Received: 26 November 2023
Revised: 4 January 2024
Accepted: 5 January 2024
Published: 12 January 2024

1. Introduction

In chelonians, aetiologies for tissue proliferations including neoplastic diseases are diverse [1–3]. Oropharyngeal swelling can generally be associated with papilloma-like lesions, which have been described several times for various reptile species [4]. Papillomaviruses have been found in skin lesions [5], also causing cutaneous papillomas in sea turtles [6]. However, the presence of papilloma-like viral sequences has not been linked to clinical symptoms in all cases. The field of reptilian virology has progressed significantly over the last two decades [4,7–10]. Standardisation of sampling methods and increased use of molecular diagnostics—particularly the use of next-generation sequencing (NGS)—has led to a marked increase in the number of described viruses in reptile medicine [4]. Also, the understanding of the viral impact in specific diseases has been enhanced in recent years. However, the interconnections between specific viruses and many extrinsic and intrinsic factors, such as the reptilian immune system, environmental conditions and coexisting infectious diseases, which contribute to a clinical illness remain unclear for a large number of viral diseases [11].

This is the first report of a Bell's hinge-back tortoise with a distinct clinical symptom alongside concurrent metagenomic determination of sequence reads corresponding to a member of the family *Adintoviridae*.

2. Material and Methods

2.1. Animals

Initially, an adult male Bell's hinge-back tortoise (*Kinixys belliana*), weighing 911 g, being approximately 35 years old and originating from Benin, West Africa, was presented (index case patient). The tortoise had a one-week history of reduced forage intake and diarrhoea. Anamnestically, the care worker indicated a former tissue enlargement in just the same oropharyngeal location that had been removed by a veterinarian four years before. At that time, the tissue sample did not undergo any further diagnostic examination. The tortoise had been kept in an enclosure (six square metres) with regular access to the suburban garden during the summer season for the last 25 years. It had been fed a variety of salads and vegetables and had received regular mineral supplementation as well as species-appropriate UVB light.

After the initial identification of adintovirus sequences in the index case, a population of Bell's hinge-back tortoises under human care were sampled to evaluate the presence of similar (adinto)virus sequences in conspecifics. The population included nine tortoises: two individuals from a German zoo and seven animals (five zoo animals and two privately kept tortoises) from Zurich, Switzerland. All animals were raised under human care and evaluated as adult tortoises in clinically healthy condition. No oral lesions or other abnormal findings were reported for any of the nine sampled tortoises.

2.2. Diagnostics

Initial diagnostic methods of the index case patient included a tissue biopsy for histopathological examination, obtaining oral and cloacal swabs for virological examination as well as oral swabs for microbiological and mycological examination, parasitological faecal examination and blood chemistry examination. Further diagnostic evaluation involved next-generation-sequencing (NGS) examinations of whole blood, cloacal and skin swabs and faeces to determine the presence of viral sequences.

Sampling of the nine conspecifics comprised cloacal and skin swabs, faeces and whole blood. The samples of these nine animals were only examined for next-generation-sequencing analysis.

For histopathological examination, the tissue sample was fixed in 4% neutral buffered formalin for at least 24 h. The formalin-fixed tissue was dehydrated and routinely embedded in paraffin wax, sectioned at 4 μm and stained with haematoxylin and eosin (HE). Slides were evaluated with a standard binocular light microscope (Carl Zeiss 670; Carl Zeiss, Jena, Germany; field of view area, 40× magnification: 0.16 mm^2). For image preparation, the specimen was digitalised using an Olympus VS200 slide scanner (Olympus Deutschland GmbH, Hamburg, Germany), and representative images were exported with the respective OlyVIA software (version 3.4.1., Olympus Deutschland GmbH, Hamburg, Germany).

For bacteriological and mycological examination, swabs were inoculated on different culture media (Columbia agar with sheep blood, Gassner agar, Staphylococcus/Streptococcus selective agar, neomycin agar, chocolate agar, Schaedler agar, Kim-mig agar) and incubated under aerobic, anaerobic or microaerophilic atmosphere at 37 °C for up to 48 h. The Kim-mig agar was incubated for 48 h at 30 °C. The swab was further placed in nutrient broth, which was also stored at 37 °C overnight. The following day, the nutrient broth was spread onto selected culture media (Columbia agar with sheep blood, Gassner agar, Staphylococcus/Streptococcus selective agar), which were incubated at 37 °C for an additional 24 h. The different colonies grown on the culture media were differentiated by mass spectrometry (MALDI-TOF, Bruker, ORT).

Blood chemistry examination was performed in a standardised manner in an inhouse laboratory (Cobas C 311, La Roche Ltd., Basel, Switzerland) 15 min after venipuncture (dorsal coccygeal vein). Haematological parameters were examined microscopically at 1000× magnification after standardised blood smear preparation with 20 uL whole blood.

For NGS examination, a section of the tumour tissue sample (20 mg) was homogenised in 500 µL PBS and centrifuged for 5 min at 12,000 relative centrifugal force (RCF). The supernatant was sieved using a 0.45 µm spin filter (Merck Millipore, Darmstadt, Germany) for 5 min at 12,000 RCF. RNA was extracted using TRIzol Reagent (Thermo Fisher Scientific, Dreieich, Germany) in accordance with the manufacturer's instructions and temporarily stored in a −80 °C freezer. Conversion of RNA to cDNA was performed using Superscript IV (Thermo Fischer Scientific, Dreieich, Germany) and nonribosomal hexamers as per protocol guidelines. Double-stranded DNA (dsDNA) was generated via the Klenow fragment (NEB), followed by random PCR amplification using a sequence-independent, single-primer amplification protocol [12]. PCR products were purified a using Monarch PCR/DNA Cleanup kit (NEB), and a DNA library was prepared with a Nextera XT DNA Library Preparation Kit (Illumina Inc., San Diego, CA, USA) prior to sequencing on an Illumina NextSeq sequencing platform with a NextSeq 500/550 High Output kit v2.5 for 150 cycles (paired-end reads, 75 base pairs). Raw sequencing data (FASTQ) were first analysed using the ID-Seq bioinformatics pipeline (https://czid.org/) [13,14]. Additional bioinformatic analyses were performed in CLC Genomics Workbench (v12) (QUIAGEN).

Further virological investigation was attempted by extracting RNA from blood (whole blood), cloacal swabs, skin swabs and faeces. Primers (Adinto_Integrase_frw 5′-TGCTCAT-GTCTGAGACACAGATATC-3′ and Adinto_Integrase_rvs 5′-CTACAGAGCTGGCATTGCTG-3′) were designed based on the recovered partial adintovirus sequence to confirm the presence of an adintovirus in the RNA which had been used for NGS analysis and to complete the sequence of the integrase gene. PCR amplification was performed using a Qiagen One Step RT-PCR kit (Qiagen, Germantown, MD, USA) and an annealing temperature of 49 °C. Additional screening of blood (whole blood), cloacal and skin swabs and faeces for the presence of adintovirus was performed according to the same primers and protocol.

3. Results

3.1. Initial Clinical Condition and Therapy

Clinical examination of the index case patient revealed an approximately 2.5 × 1.5 cm sized, irregularly shaped tissue mass with villiform projections extending from its surface in the left oropharyngeal cavity (Figure 1a). In addition, the animal showed liquid defaecation throughout the general examination. A subsequently performed faecal examination revealed flagellate overgrowth.

Initially, the index case tortoise received treatment for flagellate overgrowth (metronidazole 40 mg/kg PO; repeated after 14 days; Eradia, Virbac Tierarzneimittel GmbH, Bad Oldesloe, Germany). Furthermore, a supportive therapy was started with a liver-supporting product (50 mg/kg SID PO; Legaphyton®; Vetoquinol S.A., Lure, France) and probiotics (0.5 mL solution PO; BeneBac®, Dechra Veterinary Products Deutschland GmbH, Aulendorf, Germany). One week later, the tortoise was presented for a follow-up and a subsequent surgical extirpation of the oropharyngeal tissue enlargement. The patient's overall condition had improved. It showed regular voluntary food consumption, and the faecal quality was reported to have a more solid consistency. The animal received analgetic pain therapy with meloxicam (0.3 mg/kg SC; Metacam, Boehringer Ingelheim Vetmedica GmbH, Ingelheim, Germany) and lidocaine (1 mg/kg topical; Lidor® 2%, VetViva Richter GmbH, Wels, Austria) and was then sedated with alfaxalone (5 mg/kg IV; Alfaxan®, Jurox Pty Limited, Rutherford, Australia). The tortoise maintained spontaneous breathing during the 15 min procedure and was therefore not intubated. The oral tissue mass was removed without any macroscopic leftover tissue using forceps and a sharp spoon (Figure 1b). Only minor bleeding (approximately 0.3–0.5 mL) occurred inside the oral cavity, and the animal regained full activity within 45 min after the surgical procedure.

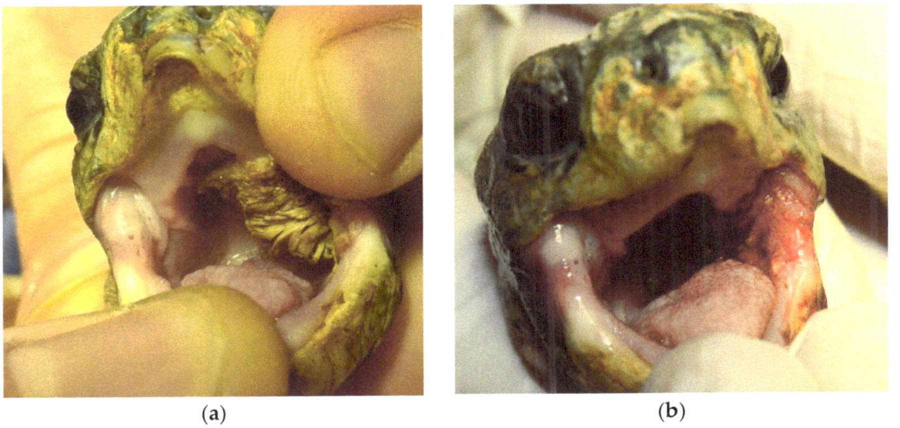

(a) (b)

Figure 1. Bell's hinge-back tortoise, oral cavity. (**a**) Upper-left side, showing an approximately 2.5 × 1.5 cm sized oropharyngeal villiform tissue enlargement on the left oropharynx. (**b**) Upper-right side, left oropharynx is now visible after complete removal of the villiform tissue enlargement.

3.2. Diagnostic Results

Blood chemistry examination revealed high levels of liver-linked enzymes GLDH (75.4 U/L), AST (521 U/L) and ALT (35 U/L). Haematological parameters were considered unremarkable.

Histopathological analysis revealed fingerlike projections of keratinising squamous epithelium with marked hyperplasia and ortho- to parakeratotic hyperkeratosis (Figure 2A,B). Regular maturation of keratinocytes was retained. Koilocytes were not present. Subepithelially, low amounts of fibrovascular stromata were present. There was mild anisocytosis and karyosis. In ten high-power fields (total field of view: 1.6 mm^2), one mitotic figure was present in the suprabasal epithelium. Furthermore, there was marked, multifocal, superficial, intraepithelial, supportive inflammation with multifocal small colonies of coccoid bacteria. Based on the histopathological findings, a squamous papilloma was diagnosed.

Microbiological assessment revealed a diverse bacterial flora including *Morganella morganii*, *Pseudomonas aeruginosa* and *Streptococcus* sp. No fungi were cultured. Microscopic faecal examination revealed a high infestation of flagellates (flagellate overgrowth). No further parasitological pathogens were found. Virological examination of dry swabs for papillomavirus and herpesvirus using family-specific PCR methods [15,16] were without positive confirmations.

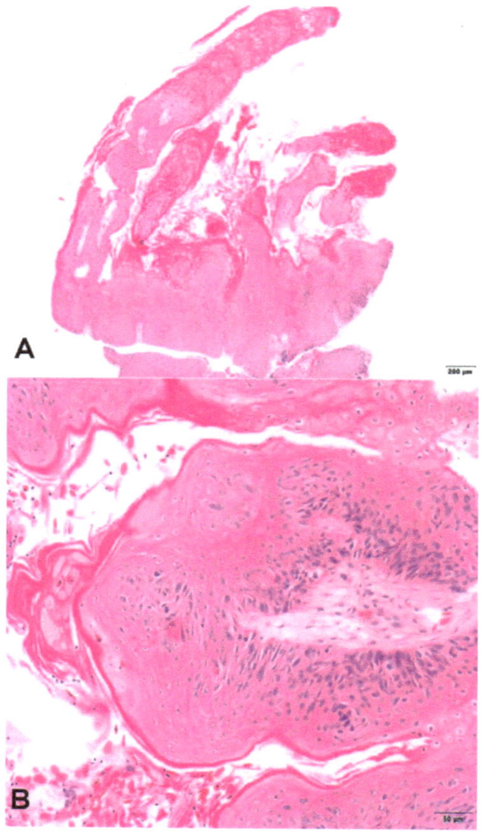

Figure 2. Bell's hinge-back tortoise (*Kinixys belliana*), oropharyngeal mass. (**A**) Squamous papilloma with thickened, exophytic, hyperplastic and hyperkeratotic epithelium arranged in fingerlike projections. Haematoxylin & eosin (H&E). Low magnification. Bar: 200 µm (**B**) High magnification of the squamous papilloma. Exophytic projections are covered by ortho- to parakeratotic hyperkeratosis. The exophytic projections of squamous epithelium are supported by subepithelial fibrovascular tissue. H&E. Bar: 50 µm.

Analysis of fastq raw reads using the CZ ID metagenomics pipeline [13,14] showed the presence of 23 non-genus-specific reads showing homology to Terrapene box turtle adintovirus (GenBank accession no. BK010890.1), a member of the *Adintoviridae* family. Reference assembly using Genomics Workbench (v12) by read mapping to the Terrapene box turtle adintovirus sequence (GenBank accession no. BK010890.1) enabled the identification of an additional 913 adintovirus reads. The complete integrase sequence was obtained by Sanger sequencing of RT-PCR amplicons generated using primers based on adintovirus sequences found in NGS analysis. A nucleotide BLAST (BLASTN) search of already known adintovirus species indicated that the adintovirus present in the clinical tortoise sample was closely related to the Terrapene box turtle adintovirus, with 95.16% nucleotide identity, followed by multiple uncharacterised mRNAs from related turtle species. To better distinguish the newly identified adintovirus from closely related viruses, we used the nomenclature 'Bell's hinge back tortoise (*Kinixys belliana*) adintovirus'. In addition to the integrase gene, analysis of NGS data enabled the recovery of a contig of 242 base pairs (bp) of the penton gene, 490 bp of the genome packaging ATPase and 457 bp from the DNA polymerase gene, with 88.84%, 94.94% and 93.44% nucleotide identity, respectively, to the homologous Terrapene box turtle adintovirus genes. An RT-PCR using specific integrase primers resulted in a detectable amplicon verified subsequently by Sanger sequencing in one additional sample from the index case patient (skin swab). Additionally, two out of nine sampled tortoises (once each for a cloacal swab and whole blood) tested positive for adintovirus RNA.

However, the complete adintovirus genome could not be recovered by Sanger sequencing and reference assembly. Furthermore, no oncogene transcripts were identified by NGS or Sanger sequencing.

3.3. Follow-up History

The index case tortoise was presented for three additional follow-ups at one, five and twenty weeks after removal of the tissue enlargement. The oral cavity remained unremarkable, and no regrowth of tissue in the affected oral region could be identified. The clinical condition continuously improved after the surgical procedure. The tortoise returned to its usual food consumption and activity within five days after surgery. The overall condition was evaluated as good at follow-ups five and twenty weeks after initial presentation. However, a regrowth of the tissue was determined 23 months after the initial presentation. At that time, the tortoise showed regular activity and unsuspicious defecation and food intake. The tissue lesion again was removed surgically according to the procedure described for the initial surgery. Macroscopically, the oral lesion strongly resembled the initial tissue mass. However, the lesion had a considerably smaller (1.0×0.5 cm) size and flatter structure. As before, the tissue mass was removed without macroscopic leftovers. The recurring papillomatous tissue was resampled and assayed by RT-PCR, but no adintovirus RNA was detected. No histological examination was conducted. The animal remained in a good clinical condition in subsequent follow-up examinations two and twelve weeks after the second surgical procedure. No further regrowth was determined.

All liver-linked parameters remained markedly elevated in each of the follow-up blood examinations.

4. Discussion

The oropharyngeal tissue mass was the primary clinical symptom observed in the tortoise and needs to be discussed as a possible recurrent finding. Based on information given by the patient's owner, a tissue enlargement in exactly the same location had been surgically removed four years previously in the absence of any further postsurgical examination. According to the owner, the shape, size and colouration of the tissue lesion were comparable to the macroscopical appearance of the initial tissue enlargement. Aetiologies for tissue proliferations, particularly of mucous membranes, are diverse in reptile medicine, including a broad variety of neoplastic diseases [1–3]. In chelonians, a plethora of neoplasia reports exist based on case reports [11]. Virus infections as causative agents need to be considered, especially since reptilian virology has progressed significantly over the last two decades by standardisation of sampling methods and increased use of molecular diagnostics [7–10]. The oropharyngeal swelling's villiform macroscopical appearance could be associated with papilloma-like lesions. These have been described for various reptile species [4]. Papillomaviruses have been found in skin lesions [5], also causing apparent cutaneous papillomas in loggerhead sea turtles and green sea turtles [6]. However, the presence of papilloma-like viral sequences has not been associated with clinical symptoms in all cases. In tortoises, papillomavirus-like particles have been detected in different species, causing papular skin lesions in one case [17]. Nevertheless, papillomaviruses have been described as being highly host- and tissue-specific [4]. The clinical status of papilloma-caused lesions (progression, resolution) has also been found to be highly variable [17]. In this case, no papillomavirus was detected, either by PCR or by next-generation sequencing (NGS). Also, reptilian herpesvirus infections need to be considered for irregular oral cavity findings in tortoises. Several herpesviruses have been described for numerous members of the family *Testudinidae* [4]. If clinical signs of an infection are present, rhinitis and conjunctivitis are oftentimes associated with indicative stomatitis and glossitis, leading to diphtheroid-necrotising alterations in advanced stages of the disease. Diphtheroid membranes covering oral cavity tissue can be seen regularly, as brownish villiform structures initially were covering the oral cavity at the patient's first presentation. However, the lack of associated lesions (rhinitis, conjunctivitis, glossitis) and the more rigid, bulging appearance of the tissue lesion did not support herpesvirus infection as a differential diagnosis.

Further analysis of a tissue biopsy of the oropharyngeal lesion using NGS resulted in the identification of sequence reads displaying homology to a member of the *Adintoviridae* family, which we termed 'Bell's hinge back tortoise (*Kinixys belliana*) adintovirus'. The name 'adintovirus' refers to a Polinton, which harbours retrovirus-like integrase proteins, and an adenovirus-like DNA polymerase [18]. The *Adintoviridae* family was first recognised in 2021 and is composed of two lineages, Alpha and Beta. A BLASTN search of the integrase gene showed that the virus is closely related to an adintovirus recovered from a box turtle, called 'Terrapene box-turtle Adintovirus'. Using data mining, a recent study found that most adintovirus genomes bear open reading frames (ORFs) with predicted structural similarity to capsid hexon and penton proteins of multiple known viruses, including adenoviruses [18]. Therefore, questions arise as to the ability of adintoviruses to produce progeny virions or rather if the adintovirus genome is maintained in the form of an integrated provirus. Although the complete genome of an adintovirus could not be recovered in this study, reference assembly of recovered sequencing reads to the Terrapene box turtle adintovirus enabled the identification of transcripts of the penton and genome packaging ATPase genes. A BLASTN search using the integrase gene of the adintovirus identified in this study showed that the most similar virus was an adintovirus from a related tortoise species, the Terrapene box turtle, followed in the same query unexpectedly by multiple uncharacterised mRNAs from related tortoises. Overall, these observations strongly suggest that adintoviruses are integrated into the respective host genome and have subsequently coevolved over time.

Another key feature of the adintoviruses genome is the presence of 'oncoid genes', which are similar to the well-characterised retinoblastoma-interacting protein and antiapoptotic proteins of small DNA tumour viruses, such as polyomaviruses, papillomaviruses, parvoviruses and adenoviruses [18]. However, adintoviruses have so far not been associated with diseases or tumours. On the contrary, transcripts from both adintovirus lineages were identified in the gill tissue of nondiseased freshwater fish by metatranscriptomic analyses [19].

In the present study, we showed that adintovirus transcripts were detectable in the initially sampled papillomatous lesion but also in nonsuspicious tissue (skin swab from the index case patient as well as each cloacal swab and plasma samples of two further clinically healthy tortoises). Related to this, adintovirus as a causative agent of the oral lesion cannot be ruled out. At this time, however, it is not possible to definitely link the presence of adintovirus sequences in the papillomatous lesion to the aetiology of the lesion itself or to the markedly increased liver enzyme elevations and the symptom of diarrhoea. Transmission electron microscopy (TEM) of sections of the papillomatous lesion would be required in order to confirm the presence of adintoviruses virions. In addition, RNA-seq analysis could also be used on appropriate tissue samples from diseased and healthy animals to look for differences in gene expression of adintovirus genes between healthy and diseased animals.

Relating to the coexisting clinical symptoms, various viruses have been described to cause liver disease in reptiles [8,20,21]. In this case, further sampling such as liver biopsy or endoscopic coelomic imaging might have been highly beneficial. However, the animal owner rejected any invasive diagnostic procedures. Although the tortoise's clinical status improved satisfyingly within five days after the initial surgery, the authors critically reflect on the perioperative management of the patient. A more multimodal pain management might have enhanced an even better postsurgical recovery. Furthermore, intubation of the tortoise might have increased intrasurgical options in case of anaesthesia incidents like major bleedings or apnea.

5. Conclusions

This report clearly shows the high potential for molecular diagnostics in reptile medicine. The routine use of advanced diagnostic methods such as NGS will facilitate the identification of new reptilian viruses and possible links to known disease syndromes. Prospective standardised study designs should be attempted, especially involving scientifically underrepresented reptile species.

Author Contributions: Conceptualization, J.H. and M.M.; validation, F.K., M.L. and M.H.-T.; resources, W.R. and I.Z.; writing—original draft and preparation, J.H.; image preparation, J.H. and W.R.; writing—review and editing: M.M., F.K., M.H.-T. and M.P.; supervision, M.P. and A.D.M.E.O. All authors have read and agreed to the published version of the manuscript.

Funding: This Open Access publication was funded by the Deutsche Forschungsgemeinschaft (DFG, German Research Foundation)—491094227 "Open Access Publication Funding" and the University of Veterinary Medicine Hannover, Foundation.

Institutional Review Board Statement: Ethical review and institutional approval were waived for this study since the study included the examination and treatment of animals with clinical disease. No animal was killed for the purpose of this study.

Informed Consent Statement: Not applicable.

Data Availability Statement: The data presented in this study are openly available in a repository at DOI:10.17605/OSF.IO/WZJ68 (https://osf.io/wzj68/).

Acknowledgments: The authors are indebted to Silvia Burgstaller, Jean-Michel Hatt and Andreas Bernhard for their guidance and support during the sampling period of this study. We wish to thank Belinda Euring and the team from the Department for Virology of the Veterinary Faculty, University of Leipzig, Germany for their advice and critical evaluation concerning the virological sampling process. We are also grateful to the staff members of the Department of Pathology and the Department of Microbiology, both at the University of Veterinary Medicine Hannover, for their diagnostic assistance.

Conflicts of Interest: The authors declare no conflict of interest.

References

1. Garner, M.M.; Hernandez-Divers, S.M.; Raymond, J.T. Reptile neoplasia: A retrospective study of case submissions to a specialty diagnostic service. *Vet. Clin. Exot. Anim. Pract.* **2004**, *7*, 653–671. [CrossRef] [PubMed]
2. Hernandez-Divers, S.M.; Garner, M.M. Neoplasia of reptiles with an emphasis on lizards. *Vet. Clin. Exot. Anim. Pract.* **2003**, *6*, 251–273. [CrossRef] [PubMed]
3. Sykes, J.M.; Trupkiewicz, J.G. Reptile neoplasia at the Philadelphia zoological garden, 1901–2002. *J. Zoo Wildl. Med.* **2006**, *37*, 11–19. [CrossRef] [PubMed]
4. Divers, S.J.; Stahl, S.J. *Mader's Reptile and Amphibian Medicine and Surgery-e-Book*; Elsevier Health Sciences: Amsterdam, The Netherlands, 2018.
5. Hellebuyck, T.; Pasmans, F.; Haesebrouck, F.; Martel, A. Dermatological diseases in lizards. *Vet. J.* **2012**, *193*, 38–45. [CrossRef] [PubMed]
6. Herbst, L.H.; Lenz, J.; Van Doorslaer, K.; Chen, Z.; Stacy, B.A.; Wellehan, J.F., Jr.; Manire, C.A.; Burk, R.D. Genomic characterization of two novel reptilian papillomaviruses, *Chelonia mydas* papillomavirus 1 and *Caretta caretta* papillomavirus 1. *Virology* **2009**, *383*, 131–135. [CrossRef]
7. Marschang, R.E. Viruses infecting reptiles. *Viruses* **2011**, *3*, 2087–2126. [CrossRef] [PubMed]
8. Gibbons, P.M.; Steffes, Z.J. Emerging infectious diseases of chelonians. *Vet. Clin. Exot. Anim. Pract.* **2013**, *16*, 303–317. [CrossRef] [PubMed]
9. Harding, E.F.; Russo, A.G.; Yan, G.J.; Mercer, L.K.; White, P.A. Revealing the uncharacterised diversity of amphibian and reptile viruses. *ISME Commun.* **2022**, *2*, 95. [CrossRef] [PubMed]
10. Nieto-Claudin, A.; Esperón, F.; Apakupakul, K.; Peña, I.; Deem, S.L. Health assessments uncover novel viral sequences in five species of Galapagos tortoises. *Transbound Emerg. Dis.* **2022**, *69*, e1079–e1089. [CrossRef] [PubMed]
11. Girling, S.J.; Raiti, P. *BSAVA Manual of Reptiles*; British Small Animal Veterinary Association: Gloucester, UK, 2019.

12. Allander, T.; Emerson, S.U.; Engle, R.E.; Purcell, R.H.; Bukh, J. A virus discovery method incorporating DNase treatment and its application to the identification of two bovine parvovirus species. *Proc. Natl. Acad. Sci. USA* **2001**, *98*, 11609–11614. [CrossRef] [PubMed]
13. Kalantar, K.L.; Carvalho, T.; de Bourcy, C.F.; Dimitrov, B.; Dingle, G.; Egger, R.; Han, J.; Holmes, O.B.; Juan, Y.F.; King, R.; et al. IDseq-An open source cloud-based pipeline and analysis service for metagenomic pathogen detection and monitoring. *Gigascience* **2020**, *9*, giaa111. [CrossRef] [PubMed]
14. Ramesh, A.; Nakielny, S.; Hsu, J.; Kyohere, M.; Byaruhanga, O.; de Bourcy, C.; Egger, R.; Dimitrov, B.; Juan, Y.F.; Sheu, J.; et al. Metagenomic next-generation sequencing of samples from pediatric febrile illness in Tororo. *Uganda PLoS ONE* **2019**, *14*, e0218318. [CrossRef] [PubMed]
15. Husnjak, K.; Grce, M.; Magdić, L.; Pavelić, K. Comparison of five different polymerase chain reaction methods for detection of human papillomavirus in cervical cell specimens. *J. Virol. Methods* **2000**, *88*, 125–134. [CrossRef] [PubMed]
16. VanDevanter, D.R.; Warrener, P.; Bennett, L.; Schultz, E.R.; Coulter, S.; Garber, R.L.; Rose, T.M. Detection and analysis of diverse herpesviral species by consensus primer PCR. *J. Clin. Microbiol.* **1996**, *34*, 1666. [CrossRef] [PubMed]
17. Jacobson, E.R.; Gaskin, J.M.; Clubb, S.; Calderwood, M. Papilloma-like virus infection in Bolivian side-neck turtles. *J. Am. Vet. Med. Assoc.* **1982**, *181*, 1325–1328. [PubMed]
18. Starrett, G.J.; Tisza, M.J.; Welch, N.L.; Belford, A.K.; Peretti, A.; Pastrana, D.V.; Buck, C.B. Adintoviruses: A proposed animal-tropic family of midsize eukaryotic linear dsDNA (MELD) viruses. *Virus Evol.* **2021**, *7*, veaa055. [CrossRef] [PubMed]
19. Perry, B.J.; Darestani, M.M.; Ara, M.G.; Hoste, A.; Jandt, J.M.; Dutoit, L.; Holmes, E.C.; Ingram, T.; Geoghegan, J.L. Viromes of Freshwater Fish with Lacustrine and Diadromous Life Histories Differ in Composition. *Viruses* **2022**, *14*, 257. [CrossRef] [PubMed]
20. Darke, S.; Marschang, R.E.; Hetzel, U.; Reinacher, M. Experimental infection of Boa constrictor with an orthoreovirus isolated from a snake with inclusion body disease. *J. Zoo Wildl. Med.* **2014**, *45*, 433–436. [CrossRef] [PubMed]
21. La'Toya, V.L.; Wellehan, J. Selected emerging infectious diseases of squamata. *Vet. Clin. Exot. Anim. Pract.* **2013**, *16*, 319–338.

animals MDPI

Article

The Role of Host Species in Experimental Ferlavirus Infection: Comparison of a Single Strain in Ball Pythons (*Python regius*) and Corn Snakes (*Pantherophis guttatus*)

Michael Pees [1,*], Annkatrin Möller [2], Volker Schmidt [3], Wieland Schroedl [4] and Rachel E. Marschang [5]

[1] Department of Small Mammal, Reptile and Avian Medicine, University of Veterinary Medicine Hannover, 30559 Hanover, Germany
[2] Tierarztpraxis Dr. Kühnel, 98527 Suhl, Germany; annkatrin_moeller@web.de
[3] Clinic for Birds and Reptiles, Veterinary Teaching Hospital, University of Leipzig, 04103 Leipzig, Germany; volker.schmidt@vogelklinik.uni-leipzig.de
[4] Institute of Bacteriology and Mycology, University of Leipzig, 04103 Leipzig, Germany
[5] Laboklin GmbH & Co. KG, 97688 Bad Kissingen, Germany; rachel.marschang@googlemail.com
* Correspondence: michael.pees@tiho-hannover.de; Tel.: +49-511-9536807

Simple Summary: Paramyxoviruses in the genus *Ferlavirus* are well-documented pathogens in snakes. Disease severity appears to depend on multiple factors which are not fully understood. In order to further understand the role of host species in ferlaviral infection and disease, a genogroup B ferlavirus that had previously been shown to be highly pathogenic in corn snakes (*Pantherophis gutattus*) was inoculated into ball pythons (*Python regius*). The pythons became infected but developed much milder disease than that observed in the corn snakes. The corn snakes also had a higher rate of bacterial involvement in the lungs as well as much weaker humoral immune responses to infection. In both species, the respiratory tract was the primary target of the virus, but systemic spread was also observed. While this study supports previous findings indicating a wide host range among squamate reptiles for ferlaviruses, it also shows that specific host species can react very differently to infection with individual virus strains.

Abstract: Ferlaviruses are a cause of respiratory disease in snakes. Four genogroups (A, B, C, and tortoise) have been described. Disease development is believed to depend on virus, host, and environment-specific factors. There is evidence of transmission of individual strains between genera and families of reptiles. A genogroup B virus previously used in a transmission study with corn snakes (*Pantherophis guttatus*) was applied intratracheally in ball pythons (*Python regius*) using the same protocol as for the corn snakes. Ball pythons became infected, with initial mild clinical signs noted four days post infection (p.i.), and the virus was detected first in the lungs on day 4 and spread to the intestine, pancreas, kidney and brain. Hematology showed an increase in circulating lymphocytes which peaked on day 28 p.i. Antibodies were detected beginning on day 16 and increased steadily to the end of the study. In comparison to corn snakes, ball pythons exhibited milder clinical signs and pathological changes, faster development of and higher antibody titers, and a hematological reaction dominated by lymphocytosis in contrast to heterophilia in corn snakes. These differences in host reaction to infection are important to understand ferlavirus epidemiology as well as for clinical medicine and diagnostic testing.

Keywords: ball python; ferlavirus; infection; pathology; detection; PCR; cell culture; immunology; IgY; IgM

Citation: Pees, M.; Möller, A.; Schmidt, V.; Schroedl, W.; Marschang, R.E. The Role of Host Species in Experimental Ferlavirus Infection: Comparison of a Single Strain in Ball Pythons (*Python regius*) and Corn Snakes (*Pantherophis guttatus*). *Animals* **2023**, *13*, 2714. https://doi.org/10.3390/ani13172714

Academic Editor: Clive J. C. Phillips

Received: 31 March 2023
Revised: 26 June 2023
Accepted: 23 August 2023
Published: 26 August 2023

1. Introduction

Paramyxoviruses in the genus *Ferlavirus* were first described in snakes in the 1970s [1]. They have since been detected in a wide range of snake species as well as in lizards and, occasionally, in chelonians [2]. They are known to primarily cause respiratory disease in infected animals but have also been associated with central nervous system signs as well as systemic disease. The severity of clinical signs can range from inapparent or mild to severe with multiple reports of severe disease outbreaks in larger collections [1,3,4].

Characterization of several ferlavirus isolates based on partial genome sequences has shown that they can be divided into at least four different genogroups, referred to as A, B, C, and tortoise [5,6]. Studies have shown some serological differences between different strains [2,7]. Viruses in genogroups A and B appear to strongly cross-react, while viruses in genogroup C cross-react less strongly with A and B viruses [8]. Differences in host range and pathogenicity between the different genogroups and different strains are not yet understood [9], although various genotypes have been found in a variety of host species [10,11].

The first transmission study with ferlaviruses was conducted using intratracheal inoculation in a group of Aruba Island rattlesnakes (*Crotalus unicolor*) [12]. The virus used in that study was not further characterized. Five animals were inoculated and were euthanized or died between 4 and 22 days post infection (p.i.). Examination of lung tissue showed progressive pathological changes in the lungs of infected animals beginning 4 days p.i. The two animals that were not euthanized earlier died suddenly on days 19 and 22, respectively, both with blood found in their oral cavity. None of the snakes in that study developed antibodies against the virus used for inoculation as determined by hemagglutination inhibition assay (HI).

A transmission study using genogroup A, B, and C ferlaviruses in different groups of corn snakes (*Pantherophis guttatus*) showed distinct differences in pathogenicity between the viruses [9]. In that study, 12 animals in each group were inoculated intratracheally with a ferlavirus genogroup A, B, or C isolates and 3 animals were each euthanized and examined 4, 16, 28, and 49 days p.i. Significant differences were found in clinical disease and pathological changes between the study groups, with the most severe disease found in animals inoculated with the genogroup B virus. In this group, snakes developed severe clinical disease starting from day 16 p.i., and the last animals remaining had to be euthanized earlier than planned on day 35 p.i. due to disease. In contrast, only two animals infected with the group A virus developed moderate clinical signs, none died, and none had to be euthanized earlier than planned. In the group inoculated with the group C virus, two developed mild clinical signs, three died, and one had to be euthanized due to disease. Analysis of the immune responses of snakes to infection using HI, serum IgM and IgY concentrations, and hematology showed a leukocytosis that was detectable beginning 16 days p.i. and antibodies detectable beginning 28 days p.i. [8]. However, there were significant differences in immune response depending on the virus used. Animals that were infected with the genogroup B isolate developed the highest white blood cell counts (WBC), but antibodies were only detectable in a single animal at a low titer shortly before death (which in this group occurred on day 35). Animals in the groups that received the group A and C viruses developed high antibody titers (by HI) over the course of the study and showed significantly fewer clinical signs and pathological findings.

Differences in clinical signs and pathological findings between different reports of ferlavirus infections and outbreaks as well as in the transmission studies available indicate that a variety of factors including host-specific factors (e.g., species, age, health status) and virus-specific factors (e.g., genogroup and strain), as well as additional factors including temperature, other environmental factors, and secondary infections, likely play a role in infection and disease development. The aim of the current study was to evaluate the effects of host species on infection and disease using a genogroup B virus previously found to be highly pathogenic in corn snakes [9]. The same virus and protocol used in those experiments was used to infect ball pythons, and clinical, pathological and immunological reactions to infection were compared between the ball pythons and the previous results from the corn snakes.

2. Materials and Methods

The overall experimental design was identical to a study conducted with three different virus strains in corn snakes [8,9,13], as it was the aim to allow direct comparison of the study results. The animal trial was approved by the national authority (Landesdirektion Sachsen, application number TVV 61/13).

The study was conducted with 12 adult ball pythons (*Python regius*). The animals were captive-bred and acquired as adult healthy snakes from a commercial company. Nine of the snakes were female, three were male. Average body length (snout–cloaca) was 125 cm (range 92 cm to 161 cm), and mean body mass was 1409 g (range 644 g to 2977 g). No further details on individual histories of the snakes were available.

All animals underwent a thorough clinical examination following established standards. They were checked for endo- and ectoparasites and samples were taken (swabs from the choana and the cloaca, tracheal washes) and checked for aerobic bacterial and fungal growth as described previously [13]. A combined sample (tracheal wash and cloacal swab) was examined for the presence of ferlaviruses by PCR following an established protocol [14]. Briefly, the PCR targeted a 566 bp portion of the L-gene using primers F5-outer and R6-outer followed by F7-inner and R8-inner in a nested format as described by Ahne et al. [14] using the specific methods described by Kolesnik et al. [15]. This PCR has been shown to be highly specific and able to detect between 5×10^{-1} and 5×10^{-3} ng/µL viral RNA [15]. RNA prepared from a ferlavirus cell culture isolate was used as a positive control and HPLC water was used as a negative control. Only snakes without any remarkable result (clinically healthy, good body condition, negative ferlavirus PCR, negative for parasites, and bacterial and fungal flora assessed to be physiological) were included in the study. The snakes were housed for at least two months before the beginning of the study period, and examinations were repeated on day −6 (before virus inoculation), at the start of the study protocol.

The snakes were kept in two terraria of approximately 140 × 78 × 65 cm, with six snakes in each group. Husbandry conditions followed general recommendations for the species, including suitable ground material (turf), a water basin and hiding places. Data loggers (microlite II, imec, Heilbronn, Germany) were used to monitor temperature and humidity. Temperature range was 20 °C to 28 °C with a 14/10 h day/night rhythm and hot spots of 35 °C, relative humidity was kept in a range of 50% to 70%. During the complete study period, snakes were fed (one mouse per snake) every seven to ten days in separate boxes.

For experimental infections, the virus strain that proved to be most virulent in corn snakes in the preceding study [13] was chosen. This strain was isolated from a timber rattlesnake (*Crotalus horridus*) during a disease outbreak in a collection of various viper species in Germany [5]. The virus was cultured on viper heart cells (VH2) according to an established protocol [6], the virus suspension used was an aliquote of the same virus passage as that used in the corn snake study [13] that had been stored frozen at −80 °C. Using a tracheal tube, 1 mL of the prepared virus suspension was inoculated into the trachea, and the virus suspension was titrated on VH2 cells [13] and confirmed to contain $10^{6.5}$ $TCID_{50}$ per mL.

No negative control group was used for this study, as the previous study with corn snakes included a group inoculated with a cell culture suspension without the virus. Since no clinical effect of this treatment was noted in the corn snakes, a similar treatment was considered unnecessary in the present study. Instead, the aim was to compare the results of both infection groups (corn snakes vs. ball pythons) directly.

The study and sampling design included a 6-day pre-inoculation period followed by the virus application and a 49-day sampling/experimental period. Over the whole study period, animals were checked daily by thorough inspection and weekly with a complete clinical examination including body mass determination. Clinical disease conditions for euthanasia were defined in the animal trial protocol as ongoing severe clinical signs including central nervous signs, severe respiratory distress leading to continuous dyspnea, or continuous signs of pain or apathy.

Samples were collected on days −6, 4, 16, 28 and 49, according to the protocol used for the corn snake study [13]. Tracheal washes and cloacal swabs and blood were collected intra vitam from all remaining snakes on each sampling day, including from those snakes that were euthanized after swabbing. On each of the days post infection (4, 16, 28 and 49), three snakes were selected randomly, euthanized, and complete necropsies were performed.

The post-mortem examination was conducted according to published standards [16]. All organs were assessed macroscopically, and the following tissues were processed for histopathological examination: lung, liver, kidney, small and large intestine, pancreas, spleen, brain, and gonads. Virus detection by PCR and virus isolation in cell culture was carried out using samples from the lungs, intestine, pancreas, kidney, and brain (see below). For details on sample collection protocols, see Pees et al. [13] (virus detection), Neul et al. [8] (immunology) and Pees et al. [9] (pathology). Lung tissue was also examined for bacteria and fungi according to Pees et al. [9].

Hematologic examination included a leukocyte count (estimation method according to Campbell and Ellis [17]) and differential blood cell counts (100 cells at 1000× magnification), which were calculated as absolute values. HI was performed as described previously [8,18] using the virus isolate used to inoculate the pythons. Plasma IgM and IgY levels were determined using indirect ELISA as described previously in corn snakes [8], also using the same virus isolate used to infect the animals. Briefly, concentrated, purified viral antigen was adsorbed onto microtiter plates. Bovine casein was used as a control. For IgM detection, plasma samples were diluted 1:200 in assay dilution buffer. A positive sample (pooled blood sample from the group) and a negative control sample (plasma sample from a confirmed ferlavirus-negative ball python) were added accordingly. Python IgM detection was carried out using 100 µL of horseradish peroxidase-conjugated rabbit-anti-IgM (isolated from corn snakes and carpet pythons)-IgGq diluted 1:10,000 in assay dilution buffer. For IgY detection, python plasma samples were diluted 1:100, and horseradish peroxidase-conjugated IgG (from rabbit)-anti-IgY (isolated from corn snakes and carpet pythons) was used. Optical densities (ODs) were used as relative values for assessment of the immunoglobulin production.

The virus was detected by PCR and virus isolation in cell culture as described previously [13] in tissue samples from euthanized animals on days 4, 16, 28, and 49. As in the study in corn snakes, after initial sampling on day −6, intra vitam samples were tested beginning on day 16. No intra vitam samples were tested for ferlaviruses on day 4 following the protocol from the corn snake study, which was designed to rule out any chance of false positive PCR results due to the persistence of viral RNA at the inoculation site.

For PCR detection, samples were cooled to 4 °C in a microtube container without preservatives and sent overnight to a diagnostic laboratory (Laboklin GmbH & Co. KG, Bad Kissingen, Germany). Further processing was carried out according to established protocols [14] and the previous study design [13], using primers targeting a portion of the ferlaviral L gene as described above.

For virus isolation in cell culture, samples were placed in microtube containers, and for intra vitam cloacal and tracheal wash samples, a transport medium was added (Remel micro test M4, Remel, Dartford, UK). Samples were immediately frozen at −80 °C. Isolation was carried out using VH2 as described [19]. Samples were thawed and prepared as described in detail elsewhere [13]. Both VH2 subculture suspensions, as well as VH2 monolayers (each in 96-well plates), were used for virus isolation attempts. Cells were incubated at 28 °C and checked daily for cytopathic effects (CPE) or cell toxicity for eight days, and up to two passages were prepared. Samples in which no CPE was observed throughout the two passages were declared negative, whereas culture samples with CPE and positive PCR confirmation were declared positive.

Statistical analysis was performed using the software SPSS 22.0 (IBM, Armonk, NY, USA). As the majority of the values were not normally distributed, all values are reported as medians and interquartile ranges (25th to 75th percentile). For comparisons between sampling days, the Friedman test for multiple samples was used, followed by the Wilcoxon signed rank test in cases in which significance was determined (hematology and immunology). A p value of 0.05 or less was considered significant.

3. Results

3.1. Clinical and Post-Mortem Findings

Overviews comparing clinical and post-mortem findings in ball pythons and corn snakes are provided in Tables 1–3. At the beginning of the ball python study, all snakes were clinically unremarkable. Following inoculation of the virus, moderate clinical signs were first observed on day 4 (Table 1). On day 16, several snakes demonstrated reddening or mild signs of stomatitis in the oral cavity, or the tracheal wash sample was flaky. Similar signs (mucous secretion, stomatitis, or flaky tracheal wash sample) were observed in a majority of the snakes up to day 49. Severe clinical signs such as respiratory sounds were found only intermittently over the course of the study. No snake died spontaneously or had to be euthanized during the study due to disease conditions.

Table 1. Changes in the respiratory tract of ball pythons (*Python regius*) and corn snakes (*Pantherophis guttatus*) [9] following infection with a genotype B ferlavirus. (n.t.—not tested, as animals died or were euthanized).

Days p.i.	Signs of Respiratory Disease	Number Affected/Observed	
		Ball Pythons	Corn Snakes [9]
1–4	Mucous secretion in the oral cavity	4/12	0/12
5–16	Flaky tracheal wash sample	4/9	1/9
	Reddening tracheal opening or mild signs of stomatitis	2/9	0/9
	Purulent secretion in the oral cavity	0/9	1/9
	Respiratory sounds	1/9	0/9
17–28	Mucous secretion in the oral cavity	3/6	0/6
	Flaky tracheal wash sample	3/6	4/6
	Respiratory sounds	2/6	0/6
	Purulent secretion in the oral cavity	0/6	1/6
	Acute death	0/6	2/6 (days 27, 28)
29–49	Mucous secretion in the oral cavity	3/3	0/3
	Flaky tracheal wash sample	3/3	n.t.
	Apathy, abnormal position	0/3	2/3
	Acute death	0/3	1/3 (day 33)
	Euthanasia due to clinical disease	0/3	2/3 (day 35)

Histological changes in the lungs of infected ball pythons ranged from mild to moderate (Table 2, Figure 1), and none of the changes were judged to be severe as determined previously during the corn snake study (Figure 1).

Other histological findings in the ball pythons included intracytoplasmic brown and Prussian blue positive pigment in the hepatocytes (8/12) and in the renal tubule cells (12/12) and melanomacrophages in the liver (12/12). Mild multifocal lymphoplasmacytic proliferation was seen in the enteric submucosa (1/12), in the renal interstitium (1/12) and perivascular in the liver (1/12). One snake presented with multifocal pancreatic necrosis (1/12) and vacuolization of adrenal cortex cells was seen in four snakes (4/12). Hemosiderosis in the liver and hemosiderotic pigment nephrosis were also detected in corn snakes (5/12). Additionally, acute reactive splenohepatitis (6/12), acute catarrhal enteritis

(1/12), epicarditis (1/12) and acute necrotizing pancreatitis (1/12) were diagnosed in corn snakes infected with the same virus isolate [9].

Microbiologically, no fungi could be isolated from the lung tissue. *Salmonella* spp. were isolated from the lungs of two ball pythons on day 49, and *Citrobacter freundii* was cultured from the lung of one snake on day 4 (Table 3).

Table 2. Gross and histological findings observed in the lungs of ball pythons (*Python regius*) and corn snakes (*Pantherophis guttatus*) [9] following infection with a genotype B ferlavirus. Findings are categorized as mild, moderate, or severe.

Day p.i.	Criteria	Findings	Number Affected	
			Ball Pythons	Corn Snakes [9]
4	Gross lesions	Mild colorless mucous	3/3	
	Histology	Mild multifocal subepithelial lymphoplasmacytic infiltrates	3/3	
16	Gross lesions	Mild colorless mucous	2/3	
		Moderate amount of colorless mucous combined with mild tissue thickening	1/3	
		Severe tissue thickening, reddish discoloration, yellow mucous		3/3
	Histology	Mild multifocal subepithelial lymphoplasmacytic infiltrates	1/3	
		Moderate diffuse heterophilic and lymphoplasmacytic interstitial infiltrates, edema	1/3	
		Severe diffuse heterophilic and lymphoplasmacytic infiltration in the interstitium, fibrin in faveolar space		3/3
28	Gross lesions	Mild colorless mucous	3/3	
		Severe tissue thickening, reddish discoloration, yellow mucous		3/3
	Histology	Mild multifocal subepithelial lymphoplasmacytic infiltrates	3/3	
		Severe diffuse heterophilic and lymphoplasmacytic infiltration in the interstitium, fibrin in faveolar space		3/3
49 (35)	Gross lesions	Mild colorless mucous	3/3	
		Severe tissue thickening, reddish discoloration, yellow mucous		3/3
	Histology	Mild subepithelial lymphoplasmacytic infiltrates	1/3	
		Severe diffuse heterophilic and lymphoplasmacytic infiltration in the interstitium, fibrin in faveolar space		3/3

Table 3. Bacteria isolated from the lungs of necropsied ball pythons (*Python regius*) and corn snakes (*Pantherophis guttatus*) [9] following infection with a genotype B ferlavirus.

Day p.i.	Ball Pythons		Corn Snakes	
	Number Affected	Bacteria Isolated	Number Affected	Bacteria Isolated
4	1/3	*Citrobacter freundii*	0/3	
16	0/3		1/3	*Salmonella* ssp. IIIa 41:z4,z23;-
28	0/3		3/3	2× *Salmonella* ssp. IIIb 48:l.v:1,5; *Citrobacter freundii*
49	2/3	1× *Salmonella* Treforest (1),51:z:1,6; 1× *Salmonella* Apeyeme 8,20:z38:-	3/3	*Salmonella* ssp. IIIb 14:z10:5; *Salmonella* ssp. IIIb 48:l,v:1,5; *Salmonella* Georgia 6,7:b:e:n:z15; 2× *Klebsiella pneumoniae*

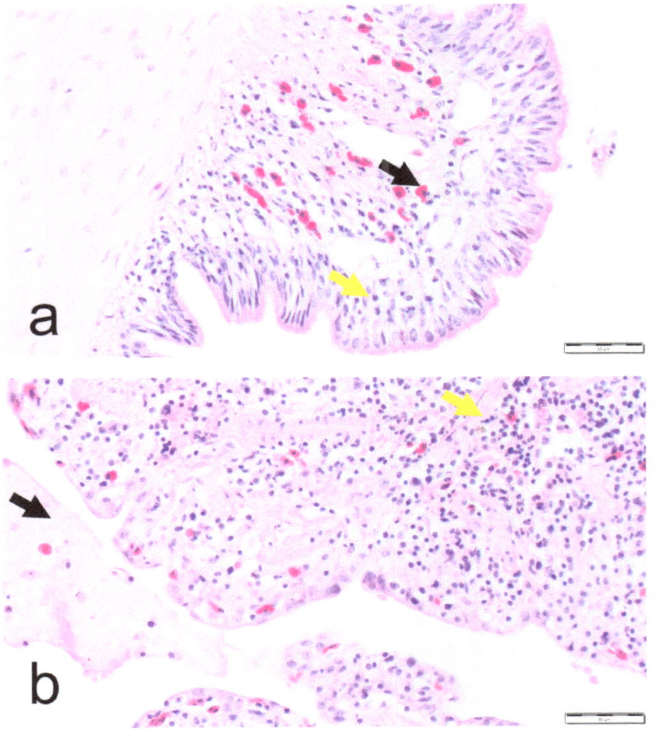

Figure 1. Histologic lung assessment: examples for (**a**) mild, and (**b**) severe changes of the lung tissue of ball pythons (*Python regius*) following infection with a genotype B ferlavirus. Magnification 400×, H&E stain. (**a**) mild subepithelial lymphoplasmacytic (yellow arrow) and heterophilic infiltrates (black arrow); (**b**) severe diffuse heterophilic and lymphoplasmacytic infiltration in the interstitium (yellow arrow), fibrin in faveolar space (black arrow).

3.2. Immune Reaction

3.2.1. Hematology

The baseline value for the total estimated white blood cell count was determined on day −6 with a median of 12,525 cells/μL, with a broad range of 7500 cells/μL to 25,400 cells/μL. Regarding the median, no increase was seen on day 4 post infection, but from day 16 to day 28, the number of white blood cells increased to a median of 17,755 cells/mL, before dropping again to almost base levels at the end of the study. Calculating the total number of heterophils and lymphocytes based on the differential blood cell count, it can be seen that this increase was mainly triggered by an increase in the number of lymphocytes (of almost 100% between day −6 and day 28), whereas the number of heterophils remained almost constant during the course of the study (Figure 2). Despite the clear increase in the median to day 28, the differences between the sampling days were neither significant for lymphocytes nor for heterophils.

In comparison with the ball pythons, the relative increase in lymphocyte counts following infection was even higher for the corn snakes (Figure 3a, comparison to day −6), and continued until the last day of the study (day 35). In contrast to the ball pythons, the heterophil count also rose starkly in the corn snakes (Figure 3b).

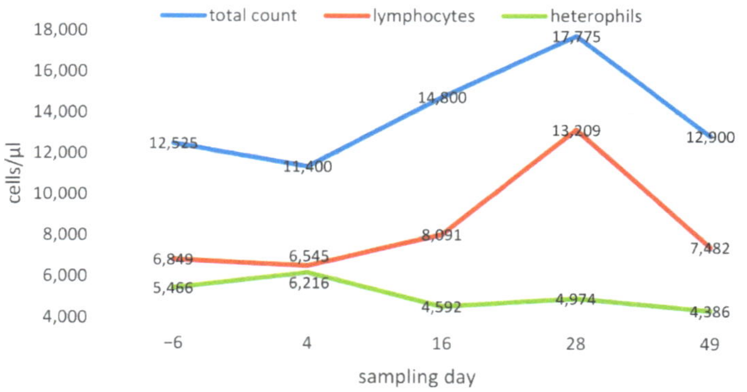

Figure 2. Haematology results in ball pythons (*Python regius*): white blood cell counts, calculated heterophil and lymphocyte counts, shown as median values per sampling day.

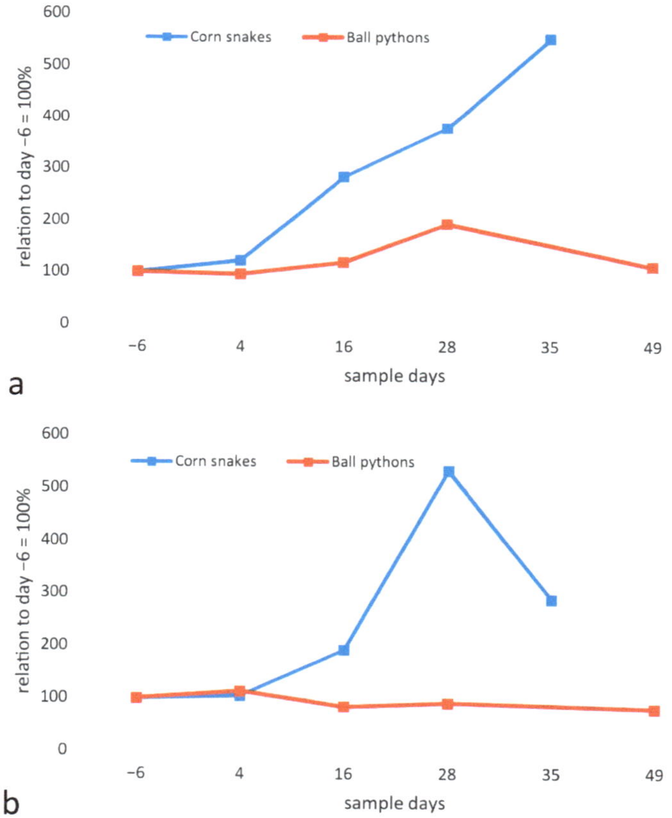

Figure 3. Blood leukocyte counts, relative changes (in percent, median values), (**a**) total lymphocyte count, (**b**) total heterophil count. Ball pythons (*Python regius*) in this study in comparison to results in corn snakes (*Pantherophis guttatus*) [8].

3.2.2. Hemagglutination Inhibition Assays (HI)

An HI titer of <2 was determined in all snakes on day −6. No antibodies were detected in any of the snakes on day 4 post infection either (all < 2). On day 16, HI titers of between 4 and 128 were detected in four snakes (4/9). On day 28, all remaining snakes were positive, with titers between 256 and 4096. On day 49, titers between 256 and 2048 were measured in the three remaining snakes (Figure 4). Statistical comparison confirmed a significant change over the sampling days (Friedman test, $p = 0.021$), and a significant increase in titers when comparing titers on days −6, 4, and 16 separately with results from day 28 ($p = 0.026$ to 0.027).

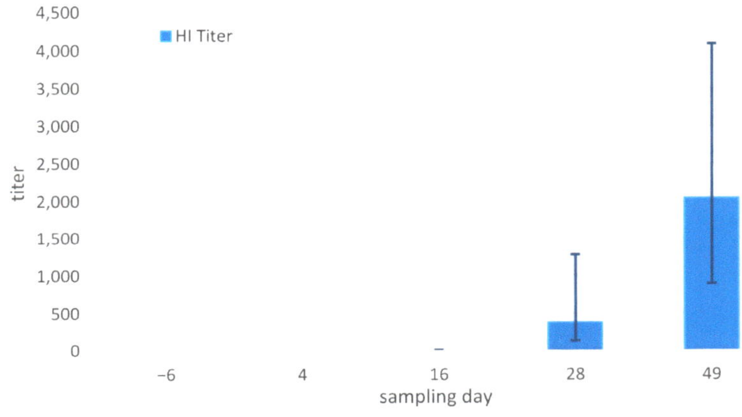

Figure 4. Hemagglutination inhibition assay (HI) titers in ball pythons (*Python regius*), median values and ranges per sampling day.

In comparison, in the corn snake study, only a single corn snake at the end of the study (on day 35) had detectable antibodies when the animal was euthanized due to clinical disease [8].

3.2.3. IgM and IgY Determination

Baseline OD median values in the ball pythons were very similar on days −6 and day 4 p.i. On day 16, only a slight increase in the ODs was measured, with a more prominent increase in IgM levels than in IgY levels. On day 28, the IgM OD median had increased by 290% and on day 49, by 548% in comparison to day −6. For IgY, the median OD had only increased by 16% on day 28, but had increased by 494% on day 49, indicating a later production of IgY (Figure 5). Comparisons in antibody detection showed significant increases for IgY over the course of the sampling days (Friedman test, $p = 0.032$, for IgM $p = 0.052$), and significant increases in the IgY concentrations from day −6 to day 28 ($p = 0.043$) and from day 16 to day 28 ($p = 0.028$).

A comparison of the relative changes in IgM and IgY levels in ball pythons and corn snakes over time showed no measurable increase in IgM levels and a much weaker IGY response to infection in corn snakes (Figure 6a,b).

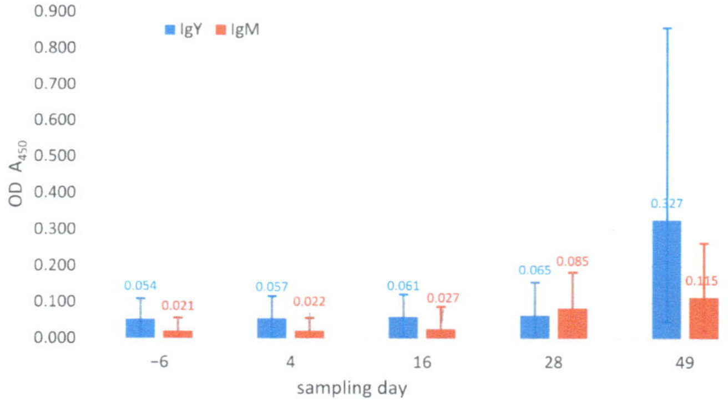

Figure 5. IgY and IgM concentrations in ball pythons (*Python regius*), optical density values, for each sampling day (median, 1st and 3rd quartile) (OD—optical density).

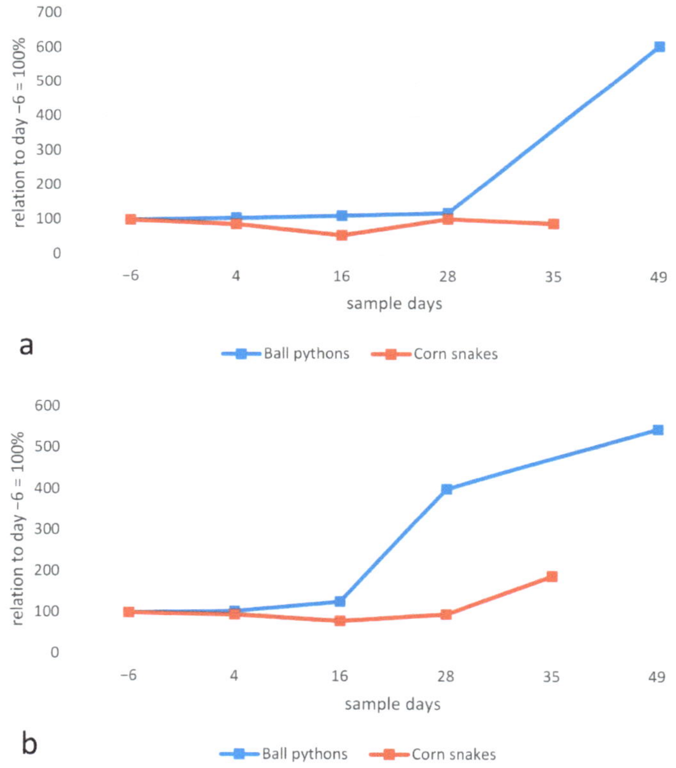

Figure 6. Relative changes in the measured optical densities reflecting concentration of IgM (**a**) and IgY (**b**) (median values) compared to day −6, ball pythons (*Python regius*) in this study in comparison to results in corn snakes (*Pantherophis guttatus*) [8].

3.3. Virus Detection

All tracheal wash samples examined were PCR positive, but the virus was only isolated in cell culture from 61% (11/18) (Table 4). PCR detected a ferlavirus in 72% (13/18) of the cloacal swabs tested, while virus isolation was successful in 61% (11/18). Regarding the sampling days, and assuming that one positive detection (PCR or cell culture) is sufficient to prove virus shedding, cloacal swabs were positive in 39% on day 16 (8/12), 67% on day 28 (4/6) and 100% on day 49 (3/3).

Table 4. Results of PCR and cell culture isolation for ferlavirus detection in intra vitam samples from ball pythons (*Python regius*) for each sampling day (days −6, 16, 28, 49). Results are shown as PCR/cell culture.

No.	Tracheal Wash				Cloacal Swab			
	day −6	**day 16**	**day 28**	**day 49**	**day −6**	**day 16**	**day 28**	**day 49**
4	−/−	+/+			−/−	+/+		
5	−/−	+/+			−/−	−/+		
6	−/−	+/+			−/−	+/+		
7	−/−	+/+	+/−		−/−	+/+	+/+	
8	−/−	+/+	+/+		−/−	+/+	+/−	
9	−/−	+/+	+/−		−/−	−/+	−/−	
10	−/−	+/+	+/−	+/−	−/−	+/+	+/+	+/−
11	−/−	+/+	+/−	+/−	−/−	−/−	+/−	+/−
12	−/−	+/+	+/−	+/+	−/−	+/+	−/−	+/+

Of the tissue samples tested following necropsy (Table 5, Figure 7), 52% were PCR positive (31/60), and the virus was isolated in cell culture from 33% (20/60). In total, 43% of the samples that were tested negative in cell culture were positive in PCR (17/40), but only 21% (6/29) of the samples that were tested negative by PCR were positive in cell culture.

Table 5. Results of PCR and cell culture isolation for the detection of ferlaviruses in tissue samples from each ball python (*Python regius*) following euthanasia on day 4, 16, 28, or 49. Results are shown as PCR/cell culture.

Day	No.	Lung	Intestine	Pancreas	Kidney	Brain
day 4	1	+/+	−/−	−/−	−/−	−/−
	2	+/+	−/−	+/−	−/+	+/−
	3	+/+	+/+	−/−	−/−	−/+
day 16	4	+/+	−/+	−/+	+/+	+/−
	5	+/+	−/+	−/−	−/+	+/+
	6	+/+	+/−	+/−	+/−	+/−
day 28	7	+/+	−/−	+/+	+/−	+/−
	8	+/−	+/+	−/−	+/−	+/−
	9	−/−	−/−	+/−	+/−	+/−
day 49	10	−/−	−/−	−/−	+/−	−/−
	11	−/−	+/−	−/−	−/−	−/−
	12	+/+	−/−	−/−	+/+	−/−
all		9/8	4/4	4/2	7/4	7/2

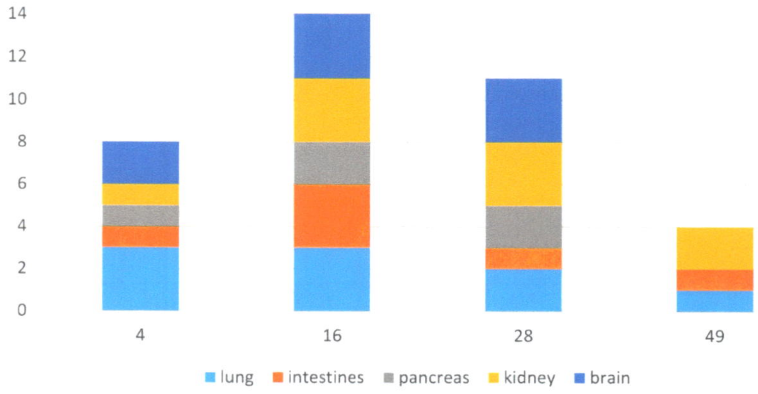

Figure 7. Ferlavirus-positive tissue samples per day (positive in PCR and/or cell culture), maximum would be 15 (three animals, five organs).

Virus detection by at least one of the methods used was successful in 75% of the lung samples, 50% of the intestinal samples, 42% of the pancreas samples, 75% of the kidney samples, and 67% of the brain samples. Regarding the sampling days, the percentage of positive detections in all organs together was 53% on day 4, 93% on day 16, 73% on day 28 and 27% on day 49 (Figure 7).

4. Discussion

This study aimed to compare the effects of an experimental ferlavirus infection in ball pythons (*Python regius*) directly to published results in corn snakes (*Pantherophis guttatus*). For the latter species defined infections with different strains were conducted and the consequences were evaluated and published in several manuscripts [8,9,13]. Previous work has shown that ferlaviruses are capable of infecting a wide range of reptile species [5,10,11,14,20–22], while some evidence has suggested that a number of factors may influence the course of clinical disease and pathological changes in infected animals. Ferlavirus infections have most often been documented in viperid species [1,3,4,23–28], often during severe disease outbreaks in collections. They have also been found in wild-caught viperids from South America [29]. Reports of infection in colubrids are slightly rarer, and include colubrid infections in mixed collections together with viperid snakes as well as in cases in which no contact with viperid snakes was reported [6,10,24,28,30,31]. There are a number of studies reporting detection of ferlaviruses in pythons. In some cases, animals were reported with clinical signs of disease, most often affecting the respiratory system [10,24,28] as well as the central nervous system [10,32], while studies screening boas and pythons in Germany for ferlaviruses by PCR reported no correlation between virus detection and disease [33,34]. While both host species and virus strain have been shown to play a role in disease development, reporting of virus characterization has been limited in many cases, making it difficult to determine what role the host species might play in disease development. The present study was designed to compare the effects of a single virus isolate found to be highly pathogenic in one species (corn snakes) on another host species (ball pythons).

In cases in which virus characterization has been reported, genogroup B viruses have been associated with respiratory or CNS disease or sudden death in boas, colubrids, pythons, and vipers in various collections in Europe [20]; with an outbreak of respiratory disease and high mortality in a mixed collection of elapids, vipers, and pythons in Croatia [35]; with colubrids or elapids experiencing respiratory disease in snake farms in

China [31]; and with viperids, colubrids, elapids, and pythons with respiratory disease (but not in clinically healthy snakes) in several collections in Thailand [28].

4.1. Clinical and Post-Mortem Results

In the pythons, clinical signs were first detected in individual animals on day 4 post infection, and only mild to moderate clinical signs were detected in any of the infected animals over the course of the study (Table 1). This is in stark contrast to the results in corn snakes, where six snakes developed severe clinical signs and the last snake died on day 35 p.i. (Table 1). This difference in the reaction to the virus is also reflected in the pathological changes in the lung (Table 2). In corn snakes, in all snakes except those from sampling day 4, severe macroscopic alterations were found, and the histology results strongly corresponded with this; in the ball pythons, macroscopic changes in the lungs were only considered severe in a single animal in the study (on day 16). These differences in reaction are also seen in the histological findings, with findings in the ball pythons considered severe only in the single animal on day 16, and all other snakes considered either unremarkable or showing mild changes. These results demonstrate that the pathogenic effects of this virus isolate in these two species were very different, and the corn snakes appear to be much more susceptible to clinical disease caused by the virus strain used. This virus isolate was originally obtained from a timber rattlesnake during what was reported to be a severe disease outbreak in a collection of viperid snakes [5]. While it is possible that passages in cell culture or other factors in the laboratory could affect the pathogenicity of the virus, the same passage was used for both the corn snake study and the ball python study described here, indicating that host-specific factors were the most likely source for the differences in clinical course noted.

In addition to the overall clinical condition of the infected animals and the impact of infection on the lungs, indications of systemic disease were found in both species. Pancreas necrosis and lymphoplasmacytic proliferation in the small intestine, liver, and kidney were all noted in snakes in both studies. The virus was also detected in the affected tissues. It is, however, not clear if all of the changes observed were directly related to the ferlavirus infection.

Another factor that could have influenced the clinical and pathological findings found in each of these studies is secondary bacterial infections of the lung. In corn snakes, bacteria, including various *Salmonella* strains, were detected in the lungs of seven of the snakes examined (all from day 28 onwards). In contrast, bacteria were isolated from a much lower number of the infected ball pythons (three animals total). These detections might be incidental findings. However, the bacteria found—*Citrobacter freundii* and *Salmonella*—have been reported previously as causes of lung disease in snakes, also in combination with ferlavirus infections [9,33,34]. Altogether, these results also confirm a lower impact but a general susceptibility of ball pythons to ferlavirus infection.

4.2. Immune Reaction

Changes in white cell counts were one of the parameters used to help evaluate the general immune response to infection. The initial values were within published reference intervals [36]. However, the values of the individual snakes in the study already varied within a broad range. This variability in clinically healthy, uninfected animals, as well as limited sample size and the uncertainties inherent in the estimation method used to determine WBC, meant that despite a clear trend of increasing WBC, the changes over the course of the study were not found to be significant when comparing the sampling days.

The absolute and relative increase in the lymphocyte count (Figures 2 and 3) indicates a strong lymphocyte reaction, which has been reported in some viral infections of reptiles [37,38], although no data are currently available on hematological evaluation of snakes naturally infected with ferlaviruses.

Within the limitations mentioned above, the change in hematological values in the individual ball pythons indicates a general immune response to the ferlavirus infection. The hematological reaction also appears to recede in the ball pythons with lower lymphocyte numbers measured on the final day of the study (day 49), which also corresponds to the histological findings. The differences in the lymphocyte and heterophil reactions in the corn snakes compared to the ball pythons could help explain the differences in the impact of the virus on each of these species. The massive heterophil reaction in corn snakes is a sign of an inflammatory response [38], possibly directly connected to secondary bacterial reactions. As discussed above, at the end of the study, bacteria were isolated from the lung tissues of all of the corn snakes, and corresponding histologic findings confirmed lung infection and inflammation. Increased numbers of heterophils and an increased heterophil:lymphocyte ratio have also been described in infections of turtles with ranaviruses, and the associated inflammation was hypothesized to have contributed to high mortality rates in that study [39].

The antibody reaction was measured according to Neul et al. [8] and focused on the standardized hemagglutination inhibition assay as well as the specific measurement of the IgM and IgY responses. The HI results show a strong antibody response in all pythons included in the study beginning on day 16, and increasing continuously to day 49. The ball pythons thereby showed a much stronger reaction, as antibodies were only detected in a single corn snake at the end of the study [8]. Interestingly, in corn snakes, the less virulent strains (genogroups A and C strains) both caused a stronger humeral immune reaction, with median titers up to 256, similar to the reaction observed against the genogroup B virus in the ball pythons. In the corn snake study, it was hypothesized that the lack of antibody production might be an important factor in the clinical course of disease and could have played a role in the severe disease caused by the genogroup B virus.

A comparison of the results of the IgM and IgY detection between the corn snake study [8] and the ball pythons showed similar dynamics to those seen for the HIs. In the ball pythons, the onset of the specific antibody production can clearly be seen from day 16 onwards, with an initial increase in IgM concentration, followed by IgY. IgM has been well documented as the initial antibody produced in response to infection in reptiles, while IgY is the long-term immunoglobulin [40]. The comparison to the values obtained for corn snakes again demonstrates the lack of an adequate adaptive immune response to this particular virus, while reactions to infection with other virus strains (of genogroups A and C) led to similar increases in both IgM and IgY to those seen against the group B virus in ball pythons, with similar timelines. Even with some limitations due to the small sample size on the last study days, the almost complete lack of a measurable antibody response in the corn snakes again demonstrates the inability of these animals to mount an adequate antibody response to this specific virus strain. As for the hematology and HI results, the comparison demonstrates a very different immune reaction in the two species, with a stronger inflammatory response in corn snakes and a stronger adaptive immune response in ball pythons. Since both species were infected with the same virus strain and aliquotes of the same virus preparation, it is most likely that these differences represent host-specific differences in reaction to infection. Why these reactions were so different remains unknown. While the immune parameters measured here do not represent the entirety of the possible immune response to infection, it is possible to hypothesize that a stronger adaptive immune response to infection in the ball pythons than in the corn snakes likely contributed to the less severe clinical disease and pathology found in the ball pythons, while inflammation likely contributed to disease development in the corn snakes.

4.3. Virus Detection

As previously reported for corn snakes, the PCR used in both studies was found to be more sensitive than virus isolation for ferlavirus detection in the ball pythons. All of the tracheal wash samples tested were PCR-positive beginning on day 16 to the end of the study. In contrast, while cell culture was also positive in all animals on day 16, almost all samples on days 28 and 49 were negative. Shedding of the virus appeared to differ between ball pythons and corn snakes. In both species, the virus was detectable by PCR in tracheal washes in most or all cases beginning on day 16. However, in the corn snakes, no virus was detectable in any of the cloacal swabs tested on day 16, and only in two out of three cloacal swabs tested on day 28 [13]. The PCR used in both studies was the same. More recent comparisons have indicated that the sensitivity of various PCRs can influence virus detection [15,31]. In a study comparing three PCRs for the detection of ferlaviruses, the PCR used in this study was found to be highly specific but less sensitive than a single-round PCR targeting a shorter segment of the same gene [15]. However, for this study, it was considered important to use the same methods as those used for the corn snake study [13] to allow direct comparison of results. Some studies have shown difficulties in detecting ferlaviruses in samples from live snakes, with negative PCR results reported from tracheal washes, while the virus was detectable in lung tissue [27]. In the present study, high infection loads may have led to better virus detection in tracheal washes than would be expected in naturally infected animals, in which time post infection, infectious dose, infection route, and immune status are all unknown.

The general virus prevalence in organs in the ball pythons (Figure 4) demonstrated an increase to day 16 with almost 100% positive samples in all organs, and then a reduction to 26% on day 49. Interestingly, this was similar in corn snakes with other virus strains, but with an identical virus, the detection rate increased continuously to 100% on the last study day (35) [13]. Together with the antibody and hematology results, this finding strengthens the hypothesis that ball pythons were better able to combat infection and therefore developed less severe disease than corn snakes infected with the same virus strain. Whether this would be true for other virus strains, or if host-specific response to infection is dependent on a combination of host species and virus strain, remains to be studied.

In both ball pythons and corn snakes, the ferlavirus was able to spread from the lungs to various other tissues, including all of the tissues tested (intestine, pancreas, kidney, and brain), although clinical signs were restricted to the respiratory tract. It is remarkable that virus detection in tissues other than the lung was possible earlier in ball pythons than in corn snakes, correlating with the higher rate of ferlavirus detection in cloacal swabs in the ball pythons. This more rapid spread throughout the body was not associated with higher pathogenicity. It appears that tissue distribution and shedding are not correlated with disease severity for this virus. However, it is possible that the more rapid spread of the virus to various tissues in the ball pythons could have contributed to the stronger adaptive immune response in this species. Receptor specificity has been studied and found to influence tissue tropism, pathogenicity, and immune response to paramyxoviruses in a wide variety of virus/host systems [41]. Systemic spread of virus has been correlated with specific motifs in the viral glycoproteins of many paramyxoviruses [40,42,43], and with higher pathogenicity in many cases. While sequence data from the genes encoding the viral surface glycoproteins F and HN have been determined for several ferlavirus isolates [44,45] including the one used in this study [9], the effects of specific changes or motifs in these receptor-binding proteins are not known, nor have the receptors in various tissues in reptiles been studied.

It is notable that while the lung samples were negative for virus detection in individual animals towards the end of the study, tracheal washes remained virus-positive in PCR tests. This is in contrast to previous reports in which virus detection in samples from tracheal washes was less sensitive than from lung samples in naturally infected vipers [27]. There are many possible reasons for this discrepancy including differences in time post infection, inoculation of a relatively high amount of virus in this transmission study, the ability of the PCR to detect RNA from viruses no longer capable of replicating, or a phase of infection in which virus has been cleared from the lung but not from the trachea.

4.4. Limitations of the Study Design

Due to the nature of experimental infections, this study has several limitations that need to be considered when interpreting the data. The study is focused on a defined group of only 12 animals, and for some sample days, the sample number was limited to three. On the other hand, the study design allowed for defined conditions and paired samples. The high virus load used in the study likely does not reflect the conditions during a natural infection. Ferlaviruses are likely spread by aerosol or direct contact between animals [1,2,4–], but the amount of virus transmitted naturally is likely much lower than that used here. This study only used one defined virus strain, and no conclusions can be drawn regarding general differences between specific species with respect to ferlavirus susceptibility.

The lack of a control group of sham-inoculated ball pythons for direct comparison with the infected animals is a further limitation of the present study. The reason no control group was used was that, since the same protocol was used as in the previous study with corn snakes [13], it was deemed unlikely to provide any additional information, since the consequences of the application of cell culture supernatant (without virus) for a placebo group had already been described and discussed. It also conformed with the RRR principle (reduce refine replace) for animal trials.

5. Conclusions

The genogroup B ferlavirus strain used in this study was clearly able to infect and cause typical disease in the ball pythons. However, differences in severity of disease between corn snakes and ball pythons infected with the same strain demonstrate species-specific differences in virus pathogenicity as well as in host reaction to infection. These differences are relevant for clinical reptile medicine, virus epidemiology, and for diagnostic testing. Additional studies to understand the molecular basis of the differences should be carried out.

Author Contributions: Conceptualization, M.P. and R.E.M.; methodology, M.P., V.S., W.S. and R.E.M.; software, M.P.; validation, M.P., W.S. and R.E.M.; formal analysis, A.M., V.S., W.S. and R.E.M.; investigation, A.M., V.S. and M.P.; resources, M.P., W.S. and R.E.M.; data curation, A.M. and M.P.; writing—original draft preparation, M.P. and R.E.M.; writing—review and editing, M.P., A.M., V.S., W.S. and R.E.M.; visualization, M.P.; supervision, M.P.; project administration, M.P.; funding acquisition, M.P. All authors have read and agreed to the published version of the manuscript.

Funding: This research was funded by the German Research Foundation (DFG), grant number Pe 877/2-2. This Open Access publication was funded by the Deutsche Forschungsgemeinschaft (DFG, German Research Foundation)—491094227 "Open Access Publication Funding" and the University of Veterinary Medicine Hannover, Foundation.

Institutional Review Board Statement: The animal study protocol was approved by the national authority (Landesdirektion Sachsen, application number TVV 61/13).

Informed Consent Statement: Not applicable.

Data Availability Statement: The data presented in this study are openly available in osf.io at https://doi.org/10.17605/OSF.IO/NRFW6 (accessed on 30 March 2023).

Conflicts of Interest: The authors declare no conflict of interest.

References

1. Fölsch, D.W.; Leloup, P. Fatal endemic infection in a serpentarium, diagnosis, treatment and preventive measures. *Tierarztl. Prax.* **1976**, *4*, 527–536. [PubMed]
2. Hyndman, T.H.; Shilton, C.M.; Marschang, R.E. Paramyxoviruses in reptiles: A review. *Vet. Microbiol.* **2013**, *165*, 200–213. [CrossRef] [PubMed]
3. Jacobson, E.; Gaskin, J.M.; Page, D.; Iverson, W.O.; Johnson, J.W. Illness associated with paramyxo-like virus infection in a zoologic collection of snakes. *J. Am. Vet. Med. Assoc.* **1981**, *179*, 1227–1230.
4. Jacobson, E.R.; Gaskin, J.M.; Wells, S.; Bowler, K.; Schumacher, J. Epizootic of ophidian paramyxovirus in a zoological collection: Pathological, microbiological, and serological findings. *J. Zoo Wildl. Med.* **1992**, *23*, 318–327.
5. Marschang, R.E.; Papp, T.; Frost, J.W. Comparison of paramyxovirus isolates from snakes, lizards and a tortoise. *Virus Res.* **2009**, *144*, 272–279. [CrossRef]
6. Abbas, M.D.; Marschang, R.E.; Schmidt, V.; Kasper, A.; Papp, T. A unique novel reptilian paramyxovirus, four atadenovirus types and a reovirus identified in a concurrent infection of a corn snake (*Pantherophis guttatus*) collection in Germany. *Vet. Microbiol.* **2011**, *150*, 70–79. [CrossRef]

7. Blahak, S. Isolation and Characterization of Paramyxoviruses from Snakes and their Relationship to Avian Paramyxoviruses. *J. Vet. Med. B* **1995**, *42*, 216–224. [CrossRef]
8. Neul, A.; Schrödl, W.; Marschang, R.E.; Bjick, T.; Truyen, U.; von Buttlar, H.; Pees, M. Immunologic responses in corn snakes (*Pantherophis guttatus*) after experimentally induced infection with ferlaviruses. *Am. J. Vet. Res.* **2017**, *78*, 482–494. [CrossRef]
9. Pees, M.; Schmidt, V.; Papp, T.; Gellért, A.; Abbas, M.; Starck, J.M.; Neul, A.; Marschang, R.E. Three genetically distinct ferlaviruses have varying effects on infected corn snakes (*Pantherophis guttatus*). *PLoS ONE* **2019**, *14*, e0217164.
10. Papp, T.; Pees, M.; Schmidt, V.; Marschang, R.E. RT-PCR diagnosis followed by sequence characterization of paramyxoviruses in clinical samples from snakes reveals concurrent infections within populations and/or individuals. *Vet. Microbiol.* **2010**, *144*, 466–472.
11. Papp, T.; Seybold, J.; Marschang, R.E. Paramyxovirus Infection in a Leopard Tortoise (*Geochelone pardalis babcocki*) with Respiratory Disease. *J. Herpetol. Med. Surg.* **2010**, *20*, 64–68. [CrossRef]
12. Jacobson, E.R.; Adams, H.P.; Geisbert, T.W.; Tucker, S.J.; Hall, B.J.; Homer, B.L. Pulmonary Lesions in Experimental Ophidian Paramyxovirus Pneumonia of Aruba Island Rattlesnakes, *Crotalus unicolor*. *Vet. Pathol.* **1997**, *34*, 450–459. [CrossRef]
13. Pees, M.; Neul, A.; Müller, K.; Schmidt, V.; Truyen, U.; Leinecker, N.; Marschang, R.E. Virus distribution and detection in corn snakes (*Pantherophis guttatus*) after experimental infection with three different ferlavirus strains. *Vet. Microbiol.* **2016**, *182*, 213–222. [PubMed]
14. Ahne, W.; Batts, W.; Kurath, G.; Winton, J. Comparative sequence analyses of sixteen reptilian paramyxoviruses. *Virus Res.* **1999**, *63*, 66–74. [CrossRef]
15. Kolesnik, E.; Hyndman, T.H.; Müller, E.; Pees, M.; Marschang, R.E. Comparison of three different PCR protocols for the detection of ferlaviruses. *BMC Vet. Res.* **2019**, *15*, 281.
16. Terrel, S.P.; Stacy, B.A. Reptile necropsy techniques. In *Infectious Diseases and Pathology of Reptiles Color Atlas and Text*; Jacobson, E.R., Ed.; CRC Press: Boca Raton, FL, USA, 2007; pp. 219–256.
17. Campbell, T.W.; Ellis, C.K. *Avian and Exotic Animal Hematology and Cytology*, 3rd ed.; Blackwell Publishing: Ames, IA, USA, 2007; pp. 3–81.
18. Rösler, R.; Abbas, M.D.; Papp, T.; Marschang, R.E. Detection of antibodies against paramyxoviruses in tortoises. *J. Zoo Wildl. Med.* **2013**, *44*, 333–339. [CrossRef] [PubMed]
19. Clark, H.F.; Lief, F.S.; Lunger, P.D.; Waters, D.; Leloup, P.; Fölsch, D.W. Fer de Lance virus (FDLV): A probable paramyxovirus isolated from a reptile. *J. Gen Virol.* **1979**, *44*, 405–418. [CrossRef]
20. Jacobson, E.R.; Origgi, F.; Pessier, A.P.; Lamirande, E.W.; Walker, I.; Whitaker, B.; Stalis, I.H.; Nordhausen, R.; Owens, J.W.; Nichols, D.K.; et al. Paramyxovirus Infection in Caiman Lizards (*Draecena Guianensis*). *J. Vet. Diagn. Investig.* **2001**, *13*, 143–151. [CrossRef]
21. Abbas, M.D.; Ball, I.; Ruckova, Z.; Öfner, S.; Stöhr, A.C.; Marschang, R.E. Virological Screening of Bearded Dragons (*Pogona vitticeps*) and the First Detection of Paramyxoviruses in This Species. *J. Herpetol. Med. Surg.* **2012**, *22*, 86–90. [CrossRef]
22. Woo, P.C.; Martelli, P.; Hui, S.-W.; Lau, C.C.; Groff, J.M.; Fan, R.Y.; Lau, S.K.; Yuen, K.-Y. Anaconda paramyxovirus infection in an adult green anaconda after prolonged incubation: Pathological characterization and whole genome sequence analysis. *Infect. Genet. Evol.* **2017**, *51*, 239–244. [CrossRef]
23. Jacobson, E.; Gaskin, J.M.; Simpson, C.F.; Terrell, T.G. Paramyxo-like infection in a rock rattlesnake. *J. Am. Vet. Med. Assoc.* **1980**, *177*, 796–799.
24. Orós, J.; Sicilia, J.; Torrent, A.; Castro, P.; Arencibia, A.; Déniz, S.; Jacobson, E.R.; Homer, B.L. Immunohistochemical detection of ophidian paramyxovirus in snakes in the Canary Islands. *Vet. Rec.* **2001**, *149*, 21–23. [CrossRef]
25. Kolesnikovas, C.K.M.; Grego, K.F.; de Albuquerque, L.C.R.; Jacobson, E.R.; Monezi, T.A.; Mehnert, D.U.; Catão-Dias, J.L. Ophidian paramyxovirus in Brazilian vipers (*Bothrops alternatus*). *Vet. Rec.* **2006**, *159*, 390–392. [CrossRef]
26. Solis, C.; Arguedas, R.; Baldi, M.; Piche, M.; Jimenez, C. Seroprevalence and molecular characterization of ferlavirus in captive vipers of costa rica. *J. Zoo Wildl. Med.* **2017**, *48*, 420–430. [CrossRef]
27. Flach, E.J.; Dagleish, M.P.; Feltrer, Y.; Gill, I.S.; Marschang, E.R.; Masters, N.; Orós, J.; Pocknell, A.; Rendle, E.M.; Strike, T.; et al. Ferlavirus-related deaths in a collection of viperid snakes. *J. Zoo Wildl. Med.* **2018**, *49*, 983–995. [PubMed]
28. Piewbang, C.; Wardhani, S.W.; Poonsin, P.; Yostawonkul, J.; Chai-In, P.; Lacharoje, S.; Saengdet, T.; Vasaruchapong, T.; Boonrungsiman, S.; Kongmakee, P.; et al. Epizootic reptilian ferlavirus infection in individual and multiple snake colonies with additional evidence of the virus in the male genital tract. *Sci. Rep.* **2021**, *11*, 12731. [CrossRef] [PubMed]
29. Junqueira de Azevedo, I.L.M.; Prieto da Silva, Á.R.; Carmona, E.; Ho, P.L. Characterization of a Paramyxovirus from a Fer de Lance viper (*Bothrops jararaca*): Partial nucleotide sequence of the putative fusion protein. *Arch. Virol.* **2001**, *146*, 51–57. [CrossRef] [PubMed]
30. Ahne, W.; Neubert, W.J.; Thomsen, I. Reptilian Viruses: Isolation of Myxovirus-like Particles from the Snake Elaphe Oxycephala. *J. Vet. Med.* **1987**, *34*, 607–612. [CrossRef]
31. Su, J.-Y.; Li, J.; Que, T.-C.; Chen, H.-L.; Zeng, Y. Detection and molecular epidemiology of ferlaviruses in farmed snakes with respiratory disease in Guangxi Province, China. *J. Vet. Diagn. Investig.* **2020**, *32*, 429–434. [CrossRef]
32. West, G.; Garner, M.; Raymond, J.; Latimer, K.S.; Nordhausen, R. Meningoencephalitis in a boelen's python (morelia boeleni) associated with paramyxovirus infection. *J. Zoo Wildl. Med.* **2001**, *32*, 360–365.
33. Pees, M.; Schmidt, V.; Marschang, R.E.; Heckers, K.O.; Krautwald-Junghanns, M.-E. Prevalence of viral infections in captive collections of boid snakes in Germany. *Vet. Rec.* **2010**, *166*, 422–425. [CrossRef]

34. Schmidt, V.; Marschang, R.E.; Abbas, M.D.; Ball, I.; Szabo, I.; Helmuth, R.; Plenz, B.; Spergser, J.; Pees, M. Detection of pathogens in Boidae and Pythonidae with and without respiratory disease. *Vet. Rec.* **2013**, *172*, 236. [CrossRef]
35. Prpic, J.; Keros, T.; Balija, M.L.; Forcic, D.; Jemersic, L. First recorded case of paramyxovirus infection introduced into a healthy snake collection in Croatia. *BMC Vet. Res.* **2017**, *13*, 95. [CrossRef] [PubMed]
36. Gibbons, P.M.; Whitaker, B.R.; Carpenter, J.W.; McDermott, C.T.; Klaphake, E.; Sladky, K.K. Hematology and Biochemistry Tables. In *Mader's Reptile and Amphibian Medicine and Surgery*, 3rd ed.; Divers, S.J., Stahl, S.J., Eds.; Elsevier: St Louis, MO, USA, 2019; pp. 333–350.
37. Allender, M.C.; Mitchell, M.A.; Phillips, C.A.; Gruszynski, K.; Beasley, V.R. Hematology, plasma biochemistry, and antibodies to select viruses in wild-caught eastern massasauga rattlesnakes (*Sistrurus catenatus catenatus*) from illinois. *J. Wildl. Dis.* **2006**, *42*, 107–114. [CrossRef] [PubMed]
38. Stacy, N.I.; Harr, K.E. Hematology of Reptiles with a Focus on Circulating Inflammatory Cells. In *Infectious Diseases and Pathology of Reptiles Color Atlas and Text*, 2nd ed.; Jacobson, E.R., Garner, M.M., Eds.; CRC Press: Boca Raton, FL, USA, 2021; pp. 267–33[illegible].
39. Rayl, J.M.; Allender, M.C. Temperature affects the host hematological and cytokine response following experimental ranavirus infection in red-eared sliders (*Trachemys scripta elegans*). *PLoS ONE* **2020**, *15*, e0241414. [CrossRef]
40. Origgi, F.C.; Tecilla, M. Immunology of reptiles. In *Infectious Diseases and Pathology of Reptiles Color Atlas and Text*, 2nd ed.; Jacobson, E.R., Garner, M.M., Eds.; CRC Press: Boca Raton, FL, USA, 2021; pp. 553–671.
41. Navaratnarajah, C.K.; Generous, A.R.; Yousaf, I.; Cattaneo, R. Receptor-mediated cell entry of paramyxoviruses: Mechanisms, and consequences for tropism and pathogenesis. *J. Biol. Chem.* **2020**, *295*, 2771–2786. [CrossRef]
42. Klenk, H.-D.; Garten, W. Host cell proteases controlling virus pathogenicity. *Trends Microbiol.* **1994**, *2*, 39–43. [CrossRef]
43. Lamb, R.A.; Kolakofsky, D. Paramyxoviridae: The viruses and their replication. In *Fields Virology*, 4th ed.; Kniepe, D.M., Howley, P.M., Eds.; Raven Press: New York, NY, USA, 2001; pp. 1305–1340.
44. Kurath, G.; Batts, W.N.; Ahne, W.; Winton, J.R. Complete Genome Sequence of Fer-de-Lance Virus Reveals a Novel Gene in Reptilian Paramyxoviruses. *J. Virol.* **2004**, *78*, 2045–2056. [CrossRef]
45. Franke, J.; Batts, W.N.; Ahne, W.; Kurath, G.; Winton, J.R. Sequence motifs and prokaryotic expression of the reptilian paramyxovirus fusion protein. *Arch. Virol.* **2006**, *151*, 449–464. [CrossRef]
46. Pasmans, F.; Blahak, S.; Martel, A.; Pantchev, N. Introducing reptiles into a captive collection: The role of the veterinarian. *Vet. J.* **2008**, *175*, 53–68. [CrossRef]

animals

MDPI

Article

Analytical and Clinical Evaluation of Two Methods for Measuring Erythrocyte Sedimentation Rate in Eastern Indigo Snakes (*Drymarchon couperi*)

James E. Bogan, Jr.

Central Florida Zoo's Orianne Center for Indigo Conservation, Brantley Branch Road, Eustis, FL 30931, USA; jamesb@centralfloridazoo.org

Simple Summary: Having a simple, reliable test to detect illness is very useful in screening animals for disease. The erythrocyte sedimentation rate (ESR) is a blood test that can detect inflammation. Although not specific for any particular disease, ESR is often used in humans as a screening "sickness indicator" due to its reliability and low cost. Little investigation of ESR in reptiles has been conducted. This study evaluates two ESR techniques in eastern indigo snakes (*Drymarchon couperi*) and found both tests performed equally. In addition, eastern indigo snakes with some inflammatory conditions had higher ESR measurements than healthy eastern indigo snakes.

Abstract: Erythrocyte sedimentation rate (ESR) is a hematological test that can detect inflammatory activity within the body. Although not specific for any particular disease, ESR is often used as a screening "sickness indicator" due to its reliability and low cost. The Westergren method is a manual ESR technique commonly used but requires special graduated pipettes and over 1mL of whole blood, precluding its use in smaller patients where limited sample volumes can be obtained. A modified micro-ESR technique has been described using hematocrit capillary tubes but is used less commonly. ESR has been reported to be a useful inflammatory indicator in gopher tortoises (*Gopherus polyphemus*) and box turtles (*Terrapene* spp.) but not in Florida cottonmouth snakes (*Agkistrodon conanti*). Having an inexpensive screening test for inflammation can help guide medical decisions within conservation efforts of imperiled species. This study evaluated the correlation between these two ESR methodologies in threatened eastern indigo snakes (*Drymarchon couperi*, EIS) and found a very strong correlation (r_s = 0.897), without constant or proportional biases and a reference interval of 0 (90% CI -1-1)–9 mm/h (90% CI 8-11) was defined. Additionally, a significant difference was found between healthy EIS and EIS in mid-ecdysis (p = 0.006) and EIS with gastric cryptosporidiosis (p = 0.006), indicating ESR as a useful inflammatory indicator in EIS.

Keywords: erythrocyte sedimentation rate; westergren method; micro-ESR; reptile; inflammation; clinical pathology

Citation: Bogan, J.E., Jr. Analytical and Clinical Evaluation of Two Methods for Measuring Erythrocyte Sedimentation Rate in Eastern Indigo Snakes (*Drymarchon couperi*). *Animals* **2023**, *13*, 464. https://doi.org/10.3390/ani13030464

Academic Editor: Tom Hellebuyck

Received: 2 January 2023
Revised: 23 January 2023
Accepted: 24 January 2023
Published: 28 January 2023

1. Introduction

Erythrocyte sedimentation rate (ESR) is a hematological test that can detect inflammatory activity within the body caused by autoimmune diseases, infections, or neoplasia [1]. Although not specific for any particular disease, ESR is typically used in conjunction with other tests to determine increased inflammatory activity. Often, ESR can be used as a screening "sickness indicator" due to its reliability and low cost [1].

When blood is placed into a vertical column, erythrocytes (RBC) will precipitate or settle at a constant rate, which is referred to as sedimentation. During acute inflammation, inflammatory proteins cause RBCs to aggregate together and precipitate faster [1,2]. This increase sedimentation rate can be measured by a variety of manual methods or with automatic machines. The most commonly used manual method to measure ESR is the Westergren method [1]. Whole blood which has been mixed with sodium citrate is placed into a 200 mm tube with a 2.5 mm internal bore. This tube is then held in a vertical position for 1 hour at which time the degree of sedimentation is measured in millimeters. The Westergren method requires 1.25 mL of blood in order to measure ESR, limiting its use in smaller patients where that volume cannot be safely collected. As an alternative, a micro-ESR method has also been described using a hematocrit tube in a similar fashion [3].

Although not uncommonly used in human medicine, ESR is relatively underutilized in veterinary medicine and especially in reptile medicine. However, a few studies have evaluated the use of ESR in reptiles. In gopher tortoises (*Gopherus polyphemus*) and box turtles (*Terrapene* spp.), ESR has been shown to be a useful indicator of inflammation [4,5]. In Florida cottonmouth snakes (*Agkistrodon conanti*), however, ESR has not been shown to be a reliable indicator of inflammation [2].

The eastern indigo snake (*Drymarchon couperi*, EIS) is a large, diurnal colubrid native to the southeastern United States and is listed as a Threatened Species through the Endangered Species Act [6]. The EIS has some unique physiological characteristics, including extremely high plasma calcium and phosphorus levels [7,8]. Determining the normal ESR of EIS would benefit conservation efforts by allowing for monitoring of inflammatory conditions in this federally threatened species. Several inflammatory conditions have been described in both free-ranging and captive EIS including ophidiomycosis [9], pentastomiasis [10], dystocia [11], and gastric cryptosporidiosis [12]. EIS collected from the wild may have been subjected to inflammation or acute phase protein related processes and ESR may be a good screening test to determine if occult inflammatory processes, such as infectious disease, are present. Screening for infectious disease is the hallmark of good biosecurity procedures either prior to an animal entering a captive collection or prior to animal release in controlled reintroductions efforts [13,14].

The purpose of this study is to (1) compare two manual ESR methods, Westergren and micro-ESR, in EIS, (2) establish an ESR reference interval for EIS, and (3) compare the established ESR reference interval with ESR results in EIS with inflammatory conditions.

2. Materials and Methods

This study was approved by the Central Florida Zoo & Botanical Gardens Research Committee (Project 2022-04) and was separated into three phases. During March 2022, 17 EIS from a captive breeding colony were used in Phase 1 of this study, 12 male and 5 female. All EIS were housed according to the Association of Zoos and Aquariums Taxon Advisory Group recommended guidelines [15]. Briefly, each snake was housed individually in an 18.4 cm × 66.7 cm × 83.8 cm (7.25 in. × 26.25 in. × 33 in.) polyvinylchloride drawer and rack system (ARS, Indianapolis, IN, USA) with newsprint substrate within a dedicated room kept at 25.5 °C. A thermal gradient was not provided, and each enclosure had a polycarbonate window on one side. Lighting was available from the room's overhead fluorescent lights and indirect sunlight through the room's glass window which was shaded by an outside awning. The fluorescent lights were on for 8 hours a day, while the indirect sunlight allowed for a natural diurnal photoperiod. The diet was offered twice weekly and consisted of thawed frozen prey items, rotated between rats (*Rattus norvegicus*), mice (*Mus musculus*), domestic chicken chicks (*Gallus domesticus*), Japanese quail chicks (*Coturnix japonica*), and rainbow trout (*Oncorhynchus mykiss*).

All 17 EIS had been previously diagnosed with gastric cryptosporidiosis. Initial diagnosis was made with a *Cryptosporidium serpentis*-specific probe-hybridization quantitative polymerase chain reaction assay (qPCR) on a cloacal swab or gastric swab sample [16]. The diagnosis was confirmed with histological and qPCR analyses of gastric mucosal biopsies obtained with an endoscope. Blood samples were collected as part of a treatment investigational study [17].

All blood samples were collected via cardiocentesis with a 23-gauge hypodermic needle attached to a 3 mL syringe. Blood was immediately placed into a lithium heparin microtainer tube (Micro tube LH, Sarstedt AG & Co., Nümbrecht, Germany) according to manufacturer's recommendations and thoroughly mixed by inversion. Measurement of ESR was completed within 2 hours of obtaining samples.

Two techniques were used to measure ESR, the Westergren method and the micro-ESR method. The blood samples were mixed thoroughly by inverting the microtainers 8 times prior to ESR measurements. The Westergren method used a commercial ESR kit (DISPETTEone, Guest Scientific AG, Cham, Switzerland) where 1.25 mL of whole blood was pipetted into the kit's plastic reservoir and a 2.5 mm × 200 mm graduated measuring pipette was then inserted into the reservoir per manufacturer's guidelines, filling the pipette by capillary action. The reservoir-pipette assembly was then placed in a rack on a level surface which ensured the tube was precisely vertical. For the micro-ESR technique, a standard non-heparinized microhematocrit capillary tube (Jorgensen Laboratories, Loveland, CO, USA) was used. The tubes were 0.9 ± 0.05 mm × 75 ± 0.05 mm. The tube was filled to the top with whole blood from the lithium heparinized microtainer then packed with 5 mm of clay, resulting in a blood volume of 0.04 mL used. The capillary tubes were placed vertically into the recesses of the hematocrit clay tray and placed on a level surface. A right angle was used to ensure the capillary tubes were precisely vertical. Both Westergren reservoir-pipette assemblies and hematocrit capillary tubes were allowed to set for 1 hour at 25.5 °C before the degree of erythrocyte sedimentation was measured in millimeters. The capillary tubes were then centrifuged (Zipocrit, LW Scientific, Lawrenceville, GA, USA) at 4400× g for 5 min to measure the packed cell volume (PCV).

During April and May 2018, Phase 2 of this study used 21 EIS (7 male, 14 female) that were considered healthy by physical examination and were slated to be released into the wild as part of a repatriation program. Blood samples were collected by a veterinarian during general anesthesia for radiotelemetry transponder placement and the Westergren method ESR was measured within 2 h of obtaining samples as previously described.

For Phase 3 of the study, medical records from the Central Florida Zoo & Botanical Gardens between February 2018 through November 2022 were reviewed for either Westergren or micro-ESR use in EIS. The EIS were separated into one of four groups based on the clinical diagnosis at the time of the ESR result: healthy, gastric cryptosporidiosis, ecdysis, or dystocia.

The results from the ESR measurements from all three study phases were then corrected for anemia and/or lymph dilution with Fabry's formula [18]:

$$ESR_{corrected} = ESR_{measured} \times (55 - PCV_{ideal})/(55 - PCV_{measured}). \quad (1)$$

The mean PCV of free-ranging EIS (32%) was used as PCV_{ideal} [7].

Analyses were computed using Excel (Microsoft 365, Redmond, WA, USA) and Real Statistics add-in software (https://www.real-statistics.com/free-download/real-statistics-resource-pack/ (accessed on 9 May 2022)). Spearman correlation coefficients (r_s) were obtained to measure the linear association of Westergren and micro-ESR methods. An r_s value of 0.80 to 1.0 was considered very strong correlation; 0.60 to 0.79 was moderate correlation; 0.30 to 0.59, fair correlation; 0.10 to 0.29, poor correlation; and 0.00 to 0.09, no correlation [19]. Passing–Bablok regression analysis was used to estimate constant and proportional bias between methods. Constant bias indicates that micro-ESR method consistently measures ESR to be higher or lower in comparison with the Westergren method. If the confidence interval for the y-intercept did not include the value 0, this was considered evidence of constant bias. Proportional bias indicates that the differences in measurements of each method are proportional to the level of measurement. If the 95% confidence interval for the slope did not include the value of 1, this was considered evidence of proportional bias. Bland–Altman plots were used for visualization and quantification of the agreement of the results between methods.

Distribution of $ESR_{corrected}$ measurements obtained from healthy EIS were evaluated for normality with kurtosis, skewness, and Shapiro–Wilk test. Outliers were determined through histogram analysis and Grubbs' test. Results between the sexes were compared with Mann–Whitney test and if a significant difference was not found the results were combined. Guidelines for establishing reference intervals from the American Society of Veterinary Clinical Pathologists (ASVCP) were followed [20]. A reference interval was calculated using Reference Value Advisor V2.1 [21] for Microsoft Excel. The Anderson-Darling test was used to evaluate the distribution of results and non-parametric methods were used to compare medians and to define the 95% reference interval.

Results of $ESR_{corrected}$ measurements between EIS groups were compared. Data distribution was evaluated with kurtosis, skewness, and Shapiro–Wilk test. Comparison between the groups was performed through a Kruskal–Wallis test and post hoc analysis with paired Mann–Whitney tests. The results of normally distributed data are reported as mean, standard deviation, minimum, and maximum values; non-normal data are reported as median, minimum, and maximum values. Statistical significance was set a $p < 0.05$.

3. Results

The Westergren and micro-ESR techniques had good agreement as all data points on the Bland–Altman plot were within 1.96 standard deviations (Figure 1). The Spearman coefficient indicated a very strong correlation ($r_s = 0.897$) between ESR methods. Passing–Bablok regression analysis of the methods resulted in a regression equation

$$y = 1.00\ (95\%\ \mathrm{CI}:\ 0.71 - 1.27) + 0.00\ (95\%\ \mathrm{CI}: -2.54\ - 1.86)\ x \quad (2)$$

while the significance of linearity was acceptable ($p = 0.799$) (Figure 2). Since $p = 0.799 > \alpha = 0.05$, it was concluded that the linearity assumption was likely to hold, without constant or proportional bias.

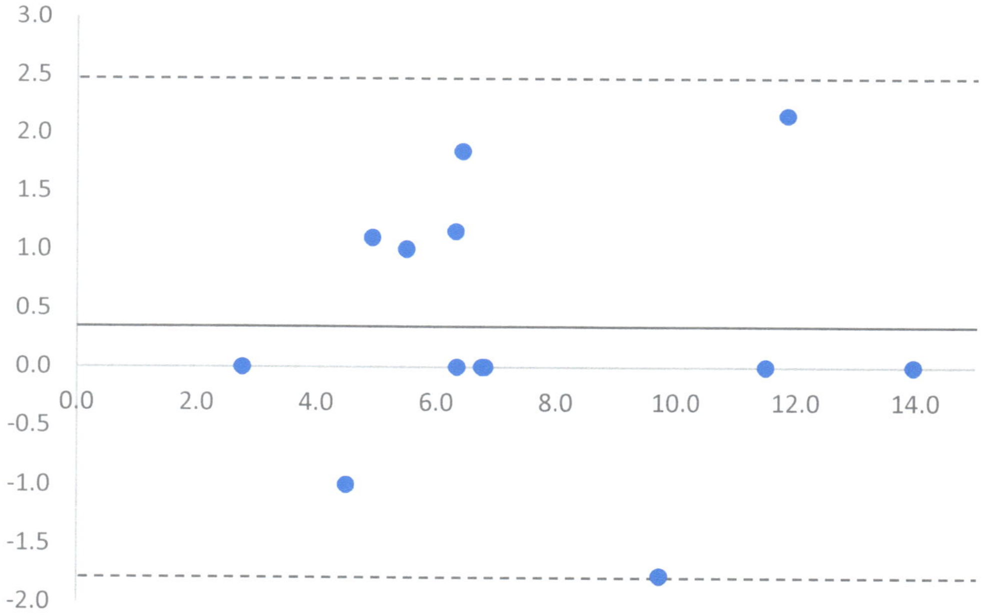

Figure 1. Bland–Altman Plot of erythrocyte sedimentation rate (ESR) measurements between the Westergren and micro-ESR methods. Dotted lines are ± 1.96 standard deviations from mean (solid black line). All data points are between 1.96 standard deviations signifying good agreement between methods.

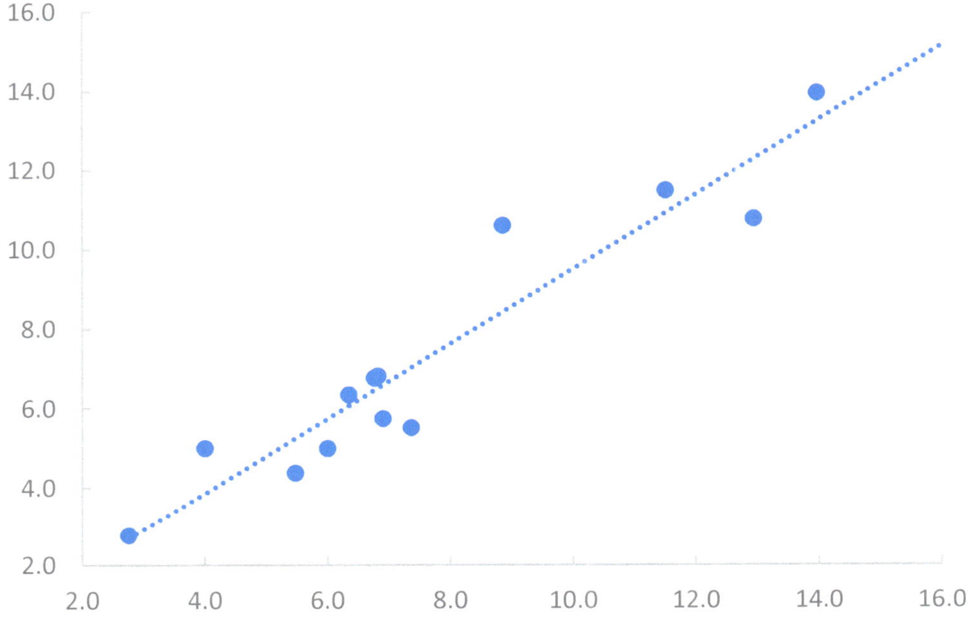

Figure 2. Correlation of erythrocyte sedimentation rate (ESR) measurements between the Westergren and micro-ESR methods. Spearman correlation coefficient (r_s) is 0.897, signifying a very strong correlation between these two test methods. The dotted line represents the line of best fit.

Two of the healthy EIS in Phase 2 were in mid-ecdysis at the time of blood collection and the results were binned in a separate cohort (mid-ecdysis). Phase 3's medical record review revealed 16 additional cases where ESR was used in EIS: 4 healthy EIS, 6 EIS in mid-ecdysis, and 6 EIS with dystocia. In the medical record review, either Westergren method or micro-ESR methods were used. Since both methods had a very strong correlation and good agreement, all results of ESR measurements were included for comparison, regardless of the method used. A total of 54 ESR values for EIS were available for comparison (23 healthy EIS, 17 EIS with gastric cryptosporidiosis, 8 EIS in mid-ecdysis, and 6 EIS with dystocia).

The results of $ESR_{corrected}$ measurements in healthy EIS were normally distributed (Shapiro–Wilk method $p = 0.233$) but one significant outlier was found, and that data point was removed (Figure 3). There was not a significant difference between the healthy male and female EIS (two-tailed Mann–Whitney method $p = 0.334$) and these 22 values were combined, ranged from 0.9 mm/h to 8.7 mm/h (mean 4.5, SD 2.2), and were normally distributed (Anderson-Darling method $p = 0.332$). Due to the small sample size of 22, the data was Cox-Box transformed and robust methods were used to calculate $ESR_{corrected}$ reference range of 0 (90% CI -1-1)–9 mm/h (90% CI 8-11).

The results of $ESR_{corrected}$ measurements in EIS in mid-ecdysis cohort were not normally distributed (Shapiro–Wilk method $p = 0.032$). There was not a significant difference between male and female EIS in mid-ecdysis (two-tailed Mann–Whitney method $p = 0.247$) and these eight values were combined and ranged from 4.4 mm/h to 19.6 mm/h (median 7.1).

The results of $ESR_{corrected}$ measurements in EIS with dystocia cohort were not normally distributed (Shapiro–Wilk method $p = 0.048$). These six values ranged from 1.8 mm/h to 21.3 mm/h (median 5.4).

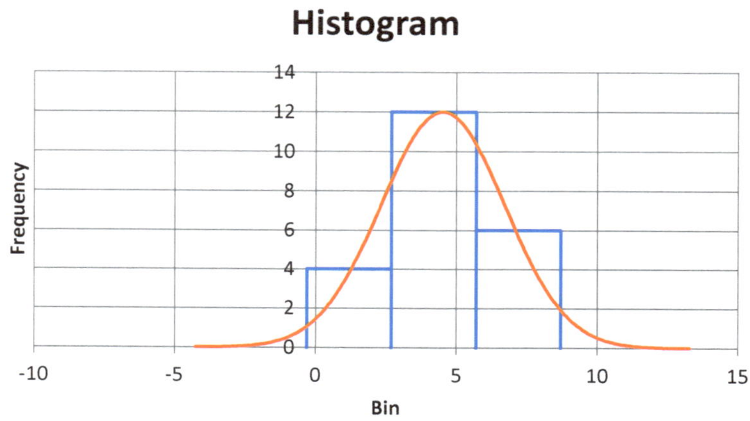

Figure 3. No outliers were found. There was not a significant difference between male and female EIS with gastric cryptosporidiosis (two-tailed Mann–Whitney method $p = 0.744$) and these 17 values were combined and ranged from 2.8 mm/h to 14.0 mm/h (mean 7.7, SD 3.6).

Reference ranges for $ESR_{corrected}$ measurements in EIS with gastric cryptosporidiosis, EIS in mid-ecdysis, and EIS with dystocia were not calculated since the number of EIS in each group was below 20 [20].

There was a significant difference in $ESR_{corrected}$ measurements between EIS cohorts (Kruskal–Wallis method, $p = 0.012$). Post hoc analyses revealed there was a significant difference in $ESR_{corrected}$ measurements between healthy EIS, EIS with gastric cryptosporidiosis, and EIS in mid-ecdysis cohorts (paired Mann–Whitney tests $p = 0.006$ and $p = 0.006$, respectively). There was not a significant difference in $ESR_{corrected}$ measurements between healthy EIS and EIS with dystocia cohorts (paired Mann–Whitney test $p = 0.614$). These results are summarized in Table 1.

Table 1. Comparison of corrected erythrocyte sedimentation rate ($ESR_{corrected}$) measurements between healthy eastern indigo snakes (EIS) and EIS with inflammatory conditions cryptosporidiosis, ecdysis and dystocia. $ESR_{corrected}$ values are reported in mm/h.

EIS	n	Mean	SD	Median	Min	Max
Healthy [1,2]	23	4.8	2.7		0.9	12.0
Cryptosporidiosis [1]	17	7.7	3.6		2.8	14.0
Ecdysis [2]	8			7.1	4.4	19.6
Dystocia	6			5.4	1.8	21.3

Normally distributed data are presented as mean, standard deviation (SD), minimum value obtained (Min) and maximum value obtained (Max) and non-normal data as median, Min, and Max. n = number of animals. $p < 0.05$ is considered significant and significant differences are denoted with a superscript number: 1 = $ESR_{corrected}$ in EIS with cryptosporidiosis was higher than in healthy EIS ($p = 0.006$), 2 = $ESR_{corrected}$ in EIS in ecdysis was higher than healthy EIS not in ecdysis ($p = 0.006$).

4. Discussion

This study demonstrates the micro-ESR method to have a high correlation and good agreement with the Westergren method for measuring ESR in EIS indicating the results of each method are comparable, allowing for the results of each testing method to be combined. The benefit of using the micro-ESR method over the Westergren method is the smaller volume needed to complete the test. Additionally, the capillary tube used in the micro-ESR can then be centrifuged to measure PCV minimizing the need for additional whole blood to perform a PCV as in the Westergren method. Combining the ESR and PCV measurements allows for correcting for anemia or lymph dilution, by using Fabry's formula.

Although mammalian ESR measurements are typically completed on whole blood preserved with either ethylene diamine tetra acetic acid (EDTA) or trisodium citrate (TSC) [22], whole blood preserved with lithium heparin is described as the standard sample used for ESR measurements in reptiles [2,4,5]. When compared to TSC samples, heparinized blood will have a higher ESR measurement [23], so samples preserved with different anticoagulants should not be compared. A benefit of using heparinized whole blood over EDTA or TSC is the increased versatility of the sample. The diminutive size of many reptile patients precludes collecting large volumes of blood for multiple microtainers. For example, the smallest TSC microtainer commercially available in the US requires 0.5 mL whole blood (Blood Collection Microtube, Blue Top, Sarstedt AG & Co. KG, Nümbrecht, Germany). Whole blood treated with TSC or EDTA may be used for additional hematology, but the plasma derived from this sample should not be used for biochemical analyses as these anticoagulants can alter some biochemical analyte values in many species [24–26]. Furthermore, erythrocytes from some reptile species will lyse when mixed with EDTA or TSC [27]. Using only one lithium heparin microtainer allows for more versatile testing options. In addition to the micro-ESR, a complete blood cell count and plasma biochemical analytes can be performed from a single microtainer if lithium heparin is used.

A reference interval for $ESR_{corrected}$ was able to be calculated based on the established guidelines set forth by American Society for Veterinary Clinical Pathology [20]. Having a reference interval for $ESR_{corrected}$ can help the clinician determine if an inflammatory condition is present. Care must be taken, however, with interpreting the results. The presence of inflammation does not necessarily indicate a disease process as healthy EIS undergoing ecdysis may have an elevated ESR compared to healthy EIS not in ecdysis. In reptiles, ecdysis is the process of renewing the skin and this occurs over five stages where the most noticeable stages are Stage 3 and Stage 4 when the spectacles have a dull, bluish hue [28]. Starting at the end of Stage 3 and continuing through Stage 4, heterophils will infiltrate the epidermis [28]. Heterophils are also the initial cells to infiltrate reptilian tissue during inflammation [29]. Additionally, cytokines are proteins associated with the reptilian inflammatory response [30] and they are also important factors during the reptilian ecdysis cycle [31]. Therefore, it is reasonable to predict ecdysis to cause an increase in ESR values. As all of the EIS placed into the ecdysis cohort had dull, bluish spectacles, it is not surprising that the ESR was also elevated for this cohort.

Similarly, a normal ESR does not rule out a disease typically associated with inflammation. Some conditions exist that can lower ESR, such as polycythemia, sickle cell disease, and spherocytosis [1]. In addition, some conditions with known significant morbidity can have normal to low ESR values. Examples include urinary tract infection, myocardial infarction, thromboembolic disease, rheumatoid arthritis, and hypoalbuminemia [32]. The EIS cohort with dystocia did not have a significantly higher ESR when compared to the ESR of healthy EIS. This may be in line with other diseases reported to have normal ESR or it may be a product of too few cases (a type II statistical error). Further investigation into ESR and dystocia is warranted.

Additional type II statistical errors cannot be completely ruled out since the overall study size is relatively small. It is possible that some of the healthy EIS may have been experiencing some form of inflammation and falsely elevating the upper range of the ESR since the one apparently healthy EIS outlier that was removed had an $ESR_{corrected}$ of 12.0 mm/hr. Likewise, some EIS with dystocia may not have inflammation at the time of phlebotomy or had some comorbidity that lowers ESR, since one EIS with dystocia had $ESR_{corrected}$ of 1.8 mm/h while another EIS with dystocia had an $ESR_{corrected}$ of 21.3 mm/h.

The EIS cohort with gastric cryptosporidiosis had an elevated ESR when compared to the healthy EIS cohort. Although humans with gastrointestinal infections will often have normal ESR values [32], this elevation is not unexpected as many snakes infected with *C. serpentis* can develop severe gastric hypertrophy with marked mucosal inflammation [33]. Further studies are needed to evaluate whether ESR can play a role in determining prognosis in these cases.

Since ESR will increase with an increase in acute phase protein elevations, additional studies combining ESR and protein electrophoresis are warranted. One particular condition described in EIS, hyperviscosity syndrome, should be investigated [34]. This condition has been associated with an increase in plasma gamma globulins implicated in increasing viscosity of the blood [34] and an increase in blood viscosity can artificially decrease ESR readings in humans [35]. Further investigation is needed to determine if hyperviscosity syndrome will increase or decrease ESR in EIS.

Several factors have been identified in mammalian samples which could affect ESR results, including room temperature, time of collection, tube orientation, and vibration. An elevated room temperature may decrease blood viscosity and artificially increase the ESR [36]. Direct sunlight can also increase ESR [1]. A tilted ESR tube and increased vibrations can also increase the ESR value. An angle of three degrees from the vertical can artificially increase the ESR by up to 30% [1]. The ESR should also be performed within two hours of blood collection as a blood sample that sits too long can cause sphering and decrease the ESR value [1]. Future studies are warranted to determine if these factors affect reptile blood in the same way. In reptiles, plasma proteins can be affected by venipuncture site and season [37]. Additional studies are warranted to investigate how these variables affect ESR results in EIS.

5. Conclusions

Erythrocyte sedimentation rate is a simple diagnostic test that can be easily performed in-house. The micro-ESR method can be readily performed using materials already present in most veterinary practices and, in EIS, the results are comparable to the Westergren method. Adding micro-ESR to the routine CBC may allow detection of occult inflammation in EIS, especially those not undergoing ecdysis. Additionally, consideration of possible *C. serpentis* infection should be investigated in EIS with an elevated ESR.

Funding: This research received no external funding.

Institutional Review Board Statement: The animal study protocol was approved by the Research Committee of the Central Florida Zoo & Botanical Gardens (protocol code 2022-04, 8 December 2022).

Informed Consent Statement: Not applicable.

Data Availability Statement: The data presented in the study are available in the article.

Acknowledgments: I would like to thank Alexandra Mason and Mark Mitchell for manuscript review, Michelle Hoffman and Sydney Seng for assistance during phlebotomy, and the entire keeper staff at the Central Florida Zoo's Orianne Center for Indigo Conservation for the excellent care of the snakes.

Conflicts of Interest: The author declares no conflict of interest.

References

1. Tishkowski, K.; Gupta, V. *Erythrocyte Sedimentation Rate*; StatPearls Publishing: Treasure Island, FL, USA, 2022. Available online: https://www.ncbi.nlm.nih.gov/books/NBK557485/ (accessed on 9 May 2022).
2. Sandfoss, M.R.; Claunch, N.M.; Stacy, N.I.; Romagosa, C.M.; Lillywhite, H.B. A tale of two islands: Evidence for impaired stress response and altered immune functions in an insular pit viper following ecological disturbance. *Conserv. Physiol.* **2020**, *8*, coaa031. [CrossRef]
3. Hashemi, R.; Majidi, A.; Motamed, H.; Amini, A.; Najari, F.; Tabatabaey, A. Erythrocyte sedimentation rate measurement using as a rapid alternative to the Westergren method. *Emergency* **2015**, *3*, 50–53.
4. Rosenberg, J.F.; Hernandez, J.A.; Wellehan, J.F.; Crevasse, S.E.; Cray, C.; Stacy, N.I. Diagnostic performance of inflammatory markers in gopher tortoises (*Gopherus polyphemus*). *J. Zoo Wildl. Med.* **2018**, *49*, 765–769. [CrossRef]
5. Adamovicz, L.; Baker, S.J.; Kessler, E.; Kelly, M.; Johnson, S.; Winter, J.; Phillips, C.A.; Allender, M.C. Erythrocyte sedimentation rate and hemoglobin-binding protein in free-living box turtles (*Terrapene spp.*). *PLoS ONE* **2020**, *15*, e0234805. [CrossRef]
6. U.S. Fish and Wildlife Service [USFWS]. Listing of the eastern indigo snake as a threatened species. *Fed. Regist.* **1978**, *43*, 4026–4029.

7. Knafo, S.E.; Norton, T.M.; Mitchell, M.; Stevenson, D.J.; Hyslop, N.; Poppenga, R.; Oliva, M.; Chen, T.; Cray, C.; Gibbs, S.E.; et al. Health and nutritional assessment of free-ranging eastern indigo snakes (*Drymarchon couperi*) in Georgia, United States. *J. Zoo Wildl. Med.* **2016**, *47*, 1000–1012. [CrossRef] [PubMed]
8. Drew, M. Hypercalcemia and hyperphosphatemia in indigo snakes (*Drymarchon couperi*) and serum biochemical reference values. *J. Zoo Wildl. Med.* **1994**, *25*, 48–52.
9. Chandler, H.C.; Allender, M.C.; Stegenga, B.S.; Haynes, E.; Ospina, E.; Stevenson, D.J. Ophidiomycosis prevalence in Georgia's eastern indigo snake (*Drymarchon couperi*) populations. *PLoS ONE* **2019**, *14*, e0218351. [CrossRef] [PubMed]
10. Bogan, J.E.; Steen, D.A.; O'Hanlon, B.; Garner, M.M.; Walden, H.D.S.; Wellehan, J.F.X. *Drymarchon couperi* (eastern indigo snake). Death associated with *Raillietiella orientalis*. *Herpetol. Rev.* **2022**, *53*, 147.
11. Bogan, J.E.; Hoffman, M.; Dickerson, F.; Antonio, F.B. A retrospective study of dystocia in eastern indigo snakes (*Drymarchon couperi*). *J. Zoo Wildl. Med.* **2021**, *52*, 618–627. [CrossRef]
12. Bogan, J.E.; O'Hanlon, B.M.; Steen, D.A.; Horan, T.; Taylor, R.; Mason, A.; Breen, T.; Andreotta, H.; Cornelius, B.; Childress, A.; et al. Health assessment of free-ranging eastern indigo snakes (*Drymarchon couperi*) from hydrologic restoration construction sites in south Florida. *J. Wildl. Dis.* 2023; *in review*.
13. Ferrell, S. Conservation. In *Mader's Reptile and Amphibian Medicine and Surgery*, 3rd ed.; Divers, S.J., Stahl, S.J., Eds.; Elsevier: St. Louis, MO, USA, 2019; pp. 1421–1428.
14. Rivera, S. Quarantine. In *Mader's Reptile and Amphibian Medicine and Surgery*, 3rd ed.; Divers, S.J., Stahl, S.J., Eds.; Elsevier: St. Louis, MO, USA, 2019; pp. 142–144.
15. Antonio, F. *AZA Snake TAG 2011. Eastern Indigo Snake (Drymarchon couperi) Care Manual*; Association of Zoos and Aquariums: Silver Spring, MD, USA, 2011; pp. 5–42.
16. Bogan, J.E.; Wellehan, J.F.; Garner, M.M.; Childress, A.L.; Jackson, B. Evaluation of a probe hybridization quantitative polymerase chain reaction assay for *Cryptosporidium serpentis* in eastern indigo snakes (*Drymarchon couperi*). *Parasitol. Res.* **2022**, *121*, 3523–3527. [CrossRef]
17. Bogan, J.E.; Mason, A.; Mitchell, M.A.; Garner, M.M.; Childress, A.; Seng, S.; Wellehan, J.F. Evaluation of paromomycin, vitamin E, selenium, curcumin, and propolis combination as a treatment for *Cryptosporidium serpentis* infection in eastern indigo snakes (*Drymarchon couperi*). *J. Herpetol. Med. Surg.* 2024; *manuscript in preparation; to be submitted.*
18. Fabry, T.L. Mechanism of erythrocyte aggregation and sedimentation. *Blood* **1987**, *70*, 1572–1576. [CrossRef]
19. Akoglu, H. User's guide to correlation coefficients. *Turk. J. Emerg. Med.* **2018**, *18*, 91–93. [CrossRef]
20. Friedrichs, K.R.; Harr, K.E.; Freeman, K.P.; Szladovits, B.; Walton, R.M.; Barnhart, K.F.; Blanco-Chavez, J. ASVCP reference interval guidelines: Determination of de novo reference intervals in veterinary species and other related topics. *Vet. Clin. Pathol.* **2012**, *41*, 441–453. [CrossRef]
21. Geffre, A.; Concordet, D.; Braun, J.P.; Trumel, C. Reference value advisor: A new freeware set of macroinstructions to calculate reference intervals with Microsoft Excel. *Vet. Clin. Pathol.* **2011**, *40*, 107–112. [CrossRef]
22. Getaneh, Z.; Ayelgn, F.; Asemahegn, G.; Geleta, H.; Yalew, A.; Melak, T. A comparison of erythrocyte sedimentation rates of bloods anticoagulated with trisodium citrate and EDTA among TB presumptive patients at the University of Gondar comprehensive specialized hospital, northwest Ethiopia. *BMC Res. Notes* **2020**, *13*, 113. [CrossRef]
23. Shallal, A.F.; Ibrahim, Z.H.; Kheder, R.; Hussein, S.H.; Hassan, S.M.; Khalil, O.A. Effect of different types of anticoagulants and storage period on the erythrocyte sedimentation rate in healthy and unhealthy people. *Biomed. Res.* **2020**, *31*, 9–12.
24. Cerón, J.J.; Martínez-Subiela, S.; Hennemann, C.; Tecles, F. The effects of different anticoagulants on routine canine plasma biochemistry. *Vet. J.* **2004**, *167*, 294–301. [CrossRef]
25. Mohri, M.; Rezapoor, H. Effects of heparin, citrate, and EDTA on plasma biochemistry of sheep: Comparison with serum. *Res. Vet. Sci.* **2009**, *86*, 111–114. [CrossRef]
26. Kamali, H.; Mohri, M. Effects of heparin, citrate, and EDTA on plasma biochemistry of cat: Comparison with serum. *Rev. Med. Vet.* **2015**, *166*, 275–279.
27. Heatley, J.J.; Russell, K.E. Hematology. In *Mader's Reptile and Amphibian Medicine and Surgery*, 3rd ed.; Divers, S.J., Stahl, S.J., Eds.; Elsevier: St. Louis, MO, USA, 2019; pp. 301–318.
28. Jacobson, E.R.; Lillywhite, H.B.; Blackburn, D.G. Overview of biology, anatomy, and histology of reptiles. In *Diseases and Pathology of Reptiles Volume 1: Infectious Diseases and Pathology of Reptiles Color Atlas and Text*, 2nd ed.; Jacobson, E.R., Garner, M.M., Eds.; CRC Press: Boca Raton, FL, USA, 2021; pp. 1–214.
29. Stacy, B.A.; Pessier, A.P.; Ossiboff, R.J. Host response to infectious agents and identification of pathogens in tissue sections. In *Diseases and Pathology of Reptiles Volume 1: Infectious Diseases and Pathology of Reptiles Color Atlas and Text*, 2nd ed.; Jacobson, E.R., Garner, M.M., Eds.; CRC Press: Boca Raton, FL, USA, 2021; pp. 375–422.
30. Origgi, F.C.; Tecilla, M. Immunology of reptiles. In *Diseases and Pathology of Reptiles Volume 1: Infectious Diseases and Pathology of Reptiles Color Atlas and Text*, 2nd ed.; Jacobson, E.R., Garner, M.M., Eds.; CRC Press: Boca Raton, FL, USA, 2021; pp. 215–266.
31. Subramaniam, N.; Petrik, J.J.; Vickaryous, M.K. VEGF, FGF-2 and TGF β expression in the normal and regenerating epidermis of geckos: Implications for epidermal homeostasis and wound healing in reptiles. *J. Anat.* **2018**, *232*, 768–782. [CrossRef] [PubMed]
32. Bray, C.; Bell, L.N.; Liang, H.; Haykal, R.; Kaiksow, F.; Mazza, J.J.; Yale, S.H. Erythrocyte sedimentation rate and C-reactive protein measurements and their relevance in clinical medicine. *Wisc. Med. J.* **2016**, *115*, 317–321.
33. Bogan, J.E. Gastric cryptosporidiosis in snakes: A review. *J. Herpetol. Med. Surg.* **2019**, *29*, 71–86. [CrossRef]

34. LaDouceur, E.E.B.; Garner, M.M. Hyperviscosity-like syndrome in reptiles. Proceedings of American Association of Zoo Veterinarians Conference, Dallas, TX, USA, 27 September 2017.
35. Kahar, M.A. Erythrocyte sedimentation rate (with its inherent limitations) remains a useful investigation in contemporary clinical practice. *Ann. Pathol. Lab. Med.* **2022**, *9*, R9–R17. [CrossRef]
36. Manley, R.W. The effect of room temperature on erythrocyte sedimentation rate and its correction. *J. Clin. Pathol.* **1957**, *10*, 354–355. [CrossRef]
37. Heatley, J.J.; Russell, K.E. Clinical chemistry. In *Mader's Reptile and Amphibian Medicine and Surgery*, 3rd ed.; Divers, S.J., Stahl, S.J., Eds.; Elsevier: St. Louis, MO, USA, 2019; pp. 319–350.

animals

MDPI

Article

The Amount of Food Ingested and Its Impact on the Level of Uric Acid in the Blood Plasma of Snakes

Miloš Halán [1], Lucia Kottferová [2,*], Karol Račka [1] and Anthony Lam [3]

1 Department of Epizootiology, Parasitology and Protection of One Health, University of Veterinary Medicine and Pharmacy in Košice, Komenského 68/73, 041 81 Košice, Slovakia
2 Clinic of Birds, Exotic and Free Living Animals, University of Veterinary Medicine and Pharmacy, Komenského 68/73, 041 81 Košice, Slovakia
3 Companion Care Vets-Folkestone, Folkestone CT19 5SY, UK
* Correspondence: lucia.kottferova@uvlf.sk

Simple Summary: The assessment of uric acid levels in snakes is an important part of the diagnosis of renal diseases. In mammals, lipemic blood from sampling too soon after an animal feeds can have substantial effects on biochemical values. However, fasting status is not routinely considered when sampling reptile blood. The investigation aims to better understand the feed-induced changes that occur and render the analysis of this parameter a more potent diagnostic tool. A study has shown that feeding snakes lead to substantial elevations in uric acid values, with postprandial concentrations significantly elevated for up to 8 days after feeding. To prevent misdiagnosis and distinguish temporary hyperuricemia from clinically relevant increases, it is recommended that sufficient data on time since the last feeding be collected, as well as repeated samples after weeks of fasting.

Abstract: In mammals, lipemic blood from sampling too soon after an animal feeds can have substantial effects on biochemical values. Plasma biochemical values in reptiles may be affected by species, age, season, and nutritional state. However, fasting status is not routinely considered when sampling reptile blood. Assessing uric acid levels in snakes is an important part of the diagnosis of the renal disease. However, the use of this biochemical indicator is undervalued without knowledge of natural uric acid fluctuations and the lack of differentiation from pathological changes. This study aimed to look at the relationship between snake feeding and uric acid concentrations. The investigation aims to better understand the feed-induced changes that occur and render the analysis of this biochemical parameter a more potent diagnostic tool. The study used ten snakes belonging to seven species, and basal uric acid values were evaluated by blood biochemical analysis before feeding. The snakes were fed in two rounds, with successive blood sampling and monitoring of uric acid changes carried out for each. The snakes were fed approximately 50% more with the second round of feeding to investigate the relationship between food supply and uric acid level. The findings show feeding led to substantial elevations in uric acid values, whereby postprandial concentrations were significantly elevated for up to 8 days after feeding. The findings show the significant changes in uric acid levels that occur after feeding and the similarities between postprandial rises in uric acid and those reported in snakes with renal disease. To minimize misdiagnosis and differentiate transient postprandial hyperuricemia from pathological increases, it is recommended that sufficient anamnestic data on time since the last feeding be collected, as well as repeated samples following weeks of fasting. This knowledge is crucial because the amount of feed in terms of intensity and volume has a significant effect on uric acid levels in the blood of snakes.

Keywords: snakes; feeding; uric acid

Citation: Halán, M.; Kottferová, L.; Račka, K.; Lam, A. The Amount of Food Ingested and Its Impact on the Level of Uric Acid in the Blood Plasma of Snakes. *Animals* **2022**, *12*, 2959. https://doi.org/10.3390/ani12212959

Academic Editor: Tom Hellebuyck

Received: 20 September 2022
Accepted: 26 October 2022
Published: 27 October 2022

1. Introduction

As most reptiles, snakes tend to mask signs of disease. It is for this reason that many reptiles are not presented to veterinarians early in the course of a disease. This presents a problem for veterinarians in their antemortem diagnosis and delays the initiation of specific therapeutic planning. Although hematology and plasma biochemistry are useful for monitoring animal health, environmental and procedural factors can have a variable effect on reported values. Biochemical analysis is a useful diagnostic tool and an important part of any veterinary diagnosis. However, there are still large gaps in understanding the physiological changes and variables affecting reptile biochemistry, leading to the devaluation of biochemical analysis as a diagnostic tool. With the exponential increase in the number of reptiles, it is becoming increasingly important that our knowledge of these species grows in parallel with their numbers [1].

Metabolic diseases are among the most common problems seen in captive reptiles. Between 1992 and 1996, McWilliams & Leeson [2] reported that 84.4% of lizard patients had a metabolic illness attributed to poor husbandry and nutritional decisions. This trend may improve with increasing reptile owner awareness and improvements in veterinary medicine, but metabolic disease remains a current issue in reptilian care. Although not a common problem in general veterinary medicine, gout is a common affliction in reptilian patients.

The renal cortex of reptiles is made up of simple nephrons (cortical nephrons) that have a tubular system lacking the loops of Henle; therefore, reptiles are unable to concentrate their urine. Variable amounts of uric acid, urea, and ammonia (nitrogenous waste) are excreted by the reptilian kidneys, depending on the animal's environment. Freshwater turtles that spend most of their lives in water excrete equal amounts of ammonia and urea, whereas amphibious reptiles excrete more urea [3]. Alligators excrete ammonia and uric acid, and sea turtles excrete ammonia, urea, and uric acid [4]. Land-dwelling or terrestrial reptiles, such as tortoises, produce more insoluble nitrogenous waste in the form of urate salts and uric acid. This is explained by the fact that these reptiles need to conserve water, and soluble forms of nitrogenous waste require large amounts of water for their excretion. Therefore, terrestrial reptiles produce more insoluble nitrogenous wastes, which are eliminated in a semisolid state [5].

In reptiles, uric acid is filtered by the renal tubules. Uric acid and urate salts are insoluble in water and are excreted by the renal tubules. When the level of circulating urate salts or free uric acid rises above the renal capacity to excrete them, hyperuricemia develops. With persistent hyperuricemia, tophi (crystals) are formed which are deposited by the blood-forming crystals in tissues, organs, and joints throughout the body—otherwise known as gout. In reptiles, tophi may precipitate throughout the body, of which the pericardial sac, kidneys, liver, spleen, lungs, subcutaneous tissues, and joints are the most common sites [6].

Gout is categorized into two types: main and secondary. The overproduction of uric acid in primary gout is caused by an innate metabolic imbalance. Secondary gout is caused by chronic diseases that affect the normal balance of uric acid production and excretion (e.g., chronic renal disease, starvation, hypertension, and the use of nephrotoxic pharmaceuticals, such as gentamycin) [1]. Although numerous factors contribute to the development of gout in reptiles, the most frequent causes leading to hyperuricemia and gout include dehydration, renal damage, and excessive consumption of proteins (purine-rich). The latter is particularly associated with herbivorous reptiles fed diets high in animal protein, as seen commonly in green iguanas on diets of cat food [7].

One study exploring the nephrotoxic effects of gentamicin in snakes found that with high doses, snakes presented extensive tubular necrosis within two weeks, and those exposed to high doses developed visceral gout. Dehydration is another factor that may impair renal function. With normal hydration, the nephron actively excretes three times the amount of urates than it would in the dehydrated state [8].

Certain species of reptiles exhibit elevated uric acid levels during hibernation, thought to be due to the reduced renal tubular blood flow at low temperatures experienced during the hibernation period. Physiological rises in uric acid are seen post-prandial in several carnivorous animals, including reptiles and birds, and are attributed to the metabolism of dietary proteins [9].

Current literature suggests the resampling of blood in healthy snakes exhibiting high uric acid levels in an attempt to rule out underlying physiological elevations in uric acid. This, however, postulates a potential problem in current reptilian biochemistry, as the reference ranges provided may be non-sensitive to potential physiological fluctuations in uric acid concentration. In an attempt to compensate, reptilian clinicians require a full patient anamnesis to identify these factors, which they must consider in the evaluation of the biochemistry panel. One, however, questions the reliability of these current methods and whether or not they may be improved [6,9].

Snakes are predominantly sit-and-wait predators, eating large meals after periods of fasting that can last for months. As a consequence of the feeding behavior and physiology in snakes we would expect more evident elevations in uric acid concentrations after feeding when compared to species that have a more regular feeding regime. Carnivorous reptiles generally have higher uric acid concentrations than herbivorous reptiles, and previous studies have shown that postprandial hyperuricemia can result in a 1.5-fold to 2-fold increase in uric acid [4]. A study run by Maixner et al. [10] on five species of reptiles showed postprandial uric acid concentrations to rise to 3-fold basal levels. Similar increases in serum uric acid concentrations have been observed in reptilian patients with renal disease. This highlights the importance of considering postprandial effects on uric acid in reptiles when evaluating plasma biochemistry [11].

Season, sex, age, size, and nutritional state are all environmental and physiological factors that can affect clinical pathology values in reptiles [10,12,13].

Seasons, hormones, and diet (depending on the time of blood collection) have all been documented to have an impact on reptile biochemistry [14,15]. Understanding differences in serum uric acid concentrations in healthy reptiles is critical for assessing hyperuricemia in gout patients. However, the existing literature's uric acid reference ranges are huge and do not account for such physiological variations, making such data less sensitive and potentially misleading interpretations [16].

The aim of the present study was to evaluate the changes of post-feeding UA levels in a limited number of individuals in 7 species.

2. Materials and Methods

2.1. Species Studied

Ten adult snakes were used in this study: Burmese python (*Python bivittatus*) Tree Red-tailed boa (*Boa constrictor*), Carpet python (*Morelia spilota*), two Beauty rat snake (*Orthriophis taeniurus friesi*), Eastern kingsnake (*Lampropeltis getula*), Rainbow boa (*Epicrates cenchria*), Four-lined snake (*Elaphe quatuorlineata*).

2.2. Feeding and Handling

Individual terrariums with glass fronts and ventilation slits in the side panels were utilized to house the snakes. Each snake was kept in a temperature-controlled environment at around 25 °C (±5 °C). The snakes each had been acclimated to the terrarium conditions for at least 6 months before the study began. Each snake was fasted for at least 14 days to avoid anomalies from prior meal preprandial sampling. Rodents (*Rattus norvegicus*) were given to the snakes, and the rats' body weights were recorded after first and second meal (Table 1) and they have free access to water.

Table 1. Weight of rats fed to each snake for both rounds of feeding.

Species	Rat Weight (g)	
	Meal 1	Meal 2
Python bivittatus individual 1′	90	130
Python bivittatus individual 2′	120	190
Python bivittatus individual 3′	200	310
Morelia spilota	160	240
Orthriophis taeniurus friesi individual 1′	60	80
Orthriophis taeniurus friesi individual 2′	80	120
Elaphe quatuorlineata	50	80
Lampropeltis getula	40	60
Epicrates cenchria	40	70

2.3. Blood Collection and Processing

Before each snake was fed, blood samples were taken and tested to determine baseline uric acid concentrations. Day 0 was the day when we took blood in the morning to determine the current value and the snakes were fed in the afternoon. From the day 1, we monitored the changes in the uric acid level during 9 days. After each meal, blood uric acid concentrations were monitored daily until they reverted to near-baseline values.

Blood was collected via the ventral coccygeal vein with a 23-gauge needle, and the samples were collected into EDTA tubes. After centrifuging the samples, a volume of 5 µL of plasma was collected using a micropipette and aliquoted for analysis. For the measurement of uric acid concentrations from the serum were performed with the Cobas c111 analyzer, which provided readings for each sample in µmol/l.

3. Results

The results in the Tables 2 and 3 show the variation in plasma uric acid concentration in the snakes before and after feeding.

Table 2. Pre-feeding (day 0) and post-feeding (day 1–9) uric acid concentrations in snake plasma after meal 1.

	Plasma Uric Acid (µmol/l)									
Day	0	1	2	3	4	5	6	7	8	9
Python bivittatus individual 1′	176.2	392.3	578.4	397.3	361.1	329.2	250.1	212.2	197.0	191.4
Python bivittatus individual 2′	349.8	535.9	738.5	891.1	784.2	730.6	866.5	697.5	611.9	536.4
Python bivittatus individual 3′	233.8	1115.8	1284.8	1030.3	897.9	526.8	370.1	314.0	270.8	253.1
Boa constrictor	370.7	653.6	1926.4	1147.5	741.9	408.3	328.5	373.1	348.7	343.5
Morelia spilota	259.6	1010.3	2524.3	1912.8	824.0	571.8	425.5	350.1	285.3	267.9
Orthriophis taeniurus friesi individual 1′	277.9	1246.4	1617.2	1050.9	602.3	343.8	333.7	299.3	289.2	284.2
Orthriophis taeniurus friesi individual 2′	254.3	404.0	1314.1	1116.5	446.2	279.1	264.4	219.3	218.4	223.6
Elaphe quatuorlineata	193.6	1286.5	1333.4	1091.2	814.6	579.3	389.4	439.6	389.4	292.5
Lampropeltis getula	238.9	622.8	508.5	417.5	221.6					
Epicrates cenchria	175.3	578.9	739.5	538.3	152.8					

Table 3. Pre-feeding (day 0) and post-feeding (day 1–9) uric acid concentrations in snake plasma after meal 2.

	Plasma Uric Acid (μmol/l)									
Day	**0**	**1**	**2**	**3**	**4**	**5**	**6**	**7**	**8**	**9**
Python bivittatus individual 1′	297.8	682.7	894.7	723.5	463.2	391.4	324.7	351.7	339.1	312.3
Python bivittatus individual 2′	396.5	650.6	1011.4	486.3	330.3	357.8	265.3	294.9	271.6	295.3
Python bivittatus individual 3′	352.2	886.0	2057.8	1255.4	682.5	435.2	358.6	458.1	412.6	395.2
Boa constrictor	370.7	653.6	1926.4	1147.5	741.9	408.3	328.5	373.1	348.7	343.5
Morelia spilota	259.6	1010.3	2524.3	1912.8	824.0	571.8	425.5	350.1	285.3	267.9
Orthriophis taeniurus friesi individual 1′	212.0	1033.5	2551.1	1957.5	1092.7	688.0	376.9	333.3	350.1	314.6
Orthriophis taeniurus friesi individual 2′	216.2	1234.3	1935.3	1951.7	1006.7	702.3	308.2	306.1	271.7	289.2
Elaphe quatuorlineata	298.1	1530.7	1388.7	1053.1	819.6	736.8	593.3	471.4	457.7	329.6
Lampropeltis getula	99.1	497.0	982.1	976.7	1011.7	921.9	495.9	431.6	400.7	308.3
Epicrates cenchria	203.5	549.5	609.2	387.3	216.9	201.6	213.9	210.8	210.7	208.2

As seen in Figures 1 and 2, the post-prandial levels of uric acid rise substantially in all snakes, with mean peak concentrations reaching 1251 88 and 1568.27 μmol/l in the first and second rounds of feeding, respectively. Plasma uric acid reached peak concentrations between 1 and 2 days in all snakes, with maximum recorded levels reaching 2551.1 μmol/l in *Orthriophis taeniurus friesi* during the second round of feeding. In the snakes studied, the majority showed similar patterns of sharp elevations in uric acid followed by a more gradual decline, returning to basal concentrations approximately nine days after feeding. The only exception to this was seen in the *Python bivittatus* individual 2′ during the first round of feeding, which showed 2 peaks of uric acid before returning to pre-prandial levels. However, during the second round of feeding, this individual followed the same pattern of an initial increase followed by a regression to basal levels, suggesting this snake may have had some internal factor delaying digestion during the first round of feeding.

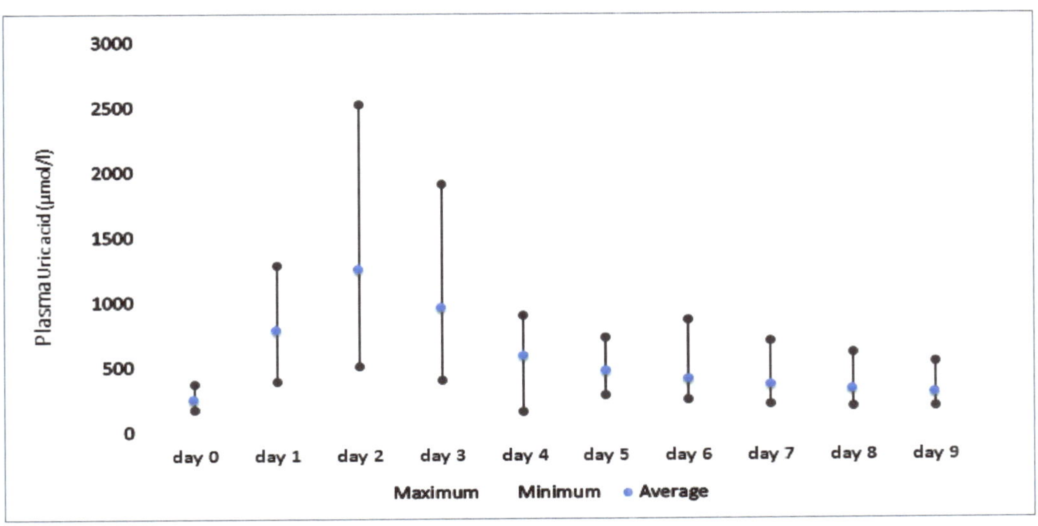

Figure 1. Pre-feeding (day 0) and post-feeding (day 1–9) plasma uric concentration range in snakes after meal 1.

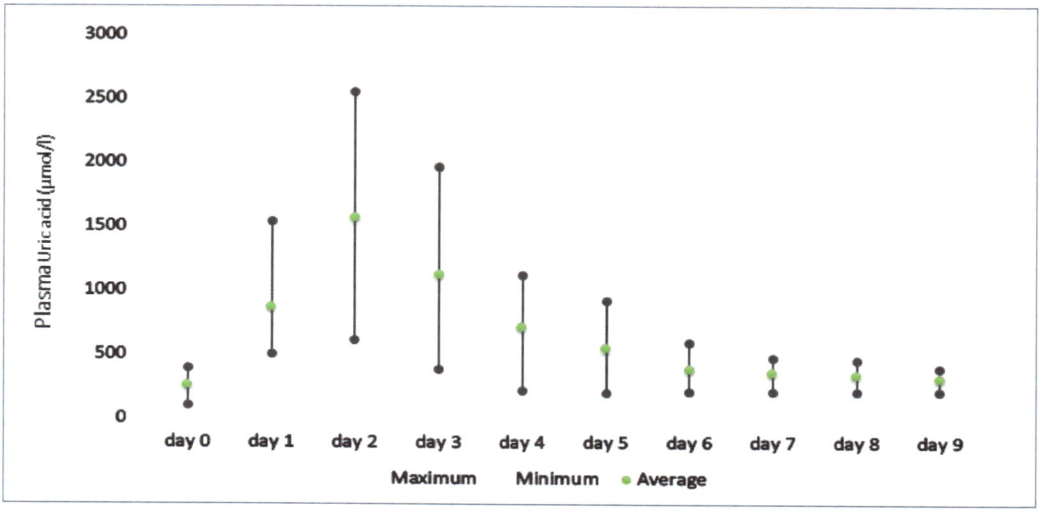

Figure 2. Pre-feeding (day 0) and post-feeding (day 1–9) plasma uric concentration range in snakes after meal 2.

With the second round of feeding, we find that the pattern of rising and falling of uric acid is similar to that seen in the first meal, but with its differences. However, a return to insignificant concentrations of uric acid did not occur until the eighth day, three more days than in the first feeding, suggesting that snakes fed larger rats took longer to return to basal concentrations. Peak concentrations of uric acid was recorded between 1 and 2 days post-prandial in all snakes, with maximum concentrations reaching 2524.3 µmol/l in the first round of feeding and up to 2551.1 µmol/l in the second.

4. Discussion

In this study, significant post-prandial increases in plasma uric acid content was observed in snakes. The levels of hyperuricemia reported in this study were similar to those found in reptiles with renal disease or gout. Uric acid levels in reptiles with gout are nearly 2-fold greater than baseline levels [3,6], similar to post-prandial concentrations observed in this study. Because of the differences in the physiologies of snake species, Finding an accurate measure of hyperuricemia is challenging. For example, Campbell [3] state that level greater than 892 µmol/l are considered hyperuricemic when compared to Mitchell [17], who gives a level exceeding 600 µmol/l to indicate hyperuricemia. The current study's post-prandial readings were significantly greater than this level, with maximum concentrations reaching 6 times higher than pre-prandial levels. These findings show that food has a significant impact on uric acid measurements, to the point that the reptile clinician would misinterpret the results as a sign of disease. Although vastly different in their physiology, finding such correlations between diet and levels of nitrogenous by-products in the blood may point out that such relationships have yet to be found in their reptilian counterparts. Maixner et al. [10] conducted a study on black rat snakes to investigate this relationship between meal size and uric acid concentrations and found that the snakes showed greater elevations in plasma uric acid concentrations e.g., when fed one instead of two prey. This has major consequences for veterinarians when it comes to evaluating biochemistry panels in snakes. It emphasizes the significance of obtaining enough anamnestic data before proceeding with further diagnostic procedures. As previously stated, post-prandial effects on serum uric acid concentrations can extend for up to 5 days or more. As a result, it is appropriate to recommend that the anamnesis include information on the time interval since the last feeding. Clinicians should also keep in mind that reptiles with hyperuricemia should be sampled and re-assessed within one week of fasting to provide a more precise picture of uric acid content and avoid misdiagnosis. Our results are representative for snakes, and the same post-prandial elevations in uric acid may not be present in other species of reptiles. Snakes are transient predators that can undergo long periods of fasting, during which basal levels of uric acid are reduced quite significantly. It is for this reason that we may attribute such dramatic changes in uric acid concentrations in these reptiles [18,19].

These findings also raise the question of why urate crystals do not form when uric acid levels rise dramatically after feeding. There is no evidence to suggest that urate crystals are deposited in the tissues of snakes under physiological conditions, apparently even under such significant post-prandial elevations. It is for this reason that we may assume that other substances or processes are involved in reducing urate solubility and facilitating deposition. A study carried out by Alvsaker [20] found that there was an absence of albumin and alpha globulin in the plasma of some people suffering from gout. Primary gout in humans, however, is often attributed to hereditary factors, whereas this link has not yet been established in snakes. The most common causes of gout in reptiles are attributed to husbandry or dietary etiologies [21].

Similar studies were conducted investigating the effects of protein supply on plasma urea and creatinine concentrations in dogs and minks [19,22]. The results presented a linear relationship between protein supply and plasma urea concentration, and Tauson et al. [23] conclude that consideration of protein intake and time of feeding should be made when interpreting plasma biochemistry. The Parkinson and Mans [24] investigated the effects of cricket ingestion on plasma uric acid concentration in 12 bearded dragons. Results suggested food should be withheld for $\geq$48 h prior to blood collection if bearded dragons are used to establish reference intervals for plasma uric acid concentration or when obtaining samples for clinical evaluation.

Although meal size did not significantly affect peak uric acid concentration. This has clinical implications when determining the time interval between re-sampling in hyperuricemic patients, and highlights the potential error if sufficient time is not taken. Based on the present results, it is evident that re-sampling after 8 days would allow for a more accurate interpretation. However, many variables that would affect these results were controlled, including temperature and access to water. Snakes are poikilothermic and are dependent upon their environmental temperature to fuel their metabolism. An investigation into the effect of temperature on metabolic rate in the *Boa constrictor amarali* found that ambient temperature had a profound effect on the duration of digestion and also digestive efficiency [25]. From this, we can assume that under conditions with more variable temperatures, in particular lower temperatures (not uncommon in captive reptile husbandry), we may see changes in the relationship between feeding and uric acid concentration. Unlike in the wild, snakes in a captive environment are feed on a more regular basis, and smaller species of snake are often fed weekly, approximately the same time interval as it took the snakes in this study to return to basal levels of uric acid. Consequently, it seems that without sufficient time intervals between feeding, basal levels of uric acid would not be reached and the individual would be in a persistent state of mild to moderate hyperuricemia. This underlines another difficulty faced by veterinarians when trying to obtain accurate representations of uric acid levels in snakes and that a period of fasting in these patients would be required to allow for more accurate interpretation.

5. Conclusions

Our research shows how feeding influences plasma uric acid content in snakes and emphasizes the need to recognize this link when evaluating clinical chemistry in snakes. This information has major consequences for veterinarians when it comes to evaluating biochemistry panels in snakes. It emphasizes the significance of obtaining enough anamnestic data before proceeding with further diagnostic procedures. As previously stated, postprandial effects on serum uric acid concentrations can extend for up to 5 days or more. As a result, it is appropriate to recommend that the anamnesis include information on the time interval since the last feeding. Clinicians should also keep in mind that reptiles with hyperuricemia should be sampled and re-assessed within one week of fasting to provide a more precise picture of uric acid content and avoid misdiagnosis.

Author Contributions: Conceptualization, M.H. and L.K.; Investigation, K.R.; Methodology, A.L.; Supervision, M.H.; Visualization, L.K.; Writing—original draft, L.K.; Writing—review & editing, M.H. All authors have read and agreed to the published version of the manuscript.

Funding: This research received no external funding.

Institutional Review Board Statement: The animal study protocol was approved by the Institutional Review Board (or Ethics Committee) of the University of Veterinary Medicine and Pharmacy in Kosice, Slovakia (EKVP/2022-12, 27 May 2022).

Informed Consent Statement: Informed consent was obtained from all subjects involved in the study.

Data Availability Statement: The data presented in this study are available on request from the corresponding author.

Conflicts of Interest: The authors declare no conflict of interest.

References

1. Anderson, E.T.; Minter, L.J.; Clarke, E.O.; Mroch, R.M.; Beasley, J.F.; Harms, C.A. The effects of feeding on hematological and plasma biochemical profiles in green (*Chelonia mydas*) and Kemp's ridley (*Lepidochelys kempii*) sea turtles. *Vet. Med. Int.* **2011**, *2011*, 890829. [CrossRef] [PubMed]
2. McWilliams, D.A.; Leeson, S. Metabolic Bone Disease in Lizards: Prevalence and Potential for Monitoring Bone Health. In Proceedings of the Fourth Conference on Zoo and Wildlife Nutrition, AZA Nutrition Advisory Group, Lake Buena Vista, FL, USA, 4–8 October 2001.
3. Campbell, T.W. Clinical pathology. In *Current Therapy in Reptile Medicine & Surgery*; Saunders/Elsevier: St. Louis, MO, USA, 2014; ISBN 9781455708932.
4. Khalil, F.; Haggag, G. Nitrogenous Excretion in Crocodiles. *J. Exp. Biol.* **1958**, *35*, 552–555. [CrossRef]
5. Moyle, V. Nitrogeno, us excretion in chelonian reptiles. *Biochem. J.* **1949**, *44*, 581–584. [CrossRef] [PubMed]
6. Appleby, E.C.; Siller, W.G. Some cases of gout in reptiles. *J. Pathol. Bacteriol.* **1960**, *80*, 427–430. [CrossRef] [PubMed]
7. Frye, F.L.; Mader, D.R.; Centofanti, B.V. Interspecific (Lizard: Human) Sexual Aggression in Captive Iguanas (*Iguana iguana*): A Preliminary Compilation of Eighteen Cases. *Bull. Assoc. Reptil. Amphib. Vet.* **1991**, *1*, 4–6.
8. Montali, R.J.; Bush, M.; Smeller, J.M. The pathology of nephrotoxicity of gentamicin in snakes. A model for reptilian gout. *Vet. Pathol.* **1979**, *16*, 108–115. [CrossRef] [PubMed]
9. Dutton, C.J.; Taylor, P. A comparison between pre-and post hibernation morphometry, hematology, and blood chemistry in viperid snakes. *J. Zoo Wildl. Med.* **2003**, *34*, 53–58. [CrossRef] [PubMed]
10. Maixner, J.M.; Ramsay, E.C.; Arp, L.H. Effects of Feeding on Serum Uric Acid in Captive Reptiles. *J. Zoo Anim. Med.* **1987**, *18*, 62. [CrossRef]
11. Smeller, J.M. Effect of Feeding on Plasma Uric Acid Levels in Snakes. *Am. J. Vet. Res.* **1978**, *39*, 1556–1557. [PubMed]
12. Samour, J.H.; Hawkey, C.M.; Pugsley, S.; Ball, D. Clinical and pathological findings related to malnutrition and husbandry in captive giant tortoises (*Geochelone species*). *Vet. Rec.* **1986**, *118*, 299–302. [CrossRef] [PubMed]
13. Lawrence, K. Seasonal variation in blood biochemistry of long-term captive Mediterranean tortoises (*Testudo graeca* and *T hermanni*). *Res. Vet. Sci.* **1987**, *43*, 379–383. [CrossRef]
14. McNab, B.K. *The Physiological Ecology of Vertebrates: A View from Energetics*; Comstock Pub. Associates: Ithaca, NY, USA, 2002; ISBN 0-8014-3913-2.
15. Lam, A.; Halán, M. Monitoring of physiological changes of uric acid concentration in the blood of snakes. *Folia Vet.* **2017**, *61*, 56–60. [CrossRef]
16. Fox, J.G.; Anderson, L.C.; Otto, G.M.; Pritchettcorning, K.R.; Whary, M.T. *Laboratory Animal Medicine*; Elsevier Academic Press: Amsterdam, The Netherlands, 2015; ISBN 9780124095274.
17. Mitchell, M.A.; Tully, T.N. *Manual of Exotic Pet Practice*; Saunders/Elsevier: St. Louis, MO, USA, 2009; ISBN 9781416001195 1416001190.
18. Kolmstetter, C.M.; Ramsay, E.C. Effects of feeding on plasma uric acid and urea concentrations in blackfooted penguins (*Spheniscus demersus*). *J. Avian Med. Surg.* **2000**, *14*, 177–179. [CrossRef]
19. Watson, A.D.J.; Church, D.B. Postprandial Increase in Plasma Creatinine Concentration in Dogs Fed Cooked Meat. *Aust. Vet. J.* **1980**, *56*, 463. [CrossRef] [PubMed]
20. Alvsaker, J.O. Uric Acid in Human Plasma: V. Isolation and Identification of Plasma Proteins Interacting with Urate. *Scand. J. Clin. Lab. Investig.* **1966**, *18*, 227–239. [CrossRef] [PubMed]

21. Rossi, J.V. *General Husbandry and Management. Reptile Medicine and Surgery*; Elsevier: Amsterdam, The Netherlands, 2006; pp. 25–41. [CrossRef]
22. Lumeij, J.T.; Remple, J.D. Plasma Urea, Creatinine and Uric Acid Concentrations in Relation to Feeding in Peregrine Falcons (*Falco peregrinus*). *Avian Pathol.* **1991**, *20*, 79–83. [CrossRef] [PubMed]
23. Tauson, A.H.; Elnif, J.; Wamberg, S. Nitrogen balance in adult female mink (*Mustela vison*) in response to normal feeding and short-term fasting. *Br. J. Nutr.* **1997**, *78*, 83. [CrossRef] [PubMed]
24. Parkinson, L.A.; Mans, C. Investigation of the effects of cricket ingestion on plasma uric acid concentration in inland bearded dragons (*Pogona vitticeps*). *J. Am. Vet. Med. Assoc.* **2020**, *257*, 933–936. [CrossRef] [PubMed]
25. Toledo, L.F.; Abe, A.S.; Andrade, D.V. Temperature and Meal Size Effects on the Postprandial Metabolism and Energetics in a Boid Snake. *Physiol. Biochem. Zool.* **2003**, *76*, 240–246. [CrossRef] [PubMed]

MDPI AG
Grosspeteranlage 5
4052 Basel
Switzerland
Tel.: +41 61 683 77 34

Animals Editorial Office
E-mail: animals@mdpi.com
www.mdpi.com/journal/animals

www.ingramcontent.com/pod-product-compliance
Lightning Source LLC
Chambersburg PA
CBHW041418210926
51850CB00013B/3310
* 9 7 8 3 7 2 5 8 2 6 5 5 1 *